Environmental Hazards and Resilience

Building resilience to the world's increasingly damaging environmental hazards has become a priority. This book considers the scientific advances which have been made around the world to enhance this resilience.

Although resilience is not new, it is through the idea of resilience that governments, organisations, and communities around the world are now seeking to address the rapidly increasing losses that environmental hazards cause so that fewer lives are lost, and damage is reduced. Alternative ideas and approaches have been helpful in reducing loss, but resilience offers a fresh and potentially effective means of reducing it further. Adopting a scientific approach and scientific evidence is important in applying the resilience idea in hazard mitigation. However, the science of resilience is at an immature stage of development with much discussion about the concept and how it should be understood and interpreted. Building useful theories remains a challenge although some of the building blocks of theory have been developed. More attention has been given to developing indicators and frameworks of resilience which are subsequently applied to measure resilience to hazards such as flooding, earthquake, and climate change.

Environmental Hazards and Resilience: Theory and Evidence considers the scientific and theoretical challenges of making progress in applying resilience to environmental hazard mitigation and provides examples from around the world – including the USA, New Zealand, China, Bangladesh and elsewhere.

The chapters in this book were originally published in the journal *Environmental Hazards*.

Dennis J. Parker is Professor at the Flood Hazard Research Centre, Middlesex University, UK. His research has focused on reducing the natural hazard losses across the globe. Dennis lives in Hertfordshire and has two daughters and six grandchildren.

Edmund C. Penning-Rowsell OBE founded the Flood Hazard Research Centre at Middlesex University in 1970. He has more than 40 years' experience of research and teaching in the flood hazard field, analysing floods and investment in flood alleviation, river management, water planning, and landscape assessment. Edmund lives in Oxfordshire and is also a member of the Oxford Water Security Network. He has two daughters and four grandchildren.

Environmental Hazards and Resilience

Theory and Evidence

Edited by
**Dennis J. Parker and
Edmund C. Penning-Rowsell**

Routledge
Taylor & Francis Group

LONDON AND NEW YORK

First published 2022
by Routledge
2 Park Square, Milton Park, Abingdon, Oxon OX14 4RN

and by Routledge
605 Third Avenue, New York, NY 10158

Routledge is an imprint of the Taylor & Francis Group, an informa business

British Library Cataloguing in Publication Data
A catalogue record for this book is available from the British Library

ISBN: 978-0-367-77449-3 (hbk)
ISBN: 978-0-367-77450-9 (pbk)
ISBN: 978-1-003-17143-0 (ebk)

Typeset in Minion Pro
by Newgen Publishing UK

Publisher's Note
The publisher accepts responsibility for any inconsistencies that may have arisen during the conversion of this book from journal articles to book chapters, namely the inclusion of journal terminology.

Disclaimer
Every effort has been made to contact copyright holders for their permission to reprint material in this book. The publishers would be grateful to hear from any copyright holder who is not here acknowledged and will undertake to rectify any errors or omissions in future editions of this book.

Contents

Citation Information

The following chapters were originally published in different issues of the *Environmental Hazards*. When citing this material, please use the original citations and page numbering for each article, as follows:

Chapter 1
Using vulnerability and resilience concepts to advance climate change adaptation
Erin P. Joakim, Linda Mortsch and Greg Oulahen
Environmental Hazards, volume 14, issue 2 (2015), pp. 137–155

Chapter 2
Foundations of community disaster resilience: well-being, identity, services and capitals
Scott B. Miles
Environmental Hazards, volume 14, issue 2 (2015), pp. 103–121

Chapter 3
Temporal and spatial change in disaster resilience in US counties, 2010–2015
Susan L. Cutter and Sahar Derakhshan
Environmental Hazards, volume 19, issue 1 (2020), pp. 10–29

Chapter 4
Assessing community resilience: mapping the community rating system (CRS) against the 6C-4R frameworks
Ajita Atreya and Howard Kunreuther
Environmental Hazards, volume 19, issue 1 (2020), pp. 30–49

Chapter 5
Research on disaster resilience of earthquake-stricken areas in Longmenshan fault zone based on GIS
Bin Liu, Xudong Chen, Zhongli Zhou, Min Tang and Shiming Li
Environmental Hazards, volume 19, issue 1 (2020), pp. 50–69

Chapter 6
Coping and resilience in riverine Bangladesh
Parvin Sultana, Paul M. Thompson and Anna Wesselink
Environmental Hazards, volume 19, issue 1 (2020), pp. 70–89

Chapter 7

Urbanisation and disaster risk: the resilience of the Nigerian community in Auckland to natural hazards
Osamuede Odiase, Suzanne Wilkinson and Andreas Neef
Environmental Hazards, volume 19, issue 1 (2020), pp. 90–106

Chapter 8

The French Cat' Nat' *system: post-flood recovery and resilience issues*
Bernard Barraqué and Annabelle Moatty
Environmental Hazards, volume 19, issue 3 (2020), pp. 285–300

Chapter 9

Stakeholder participation in building resilience to disasters in a changing climate
Paulina Aldunce, Ruth Beilin, John Handmer and Mark Howden
Environmental Hazards, volume 15, issue 1 (2016), pp. 58–73

Chapter 10

How does social learning facilitate urban disaster resilience? A systematic review
Qingxia Zhang, Junyan Hu, Xuping Song, Zhihong Li, Kehu Yang and Yongzhong Sha
Environmental Hazards, volume 19, issue 1 (2020), pp. 107–129

Chapter 11

Local government, political decentralisation and resilience to natural hazard-associated disasters
Vassilis Tselios and Emma Tompkins
Environmental Hazards, volume 16, issue 3 (2017), pp. 228–252

For any permission-related enquiries please visit:
www.tandfonline.com/page/help/permissions

Notes on Contributors

Paulina Aldunce, Department of Environmental Science and Resource Management, Facultad de Ciencias Agronómicas, Universidad de Chile, La Pintana, Santiago, Chile; Department of Resource Management and Geography, The University of Melbourne, Melbourne, Victoria, Australia; Centre for Climate and Resilience Research (CR), Santiago, Chile.

Ajita Atreya, Freddie Mac, McLean, VA, USA.

Bernard Barraqué, Centre International de Recherches sur l'Environnement et le Développement, Paris, France.

Ruth Beilin, Faculty of Science, The University of Melbourne, Melbourne, Victoria, Australia.

Xudong Chen, College of Management Science, Chengdu University of Technology, Chengdu, People's Republic of China.

Susan L. Cutter, Department of Geography and Hazards & Vulnerability Research Institute, University of South Carolina, Columbia, SC, USA.

Sahar Derakhshan, Department of Geography and Hazards & Vulnerability Research Institute, University of South Carolina, Columbia, SC, USA.

John Handmer, Centre for Risk and Community Safety, RMIT University, Melbourne, Victoria, Australia.

Mark Howden, SCIRO, SCIRO Agriculture, Canberra, Australia.

Junyan Hu, School of Management, Lanzhou University, Lanzhou, China.

Erin P. Joakim, Faculty of Environment, University of Waterloo, Waterloo, ON, Canada.

Howard Kunreuther, Decision Sciences and Business Economics and Public Policy, Wharton Risk Management and Decision Processes Center, University of Pennsylvania, Philadelphia, PA, USA.

Shiming Li, School of Management and Economics, University of Electronic Science and Technology of China, Chengdu, China.

Zhihong Li, School of Management, Lanzhou University, Lanzhou, China.

Bin Liu, School of Management and Economics, University of Electronic Science and Technology of China, Chengdu, China; College of Management Science, Chengdu University of Technology, Chengdu, China; Geomathematics Key Laboratory of Sichuan Province, Chengdu, China.

Scott B. Miles, Resilience Institute, Western Washington University, Bellingham, WA, USA.

Annabelle Moatty, Laboratoire de Géographie Physique, (UMR 8591) CNRS, Université Paris I Panthéon-Sorbonne, UPEC, Meudon, France.

Linda Mortsch, Faculty of Environment, University of Waterloo, Waterloo, ON, Canada.

Andreas Neef, Development Studies Programme, The University of Auckland, Auckland, New Zealand.

Osamuede Odiase, Department of Civil Engineering, The University of Auckland, Auckland, New Zealand.

Greg Oulahen, Department of Geography, Western University, London, ON, Canada.

Yongzhong Sha, School of Management, Lanzhou University, Lanzhou, China; Evidence Based Social Science Research Center, School of Public Health, Lanzhou University, Lanzhou, China.

Xuping Song, Evidence Based Medicine Center, School of Basic Medical Sciences, Lanzhou University, Lanzhou, China; Evidence Based Social Science Research Center, School of Public Health, Lanzhou University, Lanzhou, China.

Parvin Sultana, Flood Hazard Research Centre, Middlesex University, Hendon, London, UK.

Min Tang, School of Management and Economics, University of Electronic Science and Technology of China, Chengdu, China.

Paul M. Thompson, Flood Hazard Research Centre, Middlesex University, Hendon, London, UK.

Emma Tompkins, Geography and Environment, University of Southampton, Southampton, UK.

Vassilis Tselios, Department of Planning and Regional Development, University of Thessaly, Volos, Greece.

Anna Wesselink, Independent Scholar, Hengelo, The Netherlands.

Suzanne Wilkinson, School of Built Environment, Massey University, Auckland, New Zealand.

Kehu Yang, Evidence Based Medicine Center, School of Basic Medical Sciences, Lanzhou University, Lanzhou, China; Evidence Based Social Science Research Center, School of Public Health, Lanzhou University, Lanzhou, China.

Qingxia Zhang, School of Management, Lanzhou University, Lanzhou, China; Evidence Based Social Science Research Center, School of Public Health, Lanzhou University, Lanzhou, China; School of Public Management, Gansu University of Political Science and Law, Lanzhou, China.

Zhongli Zhou, College of Management Science, Chengdu University of Technology, Chengdu, China.

Preface

This volume owes its origins to a special edition of the *Environmental Hazards* journal entitled 'Resilience' and published by Taylor and Francis in early 2020 and edited by me. The idea behind the special edition came from Edmund Penning-Rowsell who is the Editor of *Environmental Hazards*. Taylor and Francis then invited us to turn the special edition of Environmental Hazards into a book by adding further papers published recently in the journal on the subject of disaster resilience. Edmund was instrumental in identifying the authors and articles submitted to the journal for the 2020 edition and I took on the role of editing each of the papers before they were accepted for the journal. As the invited editor of the special edition I also produced the introductory chapter which is the basis of the introductory chapter in this volume. The special edition generated considerable interest. I received more invitations to speak at conferences, write further journal articles on resilience and collaborate in making podcasts on the subject etc. than I have ever received before. Clearly there is much interest among academics, practitioners and policy-makers alike in the subject of disaster resilience and resilience in general.

Dennis Parker
St Albans, England

Disaster resilience: developing a challenged science

Dennis J. Parker

Introduction

The multi-dimensional concept of resilience has gained prominence and now dominates thinking about the management of risks facing humankind. For risks associated with climate change, environmental disasters, critical infrastructure, security and terrorism, or other kinds of risks, resilience has become an attractive notion. The concept is now embedded within international discourse and plays a key role in international, national, and local policy and development. It underpins the UN's Sendai Framework for Disaster Risk Reduction (UNDRR, 2015), its Paris Agreement on climate change (UN, 2015a), and its Sustainable Development Goals (UN, 2015b). Forty-five OECD (Organisation for Economic Co-operation and Development) countries have adopted national resilience strategies,[1] and international development is experiencing a 'resilience revolution'.[2] Over 100 world cities are in the 100 Resilient Cities network;[3] over 1,000 participate in the Making Cities Resilient Campaign (UNISDR, 2012) to make their cities resilient to disasters and local/community projects abound. Resilience citations in the Web of Science publications rose from almost zero in 1997 to nearly 30,000 in 2015 (Lovell et al., 2016).

Resilience has particularly wide applicability: hazard and disaster resilience being just one subset of the wider application of the concept. This chapter examines the origins and prominent early conceptualisations of hazard and disaster resilience and its complex dynamics. It then questions whether resilience is a rebranding of the earlier mitigation concept and considers the complex dynamics of resilience. This chapter argues that disaster resilience science is not yet a mature science because it is characterised by a theory gap and faces considerable challenges. Lastly, there is an introduction to the chapters in this volume.

Origins of resilience thinking

Resilience thinking has diverse origins. Physical, engineering, and ecological scientists first used the concept but soon social scientists became attracted to the idea. Disciplines including psychology, psychiatry, leadership studies, and regional economic analysis employed the concept. Psychologists and psychiatrists applied it to individuals and to human communities and became interested in connected ideas of self-reliance and human support systems. Alexander (2013) attributes the first serious use of the concept to Rankine who used it to characterise the mechanical strength and deformation of steel beams. Most scholars attribute the first modern adoption of resilience to Holling (1973, 1986) who analysed the stability of ecological systems which demonstrated attributes of rebounding by absorbing stresses (Folke, 2006). However, it was disaster vulnerability rather than resilience which stole the limelight during the 1990s (Blaikie et al., 1994). The year 1990 heralded the start of the International Decade for Natural

Disaster Reduction (IDNDR) which focused on the role of vulnerability in disaster reduction. Remarkably, the Final Report of the IDNDR's Scientific and Technical Committee includes a section on vulnerability and risk assessment but it makes no mention of resilience (UNIDNDR, 1999). Even so resilience emerged rapidly thereafter to dominate disaster risk management thinking (Walker et al., 2004; Cutter et al., 2008; Zhou et al., 2010; Cutter, 2010; Aldunce et al., 2014)

Resilience has some roots in earlier conceptualisations of hazard management. The notion of adjusting to hazards goes back to White's conceptualisation of human adjustment to floods which became dominant in the 1970s (White, 1945). In an influential work, White argued that "floods are largely acts of man" and adjustments were a range of structural and non-structural measures which could reduce flood losses. Subsequently hazard mitigation became virtually synonymous with this idea, although surprisingly its adoption is comparatively recent in the UK and European Union (Macdonald et al., 2011).

Is resilience just a rebranding of mitigation?
The Dutch face managing one of the most flood-prone regions of the planet but have no word for resilience in their language.[4] Instead, they think of risk consequence mitigation or management and apply this to strengthen their country against the threat of potentially devastating sea and river flooding. Risk consequence mitigation is about reducing probability and/or consequence where consequence is exposure ⊠ vulnerability. Introducing resilience into this context begs the question – is resilience simply mitigation rebranded or it is something else? On the face of it, the two concepts are quite similar. When applied to disasters, mitigation usually means lessening of hazard or disaster severity and their consequences. Resilience is similar but usually implies the ability to recover quickly from, or adjust easily to, a hazard or disaster and their consequences. Mitigation is therefore about 'lessening' or 'reducing' whereas resilience is about 'recovering' and 'adjusting'. Within White's concept of 'adjustment' the notion of recovering from floods was implicit (White, 1945). However, despite the apparent closeness of the concepts of disaster mitigation and resilience, resilience has potentially much greater reach and penetration into the opportunities that social institutions and groups inherently possess to effectively manage future scenarios of disaster risk. In this sense resilience is not at all a rebranding of mitigation but something that goes potentially well beyond it.

Mitigation is mostly about reducing loss so that the status quo can be safely maintained whereas resilience is more than about bouncing back to this status quo: the difference between static and dynamic perspectives (Cutter, 2016). Resilience as simply bouncing back was not accepted by Handmer and Dovers (1996) who viewed it as an adaptive process which emphasised the notion of development and improvement beyond simply returning to pre-disaster conditions (Aldunce et al. 2014). Walker et al. (2004) were also attracted to adaptation and learning, change and transformation in their conceptualisation of the resilience of socio-ecological systems. Martin (2012) argued that realignment and adaptation is an integral part of the concept of resilience. Recent conceptual developments of resilience view it as an attribute of social systems (Adger, 2000) or of place (Cutter et al., 2008). Some recent constructions strongly reflect Blaikie et al.'s (1994) understanding of the root causes of disaster vulnerability. These are limited access to power, institutional structures and resources within political and economic systems – causes largely ignored in the 'human adjustment' conceptualisation and subsequent work of Burton, Kates, and White (1968). So some recent conceptualisations of disaster resilience concentrate on human agency, power relations, equity, adaptive capacities, and transformation (e.g. Keck and Sakdapolrak, 2013).

The complex dynamics of resilience

Resilience to disruptive and damaging disaster events – whether they be natural hazards and associated disasters such as floods, pandemics such as Covid-19, or any other type of disaster which poses a serious threat – is dynamic (i.e. ever changing) and a function of several sets of complex, interrelated conditions and processes. Individuals, households, communities (or any other entity, e.g. businesses, governments) will possess inherent resilience (Cutter et al., 2014) stemming from their basic characteristics and qualities which either enhance or detract from resilience. If we focus just on communities, then these characteristics and qualities include socio-economic characteristics including age, ethnicity, gender and health profiles, and levels of educational attainment, employment and income, among others. Processes include everyday ones such as membership of local social networks, involvement in volunteer work, and perhaps participation in disaster preparation and training. Inherent resilience is linked to sense of place, identity, culturally meaningful practices, and social interactions (Simms, 2017). At any point in time, some community members are likely to experience declining resilience (perhaps through ill health or loss of employment) whereas others may be increasing in their resilience (perhaps by accumulating savings or joining local social networks) resulting in small incremental net changes in overall community resilience over points in time.

Communities are not isolated islands and their hazard- and disaster-related resilience will be affected by numerous processes and policies at national, regional and international scales which may also enhance or detract from their resilience. For example, national policies on welfare, health and education – all subject to change over time – are likely to have some, and possibly major, resilience implications for communities and their members, lending resilience a political dimension. Some of these policies are likely to have unintended effects, perhaps in some cases constraining or reducing a community's resilience. Others may have intended impacts including ones designed to enhance overall resilience through formal adaptive resilience-enhancing initiatives, whether they be general ones or ones targeted specifically at hazard- and disaster-related resilience. Processes, including, for example economic globalisation and regional and global scale climate change, are also likely to have changing medium to long-term resilience implications for communities leading perhaps, in the latter case, to declining resilience to sea-level rise and flooding. Economic globalisation may in some cases reinforce the structural inequalities that exist between countries thereby deepening their vulnerability to disasters. In some cases rigid social relations which support the status quo are deeply embedded in existing systems of governance preventing the development of resilience.

However, by no means all processes affecting community resilience to hazards and associated disasters are top-down ones. Indeed, communities may well organise their own adaptive resilience enhancement initiatives and campaigns so that resilience is improved from the bottom upwards.

In these examples, it is clear that resilience is a complex dynamic process which incorporates the temporal scale. It connects the past, present, and future because today's resilience initiatives are often stimulated by past disruptive events and are aimed at reducing potential future disruption. In the short term, resilience may involve bouncing back better whereas in the long term it may involve transformation in which power relations change breaking the status quo which maintains inequalities which influence resilience. Resilience also incorporates the spatial scale because it connects enhancing or detracting processes across the spectrum of local to the global scale. Here the theory of panarchy (Gunderson and Holling, 2002) is highly relevant because it describes the ways in which complex systems of people and nature are dynamically organized and structured across scales of space and time.

Is resilience science a mature science?

Given the long evolutionary history of resilience and the impressive rise in resilience citations, a number of questions arise: Are the increasing rate of publication and the obvious attractiveness of resilience within policy development and implementation signs of subject maturity? Is the disaster resilience approach to disaster risk management based on a mature science? These questions exercise the minds of some academics and to which policymakers and practitioners wishing to employ disaster resilience strategies need clear answers. Taking just one example, the Environment Agency – England's flood risk management agency – has published a national strategy aiming to move from the concept of protection to resilience (Environment Agency, 2019). This strategy mentions resilience over 100 times and is clearly dependent on a resilience approach to flood and coastal erosion risk management working properly. A flood resilience framework and measurement method, anchored in theory and practically applicable, will undoubtedly be required to ensure that the strategy is effective. The policymakers and practitioners will hope to found their strategy on a science which is sufficiently grounded and stable, well-structured and well-tested to meet their aims. However, is such a science currently available?

Recent literature on disaster resilience contains a superabundance of definitions (e.g. Zhou et al., 2010; Alexander, 2013; Sudemeier-Rieux, 2014; Cutter et al., 2014; Aldunce et al., 2014; Demoriz and Haase, 2018; Koliou at al., 2018). Published disaster resilience measurement methods and application case studies are rapidly expanding. As Zhou et al. (2010) point out, differences in understandings often stem from varied epistemological orientations. There is no single precise and universally accepted definition of resilience or means for how it should be measured, although there is some common ground (Koliou et al. 2018; Heng et al., 2018). Some degree of general consensus is emerging that resilience is about attributes (e.g. economic, social, informational) or capacities across a number of dimensions (e.g. disaster cycle stages) leading to groups of components of resilience indicators (e.g. the use of 'Cs' or capitals, as in the '5Cs' such as physical, natural, economic, social, and infrastructural capacities and the '4R's which are attributes, qualities, or properties of resilience: rapidity, robustness, redundancy, and resourcefulness). However, notwithstanding indications of convergence over some aspects of measurement, the current degree of ambiguity about the concept and its measurement is unhelpful to policymakers and practitioners. Partly because of different epistemological perspectives and contexts in which resilience is conceptualised, it will require an almost Herculean effort to somehow integrate all the different definitions and achieve any kind of consensus. Ideas on the sources of disaster resilience or variables influencing it also vary widely. For example, Burton (2015) employed 64 variables to measure community resilience whereas the Zurich Alliance community flood resilience measurement framework drew upon 88 sources of resilience (Keating et al., 2017).

The theory gap

The above differences and inconsistencies lead to a sense of disaster resilience not yet being paradigmatic[5] i.e. not possessing a settled definitional, conceptual, and theoretical basis which is widely recognised and adhered to. Miles (see Chapter 3) holds that, in community disaster resilience research, methodological and empirical progress has outpaced theory such that there is currently a theory gap. He finds a paucity of research on community disaster resilience theory in the literature. Furthermore, there is some confusing imprecision in disaster resilience research literature about how theories differ from and relate to concepts, such that concepts are sometimes confusingly referred to as theories when they are more the building blocks of theory yet to be fully constructed.

Naturally, a close relationship exists between concepts and theories but an essential difference also exists between them. Concepts are abstract ideas or general notions about circumstances or occurrences that have come to mind through experience, reasoning, or perhaps imagination. They are the bricks or walls from which the houses of theory are constructed. As Akintunde (2017) states, theory is constructed after relevant concepts have been thoroughly explored and repeatedly tested at a broader scale through a large number of empirical investigations of how circumstances and phenomena occur. Theories incorporate definitions, facts, variables, laws, and tested assumptions that are widely accepted and they predict or explain events or situations by specifying relationships between variables. Essential features of theory are generality or broad application and empirical testability and the ability to verify or confirm relationships between variables. There are plenty of widely tested and accepted mature theories, for example in environmental human behaviour, including innovation diffusion theory (Rogers, 1962), planned behaviour theory (Ajzen, 2002) and hazard perception theory (Kates, 1962), all of which have some relevance to building resilience theory. Here, in each case the references are to the original research which has subsequently been repeatedly investigated, demonstrated, and confirmed.

Related terms that are frequently employed in disaster resilience literature are 'theoretical frameworks', 'conceptual frameworks', and 'conceptual models'. Theoretical frameworks should be based on a set of ideas and existing theory which has been previously well tested and validated by other researchers. On the other hand, a conceptual framework or model need not be widely accepted and is the researcher's mental construction of relationships between variables, or concepts that reflect variables, often relating back to some element of an existing theory.

Despite Miles's criticism in Chapter 3 of community disaster resilience research having precious few theoretical foundations, as well as observations that the science of resilience lacks a unifying theory (for example, being incapable of unifying the ecological/environmental and social sciences) (Olsson et al., 2015), the picture is far from bleak. Even so a considerable amount of theory sorting and building remains to be done. To some extent it depends where one looks in the research literature for contributions to theory. For example, although there are important differences between the concepts of vulnerability and resilience, as Cutter (2008) points out, there are more than a few theoretical frameworks and conceptual models of vulnerability which have been advanced to enhance the theoretical foundations of vulnerability. Within these there are a number of elements of relevance to the advancement of resilience theory. It is also important to take a view that scientific theory advances not through a search for some kind of single, unifying, overarching, dominant theory, but to take a pluralistic approach in which having more than one – and perhaps many – well-established theories each of which can contribute something unique to our understanding of the whole. Among others, Blaikie et al.'s (1994) Disaster Pressure and Release Model; Cutter et al.'s (2008) Disaster Resilience of Place (DROP) conceptual model; Gunderson and Holling's (2001) panarchy theory and Miles's WISC framework and conceptual model (Chapter 3) are all examples of quite different but useful contributions to the building of community disaster resilience theory. Between them the emphasise the multidimensional nature of disaster resilience theory which therefore requires building blocks derived from a diverse but interconnected set of frameworks and conceptual models.

To summarise, resilience science is characterised by competing views and only a limited consensus about measurement. The science lacks a well-developed body of theory suggesting that it has yet to reach maturity. Of course, it may be argued that the lack of consensus, and indeed disunity, is not necessarily negative and indicative of immaturity because the circumstances and contexts in which disaster resilience needs to be defined and measured vary widely.

These circumstances include whether the objective is to measure resilience at the national, regional, or local/community scales and resilience to which hazard among the range of hazards is being considered. Also measuring disaster resilience in an advanced society where there is a relatively flat social structure and little social inequity presents quite different issues to measuring it in an international development context. It may be argued that somewhat different and nuanced approaches to definition and measurement are required – an argument that has some strength. Even so the current lack of orthodoxy in the definition of resilience as well as inconsistency in its conceptualisation and measurement identified by Cutter (2016) is surely a sign of scientific immaturity. However, it is not all bad news because there are signs of scientific progress, including initiatives to identify operational measures of disaster resilience, increasing research to measure resilience, and efforts to validate and verify measurement tools.

Challenges

Disaster resilience science is a challenged science, not least because resilience dynamics are complex. Challenges include developing a more mature science; effective and readily applicable. Some challenges are identified above including greater definitional consensus and assembling and assessing different theoretical contributions in order to develop a more mature body of disaster resilience theory. More research into changes in resilience over time are required as, currently, most research takes only a time snapshot of resilience. Frazier (2012) observed regarding resilience indicators that the importance of differential weighting is often unconsidered and Keating et al. (2017) report "there is currently no empirical evidence to support a larger weight for any source over others". They point out that resilience measurement frameworks often assume that ex-ante presence or absence of an indicator of resilience (representing a source of resilience) will impact ex-post resilience positively.

Keating et al. (2017) reported that they were unable to find a single disaster resilience measurement framework or method that had been verified by longitudinal study of resilience changes, as a result of different capacities and actions. However, Heng et al. (2018) identified 18 studies in which empirical validation of resilience indices and models with external reference data had been undertaken using qualitative or quantitative methods. Even so, these studies represented only about 10 per cent of their analysed articles on disaster resilience measurement. Developing verified or validated disaster resilience measurement tools for different circumstances and making them more routine remain important challenges in the quest to make disaster resilience science more mature and dependable for policymakers and practitioners. A different kind of challenge in which some progress is being made, includes overcoming the limited progress to date in modelling systemic physical interdependencies of infrastructure systems (e.g. lifelines, critical infrastructure, transportation systems) and their cascading impacts on socio-economic systems which need to be understood to improve resilience and adaptation to disruption risk (Kyriazis and Argyroudis, 2013; Koliou et al., 2018).

Numerous further challenges exist – too many to be comprehensively covered here. Resilience has strong dynamic spatial and temporal elements presenting challenges which need to be taken into account in measurement. Measurement is complicated by the three basic dimensions requiring consideration: the temporal scale which may sometimes be characterised by the disaster cycle; the spatial scale; and hazard/disaster receptors (e.g. physical, economic and social). In addition many types of hazard need to be considered. The European Commission's ENSURE project, which sought to integrate disaster vulnerability and resilience assessment, developed a matrix-based assessment framework for a range of hazards, containing vulnerability and resilience variables, indicators and parameters for the three dimensions. Although vulnerability was usefully integrated with resilience assessments, the resulting framework was unconvincing in capturing the dynamics of vulnerability and resilience (Menoni et al. 2013). Difficulties also arose in developing such a framework across the European Union nations because of the incompatibilities of resilience parameter data sets between nations. Because of these kinds of complexities, resilience measurement frameworks are often confined to a single hazard or scale (e.g. a national scale or a community scale). The Zurich Alliance flood resilience measurement framework (Keating et al., 2017) addresses both ex-ante and ex-post resilience measurement. It focuses at the community scale but takes into account that community resilience is impacted upon by the household and regional, national, and global scales. An all-singing, all-dancing disaster resilience framework and measurement system in which the appropriate resilience-building roles of local, regional, and national policymakers can be identified may not be required very often but would present a major challenge.

A final challenge to resilience science is the risk, in some cases, of its practice being partially or wholly discredited by being "all too easily … captured by neo-liberal apologists, to bolster arguments in favour of the need for 'flexibility', 'self-help' and 'competitive fitness'" (Martin, 2012). Berkes and Ross (2013) critiqued resilience thinking for simply accepting socio-economic conditions and circumstances as a given, thereby avoiding any questioning of underlying structural causes of vulnerability of the type that focused the minds of Blaikie et al. (1994). This challenge is eloquently argued by Cutter (2016) in which she questions whether we should "be more forceful in our conceptualisations of resilience and point out the inequalities in patterns, processes and perspectives".

A brief review of the papers in this Special Issue
This volume continues with a chapter (1) from Canada by Joakim, Mortsch, and Oulahen which develops a critical discussion about the "powerful" concepts of vulnerability and resilience, this time in the context of adaptation to climate change. They argue that there is currently confusion in the understanding of these concepts and how they relate to each other. They propose that by integrating them and their recognised strengths and differences, a holistic assessment of hazards may be achieved and a useful framework may be constructed for climate change adaptation. Focusing on resilience only runs the risk, they argue, of entrenching existing social structures that produce unequal risk: a risk which may be significantly reduced by integrating vulnerability with resilience.

Next comes a chapter (2) from Australia in which Miles picks up on the disaster resilience theory gap theme in the opening paper. He focuses on community disaster resilience and develops a theoretical framework called WISC which he argues provides a strong foundation for critical theory building. The name WISC reflects the four constructs of the framework: well-being, services, and capitals. Miles's intention is to stimulate research that incorporates theory.

Miles's chapter is followed by two from the USA where Cutter and Derakhshan (3) are struggling with national level quantification and measurement of resilience. Examining disaster resilience across the USA at county scale, they provide a rare monitoring of changes over time. The study is based on the DROP (Disaster Resilience of Place) theoretical framework (Cutter et al., 2008) and employs the BRIC (Baseline Resilience Index for Communities) measurement system which uses the 'capitals' approach mentioned above and focuses on measuring pre-event inherent resilience only in communities. Indicators on variables known to be influencers of resilience and others including from national databases were employed. The latter provide a degree of data quality consistency overcoming the data inconsistency faced by the international ENSURE project mentioned above. Results reveal the spatial pattern of increases and decreases in resilience over time and the main drivers of change. The authors fundamentally question the purpose of measuring and seeking to manage resilience which may lead to inequalities in outcomes which privilege some and perpetuate disadvantage in others.

In the second chapter from the USA, Atreya and Kunreuther (Chapter 4) suggest that a community's resilience to floods may be assessed using the US National Flood Insurance Program's (NFIP) Community Rating System (CRS). The CRS is an incentive system encouraging flood risk communities to improve their flood resilience through creditable activities in return for discounted flood insurance premiums. They also employ the 'capitals' approach to describe a community's strengths together with the '4Rs' (properties of resilience), and find a positive correlation between the creditable activities and at least one of the 6Cs and 4Rs. However, additional information is required because some dimensions of resilience, for example social vulnerabilities, are not covered when using the CRS. Benefit–cost analysis of resilience enhancing measures would also be required to make the best use of communities' limited financial resources – a point that has wider applicability to resilience-building projects.

Bin, Xudong, Zhongu, Min, and Shimming's analysis (Chapter 5) of earthquake stricken areas in the Longmenshan fault zone of China's Sichuan Province uses an entirely different, macro-economic, quantitative modelling approach to measuring disaster resilience. Characteristics of a region's industrial structure before and after an earthquake are used to measure of resilience or resilient recovery. Employment and industrial structures usually coincide but fail to do so when plant and equipment is destroyed. The authors employ a measure of this degree of coincidence as a resilience indicator together with per capita GDP growth rate which disasters slow. The effects of earthquakes on resilience decline over time but seismic intensity and spatial variations in topographic complexity differentially constrain the spatial pattern of resilience improvement. Unless they can find ways of enhancing it, the resilience of those occupying high risk zones may be progressively reduced by disasters in successive years, especially if they are poor. This is precisely the case in riverine northern Bangladesh where the coping mechanisms and resilience of rural households to flooding are investigated by Sultana, Thompson, and Wesselink in Chapter 6. Using household surveys they discovered ways in which vulnerable households employ 'hydro-social landscapes', such as evacuating to flood protection embankments, to boost their resilience. Vulnerable households' sources of resilience include use of community organisations to enhance livelihoods, for example by providing access to relief, and male seasonal migration to urban areas where employment prospects exist. These examples illustrate a key point made in the earlier discussion above – that resilience is very context-specific. Even so, in this case, governmental policies familiar with the context fail to support the vulnerable households.

Odiase, Wilkinson, and Neef's chapter (7) is an example of ex-ante community resilience assessment employing the Community Resilience Index to investigate, measure and calculate a baseline resilience of the Nigerian community in urban Auckland, New Zealand to multi-hazards. They use a variant of the 'capitals' methodology by identifying six 'domains', each represented by resilience indicators the data for which are from primary and secondary sources. As explained above, weighting of indicators is a contentious issue in disaster resilience measurement but Odiase and co-authors opt for equal weighting for all indicators because differential weighting undermines interdependence among resilience domains.

The post-disaster recovery process is a potentially very important dimension of disaster resilience building, and one which is somewhat neglected in the research literature. In Chapter 8 Barraqué and Moatty examine the link between recovery and resilience within the context of France's *Catastrophe Naturelle* (or Cat' Nat') insurance fund system which requires insurance companies to provide compensation to disaster victims funded by an amount added to all insurance premiums across the nation. The authors question whether the insurance-based recovery funding enhances resilience and how landowners may better participate in vulnerability reduction – and they do this within the context of the decentralisation of flood control in France and whether or not recovery has improved and its costs. They conclude that the Cat' Nat' system has improved post-disaster recovery but an unintended outcome is that it reduces resilience because flood victims are reimbursed for the damages they experience and more resilient reconstruction which could reduce vulnerability is not encouraged. Risk zone maps are legally required in France, although they are sometimes viewed negatively by property-owners, and these encourage vulnerability reduction but this is not a part of the recovery process.

In Chapter 9 Aldunce, Beilin, Handmer, and Howden argue that stakeholder participation is pivotal to the building of resilience to disasters in a changing climate. They observe that, unfortunately, the importance of stakeholder participation is not well informed or underpinned by empirical evidence or theory. Employing social-interactive discourse theory, these authors study the frames and subsequent practices developed for a disaster policy initiative in Queensland, Australia. They discover that stakeholder participation, especially participants from representing local government and communities, require urgent attention. They also found that moving from experiental learning to social learning is also critically important.

Learning through experience and acquisition of disaster knowledge plays a vital role in adapting to hazards and disasters and, as we have seen above, adaptation is central to the concept of disaster resilience. Qingxia, Junyan, Xuping, Zhihong, Kehy, and Yongzhong's chapter (10) focuses on 'social learning' at the individual, community and organisation levels as a policy option for managing urban disaster risks. They employ thematic content analysis on carefully selected studies from around the world and investigate evidence of how, through many different types of social learning (e.g. networking, stewardship, use of memories) and a mix of cognitive change, self-organisation, civil responsibility, and open communication and deliberation, social learning can promote urban disaster resilience. The results reveal how social learning is operationalised to promote disaster resilience.

The final chapter (11), by Tselios and Tompkins, examine the effect of political decentralisation on natural hazard-associated disasters by undertaking a statistical analysis of disaster outcomes and, for example, indices which provide a proxy for political decentralisation. They focus on storms and earthquake hazards and disasters. Although they barely mention resilience (exceptions include the title of the paper and the final paragraph of the conclusions), political decentralisation in the form of more locally representative democracy is very likely to be an important factor in community resilience to hazards and disasters. Aldunce et al. in Chapter 9 make a similar point when they call for greater stakeholder participation and involvement of local government in disaster policy initiatives. Tselios and Tompkins also examine the relationship between disaster outcomes and political orientation (i.e. left-wing and right-wing) and the extent of local representation. Among their findings is that decentralised countries experience lower numbers of people affected and to some degree the numbers killed by storms. This leads them to conclude that there is a need to further decentralise disaster risk reduction because local government can be more responsive to local needs. They find similar results for earthquakes although more decentralised countries have higher economic damage which was also the case for storms.

Notes

1. www.oecd.org/cfe/regional-policy/national-policy-resilience-frameworks.pdf
2. www.odi.org/projects/2864-resilience-scan
3. www.100resilientcities.org/
4. Personal communication with Jaap Flikweert, Royal HaskoningDHV (the Netherlands-based international engineering company).
5. Here a paradigm means a settled and widely accepted, and indeed dominating, framework containing all of the commonly established and accepted perspectives on a subject, including theories, common methods, conventions, and standards (Kuhn, 1970).

References

Adger, W.N. (2000) Social and ecological resilience. Are they related? *Progress in Human Geography*, 24 (3): 347–364

Ajzen I. (2002) Perceived behavioral control, self-efficacy, locus of control, and the theory of planned behavior, *Journal of Applied Social Psychology* 32 (2002): 665–683.

Akintunde, E.A. (2017) Theories and concepts for human behavior in environmental preservation. *Journal of Environmental Science and Public Health* 1 (2): 120–133. DOI: 10.26502/JESPH.012.

Aldunce, P., Beilin, R. Handmer, J. and M. Howden (2014) Framing disaster resilience. The implications of the diverse conceptualisations of "bouncing back". *Disaster Prevention and Management* 23 (3): 252–270.

Alexander, D.E. (2013) Resilience and disaster risk reduction: An etymological journey. *Nat. Hazards Earth Syst. Sci.* 13: 2707–2716. DOI:10.5194/nhess-13-2707-2013.

Berkes, F. and H. Ross (2013) Community Resilience: Toward an Integrated Approach. *Society & Natural Resources* 26 (1): 5–20. DOI:10.1080/08941920.2012.736605.

Blaikie, P., Cannon, T., Davis, I. and B. Wisner (1994) At Risk Natural Hazards, People's Vulnerability, and Disasters. Routledge, London and New York.

Burton, C.G. (2015) A Validation of Metrics for Community Resilience to Natural Hazards and Disasters Using the Recovery from Hurricane Katrina as a Case Study. *Annals of the Association of American Geographers* 105 (1): 67–86. DOI:10.1080/00045608.2014.960039.

Burton, I., Kates, R.W. and G.F. White (1968) The Human Ecology of Extreme Geophysical Events. *Natural Hazard Research Working Paper* No. 1, Department of Geography, University of Toronto: Toronto.

Cutter, S.L. (2010) Resilience to natural hazards: A geographic perspective, *Natural Hazards* 53: 21–41. DOI:10.1007/s11069-009-9407-y.

Cutter, S.L. (2016) Resilience to What? Resilience for Whom? *The Geographical Journal*, 182 (2): 110–113. DOI:10.1111/gej.12174.

Cutter, S.L., Ash, K.D. and C.T. Emrich (2014) The geographies of community disaster resilience. *Global Environmental Change* 29: 65–77. DOI:10.1016/j.gloenvcha.2014.08.005.

Cutter, S.L., Barnes, L., Berry, M., Burton, C., Evans, E., Tate, E. and J. Webb (2008) A place-based model for understanding community resilience to natural disasters. *Global Environmental Change* 18: 598–606.

Demiroz, F. and T.W. Haase (2018) The concept of resilience: A bibliometric analysis of the emergency and disaster management literature. *Local Government Studies*, DOI: 10.1080/03003930.2018.1541796.

Environment Agency (2019) Draft National Flood and Coastal Risk Management Strategy for England. Environment Agency, Bristol.

Folke, Carl. (2006) Resilience: The emergence of a perspective for social–ecological systems analyses. *Global Environmental Change* 16 (3): 253–267.

Frazier, T.G. (2012) Selection of Scale in vulnerability and resilience assessments. *Journal of Geography and Natural Disasters* 2 (3). DOI: 10.4172/2167-0587.100e10.

Gunderson, L.H. and Holling, C.S. Eds. (2002) Panarchy: Understanding transformations n human and natural systems. Island Press, Washington D.C.

Handmer, J. and R. Dovers (1996) A typology of resilience: Rethinking institutions for sustainable development. *Organisation and Environment* 9 (4): 482–511.

Heng, C., Lam, N.S.N, Qiang, Y., Lei, Z., Correll, R.M. and V. Mihunov (2018) A synthesis of disaster resilience measurement methods and indices. *International Journal of Disaster Risk Reduction* 31: 844–885. DOI: 10.1016/j.ijdr.2018.07.015.

Holling, C.S. (1973). Resilience and stability of ecological systems. *Annual Review of Ecology and Systematics*, 1–23.

Holling, C.S. (1986). "The Resilience of Terrestrial Ecosystems." In W.C. Clark and R.E. Munn, Editors. Sustainable Development of the Biosphere. Cambridge University Press, Cambridge, UK, 292–317.

Kates, R.W. (1962) Hazard and choice perception in flood plain management. Dept. of Geography, University of Chicago Working Paper 78, Chicago.

Keating, A., Campbell, K., Szoenyi, M., McQuistan, C., Nash, D. and M. Burrer (2017) Development and testing of a community flood resilience measurement tool. *Natural Hazard Earth Systems Science* 17: 77–101.

Keck, M. and P. Sakdapolrak (2013) What is social resilience? Lessons learned and ways forward. *Erdkunde*, January–March, 5–19.

Koliou, M., van de Lindt, J.W., McAllister, T.P., Ellingwood, B.R., Dillard, M and H. Cutler (2018) State of the research in community resilience: Progress and challenges. *Sustainable and Resilient Infrastructure*.

Kuhn, T.S. (1970) The Structure of Scientific Revolutions, Vol. 2 of International Encyclopedia of Unified Science, 5th Edn. The University of Chicago Press, Chicago, IL.

Kyriazis, P. and S. Argyroudis (2013) Systemic seismic vulnerability and loss assessment: Validation studies. Systemic Seismic Vulnerability and Risk Analysis for Buildings, Lifeline Networks and Infrastructures Safety Gain (SYNER-G) D8.12 - Report 6, Seventh Framework Programme. Brussels: European Commission.

Lovell, E., Bahadur, A., Tanner, T. and H. Morsi (2016) Resilience: The Big Picture. Overseas Development Institute, London, UK.

Macdonald, N., Chester, D., Sangster, H., Todd, B. and J. Hooke. (2011) The significance of Gilbert F. White's 1945 paper 'Human adjustment to floods' in the development of risk and hazard management. *Progress in Physical Geography* 1–9. DOI: 10.1177/0309133311414607.

Martin, Ron. (2012). Regional economic resilience, hysteresis and recessionary shocks. *Journal of Economic Geography* 12 (1): 1–32.

Menoni, S., Modaressi, H., Schneiderbauer, S., Kienberger, S. and P. Zeil (2013) Risk Research ENSUREingtoMOVEahead. A cooperative paper based on the results of the projects ENSURE and MOVE, Bruxelles: European Commission.

Olsson, L., Thoren, H., Persson, J. and O'Byrne, D. (2015) Why resilience is unappealing to social science: Theoretical and empirical investigations of the scientific use of resilience. *Science Advances*, 22 May, 1 (4) e1400217 DOI: 10.1126/sciadv.1400217.

Rogers, E. (1962) Diffusion of Innovations. The Free Press, New York.

Simms, J.R.Z. (2017) "Why would I live anyplace else?": Resilience, sense of place, and possibilities of migration in coastal Louisiana. *Journal of Coastal Research*. Coastal Education and Research Foundation 332, 408–420. DOI: 10.2112/jcoastres-d-15-00193.1.

Sudemeier-Rieux, K. (2014) Resilience – an emerging paradigm of danger or of hope? *Disaster Prevention and Management* 23 (1) 67–80.

UN (United Nations) (2015a) Framework Convention on Climate Change. The Paris Agreement.

UN (United Nations) (2015b) Transforming our world: The 2030 Agenda for Sustainable Development. Resolution adopted by the General Assembly on 25 September 2015. UN, New York.

UNDRR United Nations Office for Disaster Risk Reduction) (2015) Sendai Framework for Disaster Risk Reduction 2015 – 20130. UN, New York.

UNISDR (2012) Making Cities Resilient. My city is getting ready! UNISDR, New York.

UNIDNDR (1999) Final report of the Scientific and Technical Committee of the International Decade for Natural Disaster Reduction, Report of the Secretary-General, Addendum. General Assembly Economic and Social Council. UN, New York.

Walker, B., Crawford, S., Holling, S., Carpenter, R. and A. Kinzig. (2004) Resilience, adaptability and transformability in social–ecological systems. *Ecology and Society* 9 (2): 5.

White, G.F. (1945) Human Adjustment to Floods. Research Paper 29. Chicago: Department of Geography, University of Chicago.

Zhou, H., Wang, J., Wan, J. and J. Huicong (2010) Resilience to natural hazards: A geographic perspective. *Natural Hazards* 53: 21–41. DOI:10.1007/s11069-009-9407-y.

Using vulnerability and resilience concepts to advance climate change adaptation

Erin P. Joakim, Linda Mortsch and Greg Oulahen

Adaptation is necessary if we are to minimize risks associated with climate change impacts. Vulnerability and resilience are two important concepts in the literature on hazards and climate change but have been used in a variety of ways to investigate human interaction with a hazardous environment. The result is widespread adoption of the terms but confusion about their relationship and how best they can advance work on climate change adaptation. This paper critically reviews the different understandings of the concepts and how they relate, and then proposes a framework that integrates vulnerability and resilience in order to advance adaptation thinking, planning and implementation. The paper concludes with a description of how the framework will apply findings on unequal social vulnerability to inform adaptation options that increase resilience in coastal cities.

1. Introduction

Climate change adaptation refers to the discrete actions taken to adjust to actual or expected climate and its effects, and seeks to moderate harm or exploit beneficial opportunities in human systems (Agard et al., 2014). Along with mitigation, adaptation is a cornerstone of responding to the issue of climate change. It is a fundamental and necessary response if we are to ameliorate the impacts associated with a changing climate and a future that is likely to have increasingly frequent and severe climate-related hazards. The overall aim of adaptation is ' ... to maintain and increase the resilience and reduce the vulnerability of ecosystems and people in the face of the adverse effects of climate change' (Agard et al., 2014, p. 11). From this perspective, adaptation actions contribute to reducing vulnerability and building resilience within the context of a changing climate (IPCC, 2014).

While there is a significant amount of planning being conducted to develop and mainstream adaptation policy, an emerging challenge is that this planning can often be considered incremental, primarily local and lacking a broader strategic approach and consideration of ideas of transformation and change (IPCC, 2014). While the concepts of vulnerability and resilience may be considered during adaptation planning, an overarching framework to guide adaptation thinking is often neglected. Although resilience may be used to describe or frame adaptation activities (i.e. that the adaptation option helps move the system toward resilience), a description of what resilience means and what it entails is often lacking. At a time when there is increasingly popular use of the term resilience, it is important to critically examine how this concept can be incorporated in a framework to advance climate change adaptation thinking and help make it a powerful agent for change. While there is a strong definitional, methodological and empirical understanding of vulnerability, the relationship between vulnerability and resilience, and how it contributes to framing adaptation, is unclear (Cannon & Müller-Mahn, 2010). Furthermore, there is often confusion about what these terms mean; this may stem from the fact that the terms have been defined and used across disciplines and stakeholders, leading to ambiguity about what they represent and how they can be applied (Cardona et al., 2012; Hewitson et al., 2014).

This paper critically examines how integrating the concepts of vulnerability and resilience may be useful for developing a framework for guiding climate change adaptation activities. To begin, the paper reviews how vulnerability and resilience have been interpreted in both the hazards and climate change literature, followed by an exploration of the nature of the relationship between the two concepts. The paper then moves on to a discussion of how these concepts can frame adaptation thinking, planning and implementation. The overall argument of the paper is to suggest that incorporating both concepts of vulnerability and resilience can be useful for framing adaptation; vulnerability concepts are useful for defining existing political–economic structural problems that contribute to unequal risk, whereas resilience offers the potential to identify and clarify solutions and move adaptation forward.

2. Vulnerability to climate-related hazards and stressors

The concept of vulnerability has been used in a variety of disciplines; thus there are different understandings and approaches that have been used over the past 40 years (Birkmann, 2006). The term was historically used by the Romans to describe a wounded soldier on the battlefield – indicating that the soldier was at risk for future attack (Adger, 2000). This implies that the vulnerable person is defined by their existent state, regardless of any future threat they may face or any adaptations they may take. While this may have been the historical meaning, in the context of climate change hazards and adaptation, the term vulnerability has not always been used in this way. Various disciplines have used and understood the concept in different ways; due to the various uses and understandings of the term, Füssel and Klein (2006, p. 305) have identified several ambiguities that have arisen from epistemic and semantic differences, including:

(1) Whether vulnerability is the starting point, intermediate element or outcome when conducting an assessment;
(2) Whether vulnerability should be defined in relation to an external stressor or to an undesirable outcome;
(3) Whether vulnerability is an inherent property of a system or dependent upon a specific circumstance, scenario and responses;
(4) Whether vulnerability is a static or dynamic concept.

A number of academics have reviewed the vulnerability literature and outlined a variety of approaches. Villágran's (2006) review organized vulnerability conceptualizations into three categories: physical exposure, pre-existing condition and benchmark vulnerability. Eakin and Luers (2006) similarly highlighted three categories of vulnerability conceptualizations, focusing on disciplinary approaches: the risk/hazard perspective, the political economy/ecology approach and the ecological resilience approach. Füssel (2007) divides vulnerability approaches into five categories, including the risk hazard perspective, political economy approaches, the pressure-and-release (PAR) model, the integrated approach and the resilience perspective. More simplistically, Kelly and Adger (2000) divided approaches into starting point and end point understandings of vulnerability. This is similar to O'Brien, Eriksen, Nygaard, and Schjolden's (2007) dual approach of outcome and contextual vulnerability categorizations.

Each of these reviews focuses on a different approach for categorizing the diverse ways in which vulnerability has been defined and used, with some authors reviewing vulnerability conceptualizations from a broad disciplinary perspective, while others focus almost exclusively on the climate change literature. While each of these pieces offers important understanding and synthesis of the vulnerability literature, the discussion in this paper provides a bridge among these different reviews. As vulnerability conceptualizations in the climate change literature have been influenced by other disciplines, this review includes a broad approach for classifying vulnerability approaches, including disciplinary perspectives from the climate change, hazards and other related literatures. The vulnerability categorizations mentioned above are organized into four main groups in order to synthesize the variety of vulnerability reviews. This synthesized review will then be useful for outlining how vulnerability concepts link to climate change and can be used to frame adaptation thinking.

2.1. *Vulnerability as a threshold*

The first understanding of vulnerability explores the concept as the probability of a person, community or system reaching or surpassing a certain benchmark or threshold, more commonly found in the food security literature (Villagrán, 2006). This approach is also commonly found in the engineering sciences, with vulnerability viewed as the threshold at which physical structures are likely to fail (e.g. dike failure and building collapse) or protective measures exceeded (e.g. overtopping of flood protective measures). This approach is less-often used within the climate change community, although there is a relation to concepts of tipping points; tipping points are defined as the threshold at which abrupt and irreversible harm occurs and is more commonly associated with natural as opposed to human systems (Agard et al., 2014).

2.2. *Vulnerability as exposure to hazards*

The second approach defines vulnerability in relation to exposure to particular hazards or stressors. This is highlighted as the traditional risk and hazards approach identified by both Eakin and Luers (2006) and Füssel (2007). In this approach, hazards and disasters, including climate-related hazards and stressors, are viewed as purely physical events, where impacted populations are seen as passive actors in the risk process. Vulnerability is defined specifically as a direct consequence of exposure, and there is often little differentiation between the social and spatial characteristics of impacted areas (Villágran, 2006). Within this approach, emphasis is predominantly related to physical systems, such as the built environment (Füssel, 2007). Furthermore, this approach tends to focus on the immediate impact of the hazard or stress, as opposed to issues that may arise after exposure. Cannon (2000) further argues that this approach has a tendency toward technocratic, 'hard science' strategies for responding and adapting to hazards and climate-related stressors, with a focus on engineering or management solutions. As this understanding of vulnerability relates only to exposure, there is limited recognition of the role of socio-economic systems that differentially allocate risk and contribute to populations that have a lowered ability to cope and respond.

2.3. *Vulnerability as pre-existing condition*

The third approach conceptualizes vulnerability as a particular condition or state of a system before a hazard or climate-related stressor occurs, often described in terms of criteria such as susceptibility, limitations, incapacities or deficiencies, for example, the incapacity to resist the impact of a hazard or climate change (resistance) and the incapacity to cope (coping capacity) (Villagran, 2006). This approach is the closest in line with the historical meaning of the word identified by Kelly and Adger (2000) above, whereby vulnerability is seen as a pre-existing condition.

This understanding of vulnerability is defined by Eakin and Luers (2006) and Füssel (2007) as the political economy/political ecology approach. Here, vulnerability is a pre-existing condition that is an outcome of the social, political and economic processes that create different levels of capacity among individuals, groups and communities to resist, respond and recover from environmental stresses and shocks, including climate change (Anderson & Woodrow, 1998; Blaikie, Cannon, Davis, & Wisner, 1994; Cannon, 2000; Cutter, 1996; Hewitt, 1997). This approach builds on the work of development and hazards researchers such as Hewitt (1983, 1997), Sen's entitlement theory (1981), Blaikie et al. (1994), Pelling (2003), and Wisner, Blaikie, Cannon, and Davis (2004). One of the most well-known vulnerability models, the PAR model developed by Blaikie et al. (1994) and refined by Wisner et al. (2004) epitomizes the political ecology approach for conceptualizing vulnerability as a pre-existing condition. This model focuses on examining the root causes of vulnerability, emphasizing the role of widespread, large-scale processes that impact the distribution of resources and are a reflection of the distribution of power in a society (Blaikie et al., 1994). Although Füssel (2007) has identified the PAR model as a separate approach for understanding vulnerability, we believe this model is more appropriately categorized within the 'vulnerability as pre-existing condition' category.

The 'vulnerability as a pre-existing condition' approach is also in line with the starting point understandings of vulnerability outlined in the climate change literature by Kelly and Adger (2000) and O'Brien et al. (2007). Starting point approaches view vulnerability as the lack of capacity to adapt, and are, therefore, also inherently related to aspects of resilience. Thus, this approach could also be seen as incorporating the resilience perspectives highlighted by Eakin and Luer's (2006) ecological resilience classification as well as Füssel's (2007) resilience perspective. Although these resilience approaches to understanding vulnerability are included here under the pre-existing condition category, the nature of the relationship between vulnerability and resilience is discussed in further detail below.

2.4. *Vulnerability as an outcome*

The final approach sees vulnerability as an outcome or residual generated after any adaptation has taken place (O'Brien et al., 2007). This approach is more specifically found in the climate change literature and defined as 'end-point' vulnerability by Kelly and Adger (2000). End-point approaches see vulnerability as the adverse consequences that remain after scenarios of future climate change and the biophysical impacts have been established along with the identification of adaptive strategies. Thus, vulnerability is understood as the final outcome, or the net impact, after a series of projections and estimates of changes and adaptations have taken place (Kelly & Adger, 2000). In this sense,

> vulnerability here summarizes the net impact of the climate problem, and can be represented quantitatively as a monetary cost or as a change in yield or flow, human mortality, ecosystem damage or qualitatively as a description of a relative or comparative change. (IPCC, 2007; O'Brien et al., 2007, p. 2)

According to O'Brien et al. (2007), this approach highlights 'outcome vulnerability', where vulnerability is seen from a linear and scientific framing, progressing from expected impacts of climate change, offset by the adaptation measures and strategies implemented (O'Brien et al., 2007). This approach is inherently focused on projections of the future, with limited emphasis on current vulnerability conditions.

2.5. *Summary of vulnerability approaches*

Through a review of the various uses and understandings of the concept of vulnerability, four different categorizations of vulnerability have been presented to demonstrate how various disciplines have used the term. These approaches are highlighted in Figures 1(a)–(d); these figures are meant to capture the essence of each interpretation of vulnerability and are not meant to depict how vulnerability levels will change over time.

Although each conceptualization of vulnerability has strengths and weaknesses, and the suitability of each approach is likely dependent on the context of the research and questions asked, we

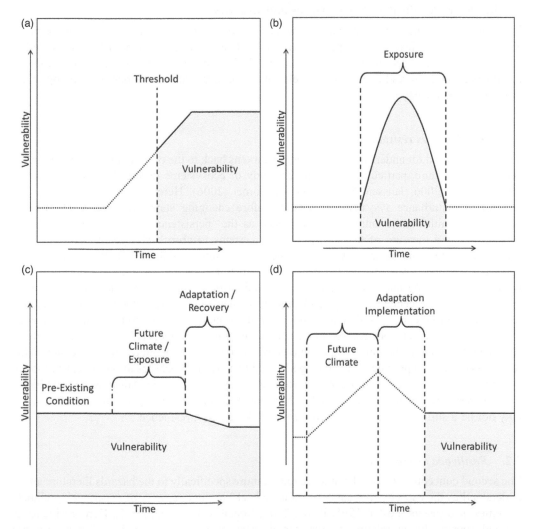

Figure 1. (a) Vulnerability as threshold. (b) Vulnerability as exposure. (c) Vulnerability as pre-existing condition. (d) Vulnerability as outcome after adaptation.
Source: Author(s).

suggest that understanding vulnerability as a pre-existing condition is most suitable for framing climate change adaptation. Incorporation of this understanding of vulnerability in adaptation thinking, planning and implementation allows for a 'socio-institutional' approach which incorporates both small- and large-scale processes that contribute to vulnerabilities in specific places (see Downing, 2012). An in-depth understanding of the vulnerabilities of the area under study can help to highlight the complexity of the accumulation of risk and how social, economic and political processes, including institutional and decision-making processes and the policy context, can contribute to the development of vulnerability to climate change. As noted in Lavell et al. (2012, p. 27), 'effective management of climate change requires an understanding of the diverse ways in which social processes and development pathways shape risk'. An in-depth understanding of vulnerability can help to achieve this, which in turn can help define key areas to target for adaptation strategies.

3.   Resilience to climate-related hazards and stressors

Resilience has recently gained popular traction as a response to both climate change and hazards. Similar to vulnerability, the concept of resilience has been used in a variety of literatures, and as such, has a variety of definitions (Lavell et al., 2012). This section outlines three main approaches or understandings of resilience and their relevance to climate change adaptation thinking, planning and implementation.

3.1.   *Resilience as resistance*

The first approach for understanding resilience hearkens back to the original use of the term in the ecological literature, particularly through the study of ecosystems during the 1960s and early 1970s (Folke, 2006; Janssen, Schoon, Ke, & Borner, 2006). Here, resilience focused on the amount of disturbance a system can absorb before changing state (Maguire & Cartwright, 2008). This initial understanding saw resilience as the 'persistence of relationships within a system and is a measure of the ability of these systems to absorb change of state variables, driving variables, and parameters, and still persist' (Holling, 1973, p. 17). In the ecosystems approach, resilience was seen as the capacity of a natural system to endure stress and change without transforming into a new state or new equilibrium. Applying this approach to human systems within the context of climate change, resilience can be understood as the capacity to resist and absorb the impacts of climate shifts, variations and extremes without leading to significant impacts on human systems. Yet, in the context of climate change, ecological systems may be significantly altered in the future and traditional ways of life and economic activities reliant on the resources the system provides may no longer be viable. Critical thresholds in ecological and social systems may be reached, requiring modification and adaptation (Nelson, Adger, & Brown, 2007). This suggests that defining resilience solely based on resistance to the impacts of climatic changes may not be a sufficiently holistic approach for thinking about resilience.

3.2.   *Resilience as recovery*

The second conceptualization of resilience relates more specifically to the hazards literature and is understood as the capacity to recover or 'bounce back' in the aftermath of experiencing climate extremes or disasters (Foster, 1995; Paton, 2006; Ronan & Johnston, 2005). Here, resilience is defined specifically in relation to the post-disaster or post-crisis conditions and response (Rose, 2007). This approach originally assumed that systems could recover and return to their original state (Alesch, 2004; Paton, 2006). More recently, it has been recognized that communities rarely return to the pre-disaster state, as a disaster results in changes to the physical, economic, social and psychological reality of societal life (Kenny, Clarke, Fanany, & Kingsbury, 2010). As households and communities recover from extreme events, a new understanding and meaning of 'normal' is continuously remade and reworked throughout this process (Ride & Bretherton, 2011). Thus, this approach would be better understood as returning to a desirable level of functioning. This recognition is increasingly relevant for long-term climate change, as it is expected that communities will not return to their previous form.

Within this perspective, Tierney (2014) distinguishes between inherent and adaptive resilience to help cope with the aftermath of a crisis situation and facilitate returning to some degree of functioning. Inherent resilience refers to the 'ordinary' and pre-existing capacities that individuals, households and communities possess that can be implemented in the post-disaster or post-crisis situation. Adaptive resilience, on the other hand, refers to 'the ability in crisis situations to maintain function on the basis of ingenuity or extra effort' (Rose, 2007, p. 385). This suggests that within the recovery approach for thinking about resilience, adaptive actions can help facilitate the development of resilience.

While this resilience approach often focuses on the speed of recovery, ideas related to the capacity to absorb the impacts of stresses and shocks are often incorporated as well. An example is the United Nations International Strategy for Disaster Reduction definition, where resilience is seen as

the ability of a system, community or society exposed to hazards to resist, absorb, accommodate to and recover from the effects of a hazard in a timely and efficient manner, including through the preservation and restoration of its essential basic structures and functions. (UNISDR, 2009)

This suggests that resilience incorporates some level of capacity to absorb impacts, while at the same time being able to return to functioning quickly when impacts are experienced (National Research Council, 2011).

Many definitions of resilience within the hazards literature focus on the ability to quickly return to 'normal' or functional operations, although focusing solely on the ability to bounce back often assumes that resilient systems can achieve a state of equilibrium, whereas human and natural systems are more accurately seen as chaotic and non-equilibrating (Birkmann & Wisner, 2006). Without consideration of adaptive resilience, returning communities to their pre-disaster level of functioning is recognized as increasingly insufficient as it is likely returning them to a condition of vulnerability to future events (Lavell et al., 2012). Further, while the recovery approach to resilience is important to consider within the context of extreme events that are projected to become more frequent and severe due to climate change, this understanding is less useful when considering longer-term, slow onset and chronic hazards associated with a changing climate.

3.3. *Resilience as creative transformation*

This leads to the final approach for understanding resilience, where the concept relates to the idea of transformation, and increasing the functionality of the community after a climatic shift or extreme event. In this sense, resilience is the process of 'adapting to new circumstances and learning from the disaster [or climate] experience' to create communities that have achieved greater resiliency and functionality (Adger, 2000; Maguire & Hagan, 2007, p. 17). In this approach, resilience concepts incorporate a measure of the adaptive and transformational capacity of individuals, groups and communities (Folke et al., 2010; Magis, 2010).

This approach leads to a growing body of literature which focuses not only on recovery resilience, or returning the community to its previous level of functionality, but also as a tool for promoting positive growth and improving overall well-being (Kulig, 2000; Kumpfer, 1999; Munich Re, 2007; Paton, 2006). Folke (2006, p. 253) focuses on the positive aspects of climatic shifts and extremes, viewing them as having the 'potential to create opportunity for doing new things, for innovation and for development'. In other words, climate change can be viewed as a catalyst for learning, transformation and growth in the community (Berkes, 2007; Kumpfer, 1999). This view of resilience 'accepts that change is inevitable, rather than seeing change as a "stressor" from which a community needs to recover its original state' (Maguire & Cartwright, 2008, p. 5).

While some resilience scholars argue that this transformative resilience approach is more effectively considered under the umbrella of adaptation and adaptive capacity (see Rose, 2011), many scholars incorporate some aspect of the capacity to learn, adapt and change within their resilience definitions. This suggests that although adaptation actions are often viewed as promoting resilience, a more holistic interpretation recognizes the complex inter-linkages between adaptation and resilience; implementation of adaptation requires some degree of resilience, while adaptation also helps promote and develop resilience (MacClune & Optiz-Stapleton, 2012). Incorporating ideas of adaptation and transformation when defining resilience is also useful in the context of climate change, particularly for expected long-term, chronic hazards such as sea-level rise; here, it is less likely communities will need to 'recover', instead requiring some degree of transformation.

3.4. *Summary of resilience approaches*

The above sections have reviewed how the concept of resilience has been characterized in the literature. Figure 2 summarizes the various approaches to thinking about resilience. These figures depict the differences in the three conceptualizations of resilience as opposed to clearly delineating how resilience levels will change over time.

There is a growing literature that conceptualizes resilience using all three of the dimensions outlined above. For example, the Resilience Alliance understands the concept of resilience along three similar dimensions: the ability to absorb, the degree of self-organization, and the capacity for

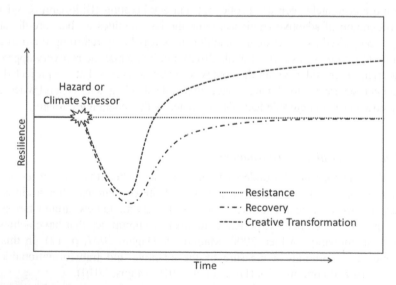

Figure 2.   Conceptualizations of resilience.
Source: Author(s).

learning and adaptation (Kuhlicke, 2010). Similarly, Pelling (2011) characterizes resilience as maintaining the status quo by refining actions, transition as incremental change to achieve some goals and transformation as radical regime change. In the IPCC fifth Assessment Report, Working Group II has adopted the Arctic Council (2013) definition of resilience, whereby it is defined as 'the capacity of a social-ecological system to cope with a hazardous event or disturbance, responding or reorganizing in ways that maintain its essential function, identity, and structure, while also maintaining the capacity for adaptation, learning, and transformation' (Agard et al., 2014, p. 23). This represents a shift from the IPCC third Assessment Report approach to resilience that focused almost exclusively on the capacity to absorb the impacts of climatic shifts, stressors and extremes, and the IPCC fourth Assessment Report which further integrated a focus on adaptation, but not transformation (IPCC, 2007; Levina & Tirpak, 2006). This increasingly comprehensive understanding of resilience recognizes that responding to climate change requires a broad range of approaches and strategies in order to prepare for the uncertainties of the future. Thus, we suggest that adopting an understanding of resilience that incorporates all three perspectives allows for a sufficiently holistic framework for adaptation thinking, planning and implementation.

While resilience thinking has gained considerable interest over the past few years, there has been some criticism related to transferring an ecological, systems-based concept onto socially, politically and economically constructed contexts. The systems approach typically used to explore resilience has a tendency to ignore power relations in society and views many climate change adaptation interventions as neutral processes. This results in a de-politicization of the processes that create vulnerability and risk (Kuhlicke, 2010). Cannon and Müller-Mahn (2010, p. 623) warn that the shift toward using the term resilience 'is dangerous because it is removing the inherently power-related connotations of vulnerability and is capable of doing the same to the process of adaptation'. They argue that a focus on resilience overlooks theorizations of how socioeconomic structures produce unequal risk. Policy measures are thus prone to be technocratic and depolitical. Climate change adaptation work has likewise been critiqued for lacking a critical analysis of political economic structures that dictate unequal access to resources, instead focusing on 'conditions' necessary for adaptive governance, like 'policy will' (Peet, Robbins, & Watts, 2011, p. 10). Miller et al. (2010), however, believe that the understandings of structural mechanisms that the vulnerability approach brings can effectively add nuance to the view of socio-ecological systems when integrated with the concept of resilience.

4. Understanding the relationship between vulnerability and resilience

In order to overcome the criticisms associated with both vulnerability and resilience perspectives, there is a need to integrate concepts of vulnerability with resilience in order to frame adaptation thinking, planning and implementation. From the interpretations previously recommended, the vulnerability perspective provides an understanding of the social, economic, historical, cultural and political processes that lead to increased risk, whereas the resilience perspective explores the opportunities for moving forward and reducing the impacts of shocks and stresses associated with climate change. Thus, vulnerability and resilience concepts complement each other, in that vulnerability explores 'the nature and extent of the adverse effects that may occur ... [while resilience provides] a consideration of the characteristics or conditions that help ameliorate or mitigate negative impacts' (Lavell et al., 2012, p. 33).

To this point, vulnerability and resilience have been discussed separately, although an emerging literature recognizes the complexity and similarities that exists between these two concepts:

> Vulnerability research and resilience research have common elements of interest – the shocks and stresses experienced by the social-ecological system, the response of the system, and the capacity

for adaptive action. The points of convergence are more numerous and more fundamental than the points of divergence. (Adger, 2006, p. 269)

Integrating resilience and vulnerability concepts is important for three key reasons:

> it helps assess hazards holistically in coupled human-environment systems; it stresses the ability of a system to deal with a hazard, absorbing the disturbance or adapting to it; and it helps explore policy options for dealing with uncertainty and future change. (Berkes, 2007; Haque & Etkin, 2007, p. 279)

Although the interconnections between vulnerability and resilience have been identified, there is some discrepancy regarding the nature of the relationship between the two concepts.

There are three main ways in which the relationship between vulnerability and resilience has been defined that are important within the context of climate change adaptation. Each of these approaches is outlined below, followed by an overview of the interpretation used to inform our proposed adaptation framework.

4.1.   *Vulnerability and resilience as a continuum*

The first approach defines vulnerability and resilience as the positive and negative ends of a singular concept that can be represented along a continuum (Adger, 2000; Berkes, 2007; Birkmann & Wisner, 2006). This is particularly true for the end-point definitions of vulnerability, where coping and adaptive capacity is viewed as an inherent or inverse part of vulnerability (O'Brien et al., 2007; Yohe & Tol, 2002). This approach indicates that an increase in resilience would reduce vulnerability and vice versa (Figure 3).

This continuum approach is more commonly highlighted in the climate change literature, through the definitions of vulnerability previously adopted by the IPCC. In this approach, vulnerability is seen as a function of exposure, sensitivity and adaptive capacity (Yohe & Tol, 2002). From this perspective, Smit and Wandel (2006) suggest that the concept of vulnerability explicitly incorporates, or is reflective of, the resilience of that system: vulnerability is not viewed as separate from resilience, but as an inherent part of vulnerability (Joakim, 2008; King & MacGregor, 2000; Yohe & Tol, 2002). For example, Turner et al. (2003) identify resilience as one contributor, along with exposure and sensitivity, to vulnerability in their conceptual framework for vulnerability analysis. This is similar to Adger (2000), who noted that resilience is a 'loose antonym' for vulnerability, in that resilience increases capacity to cope with stress, whereas vulnerability defines the level of exposure to stress. Thus, the IPCC approach has often viewed vulnerability and resilience as the positive and negative aspects of a singular concept that can be represented along a continuum (see Barnett, Lambert, & Fry, 2008; Berkes, 2007; Birkmann & Wisner, 2006): when vulnerability is reduced, the level of resilience automatically increases (Kasperson & Kasperson, 2001).

4.2.   *Vulnerability and resilience as background and action conditions*

A more nuanced interpretation of the relationship between vulnerability and resilience views vulnerability as background conditions and resilience as actions taken to dampen losses by using

Vulnerability ⟵──────────────────────────────⟶ Resilience

Figure 3.   Vulnerability and resilience as opposite ends of one spectrum.
Source: Joakim (2013).

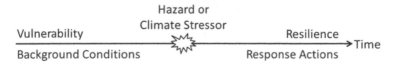

Figure 4.   Vulnerability and resilience as background and responsive actions.
Source: Author(s).

resources more efficiently or recovering more quickly (Rose & Krausmann, 2013). Conceptualizing vulnerability as background conditions acknowledges that vulnerability is a pre-existing condition, in line with our recommended interpretation of vulnerability. Understanding resilience as actions focuses on the resistance and recovery perspectives of resilience, thereby highlighting a temporal dimension to the relationship between vulnerability and resilience; vulnerability represents the pre-disaster conditions, whereas resilience represents the post-disaster response (Figure 4).

Understanding vulnerability and resilience as background and action conditions is more commonly used within the hazards and disaster discipline although applying this concept to chronic hazards expected with climate change could be useful for thinking about how vulnerability conditions influence resilient responses. Inclusion of the temporal component is an important consideration that is often missed in conceptualizations of the relationship between vulnerability and resilience (Christmann, Balgar, & Mahlkow, 2014). However, by relegating vulnerability to 'background conditions', the approach de-emphasizes the social, political and economic processes that contribute to the generation and perpetuation of vulnerabilities, instead focusing on neutral and de-politicized response actions.

4.3.   *Vulnerability and resilience as distinct concepts*

Conversely, the relationship between vulnerability and resilience can be defined as complex and 'process-oriented', where the concepts are seen as inherently linked, although distinct (Cutter et al., 2008; Maguire & Cartwright, 2008; Sapountzaki, 2012). As Mayunga (2007) and Klein, Nicholls, and Thomalla (2003) note, defining resilience as the opposite of vulnerability results in circular reasoning and provides limited new knowledge. The complexity of the relationship between vulnerability and resilience can also be observed in contexts where increased resilience can also lead to increased vulnerability. In the case of flood management, protective structures such as levees and insurance can limit the number of extreme events and help during recovery, although communities may become more vulnerable to future flood events through increased sense of security and further development on flood plains (Burby, 2001; Gunderson, 2010). Thus, as Buckle, Mars, and Smale (2000, p. 13) argue, resilience is not 'just the absence of vulnerability'. Instead, resilience should be understood through a multi-structural approach that encompasses broader concepts of adaptive capacity, exposure and the coupled human–environment system. In this way, resilience concepts focus on learning, re-organization and self-change, and it is understood that 'vulnerability features may co-exist with characteristics that improve adaptive capacity' (Sapountzaki, 2012, p. 1268).

This approach for understanding the relationship between vulnerability and resilience takes a process-oriented approach, conceptualizing the relationship between vulnerability and resilience along two separate continuums (*X–Y* axis), thereby creating a set of quadrants in which communities experience differential levels of resilience and vulnerability over time (Figure 5). In Figure 5, if all four hypothetical communities experienced similar exposure to a climatic extreme, community three would likely experience the highest level of impacts and the most

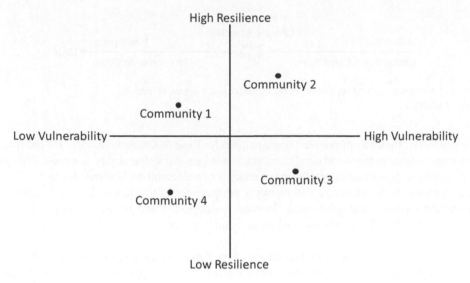

Figure 5.   Vulnerability and resilience as separate but interrelated concepts.
Source: Joakim (2013).

difficulty to recover and/or adapt due to high vulnerability and low levels of resilience. On the other hand, community two, although experiencing high levels of vulnerability, may have less difficulty recovering and adapting due to the ability to cope, adapt and transform following the extreme event. Community one approaches the ideal condition, with low levels of vulnerability and higher levels of resilience. While identifying characteristics of low vulnerability and high resilience communities is beyond the scope of the article, and likely highly contextual and scale-dependent, some commonly identified characteristics of resilience related to climate change include: flexible, diverse and redundant systems; responsive, resourceful and adaptable actors and stakeholders; and accessible, transparent, accountable institutions that facilitate the exchange of information (MacClune & Optiz-Stapleton, 2012).

This interpretation can also be applied to communities and households experiencing chronic or longer-term climate change impacts – communities may have some resilience and capacity to adapt and transform, although they could also be experiencing vulnerabilities simultaneously. This relates to Pelling's (2011) conceptualization of vulnerability and resilience as separate but related terms of adaptation pathways. A lack of adaptation results in vulnerability whereas positive adaptation pathways lead to resilience, transition or transformation. These adaptation pathways serve to change the level of risk experienced by individuals, households, communities and populations.

4.4.   *Summary of the relationship between vulnerability and resilience*

While we have outlined three different approaches for conceptualizing the relationship between vulnerability and resilience, we suggest that viewing them as separate and distinct concepts is the most useful approach for framing climate change adaptation. Understanding vulnerability and resilience as separate concepts explicitly recognizes the resilience of individuals, households and communities, highlighting the intrinsic capacities that may exist in areas impacted by climate change. Although the two-dimensional model suggests a somewhat static understanding of vulnerability and resilience, in reality, levels would continuously fluctuate over time, reflecting the dynamic nature of social, political and economic processes, as well as individual, household and community choices that impact vulnerability and resilience.

Viewing resilience and vulnerability as separate concepts has been supported by other research as well. In some cases, groups identified as vulnerable by planners and decision-makers were actually found to be more resilient during crisis events (Buckle et al., 2000; Ferrier, 2008; Handmer, 2003). Handmer (2003) noted that the elderly, generally identified as vulnerable due to age, mobility and health issues, were able to cope and adapt more effectively during the 1998 gas crisis in Victoria, Australia, than was expected by emergency managers. This was also true of some immigrant populations: their collective experiences, gained through life experiences such as the Great Depression and the Second World War, as well as refugee camps, provided them with the experience and coping strategies that other groups did not have (Buckle, 2001/2002; Buckle et al., 2000). This implies that government officials and researchers should assume that all populations (individuals and groups) have an inherent ability to cope, adapt and potentially transform on some level (Rose, 2007).

Although the relationship between vulnerability and resilience is presented along an axis, there may be some difficulty in the practical implementation of this conceptualization. As the relationship between these two concepts is rather complex, elements or themes of vulnerability and resilience may be difficult to separate along an X–Y axis. Furthermore, the contextualization of factors influencing these concepts suggests that factors selected for one context may not be suitable for other contexts (Birkmann, 2007). While there may be some difficulty in the implementation of this framework, we argue that it is a useful approach for framing adaptation thinking, planning and implementation.

5. Toward an adaptation framework

The argument of this paper is that integrating concepts of vulnerability and resilience provides a useful framework for climate change adaptation. Understanding vulnerability as a pre-existing condition generated through both larger-scale structural processes and smaller-scale agency provides an understanding of the processes, policies and actions that contribute to the development of risk. Incorporating a holistic resilience perspective focuses on how individuals, households and communities can build their capacity to resist, recover and transform, thus focusing on the elements of responding to climate stressors and perturbations. Further, viewing vulnerability and resilience as separate, but interrelated concepts recognizes the inherent complexities and potentially unanticipated feedbacks likely to occur within socio-ecological systems within the context of a changing climate. Recognizing these three key approaches provides an overarching framework for adaptation thinking, planning and implementation.

One important consideration for integrating the concepts of vulnerability and resilience into an adaptation framework includes the negativity often associated with the term vulnerability. While we have noted the importance of identifying vulnerabilities, much of the research on vulnerability has focused on negative qualities and characteristics. This has led some to argue for a recognition of the inherent resiliencies or capacities that individuals and groups can access when coping with climatic change and extreme events (Cannon, Twigg, & Rowell, 2003; Joakim, 2008). Cannon et al. (2003, p. 6) argue that the exclusive focus on vulnerability implies that the impacts of climate change 'always produce victims who have no strengths or capacities to resist and recover'. There may be a tendency to view impacted populations as 'helpless victims' who require the assistance of 'skilled' outsiders, particularly in a South–North context. Through the movement toward a resilience paradigm, this stereotype can be negated as individuals and communities are recognized as having capacities on which programs and resources can be built (IFRC, 1996). Furthermore, Buckle et al. (2000) note that lists of vulnerable populations only examine one dimension of vulnerability and lead to an understanding of vulnerability outside of place (i.e. the unique and complex interaction of various climate impacts), community facilities (actions that may have been taken to reduce vulnerabilities), time (ignoring variations that may occur seasonally, or through repeating events such as droughts) and independent of social and economic trends (economic downturns or recessions and/or political circumstances/upheavals). Thus, it is important to characterize the resilience aspects of individuals and communities in order to achieve a deeper understanding of coping and adaptive capacities.

Integrating the concepts of vulnerability and resilience also highlights the complexity and dynamism of the relationship between both concepts. In one of the author's research on disaster recovery after a large earthquake in Indonesia, it was found that the relationship between vulnerability and resilience was neither linear nor simple (see Joakim, 2013). In fact, the relationship between vulnerability and resilience was found to be highly complex, dependent on context, place and time. Various attributes of vulnerability had both positive and negative feedback on aspects of resilience. Approaching the relationship as two separate concepts explicitly recognized that even though at-risk and affected populations may experience some forms of vulnerability, they also have other forms of resilience to support coping, adaptation and transformation.

Examination of the case study villages used in the research in Indonesia highlights how the adaptation framework outlined can be applied. In one case study village, households experienced high vulnerability through poor housing construction, low income levels and low levels of education, all of which can be linked back to larger-scale structural processes, including political and economic ideologies that contribute to ongoing marginalization (Joakim, 2013). This decreased the capacity of households to resist the impacts of the earthquake, although reconstruction was facilitated through high levels of recovery and transformational resilience. Recovery resilience took the form of construction skills, community organizational capacity and cultural capitals that contributed to sharing of resources and mutual cooperation. The village also demonstrated transformational resilience through the use of social capital networks to support the implementation of a strategic five-year development plan in the aftermath of the earthquake disaster. Thus, this village typifies community two as highlighted in Figure 5 due to high levels of experienced vulnerability as well as high levels of resilience. While the example provided is from an earthquake study, a similar approach can be taken in responding to climate change hazards and stressors; the development of adaptation options can be implemented in conjunction with existing resiliencies in order to facilitate adaptation and transformation, resulting in improvements to overall welfare and living conditions.

A second example can be found in one of the other case study villages in Indonesia; here, vulnerability characteristics were similar to the first village, although there was a lack of social capital networks and capacity to implement learning and adaptation within the village. This village particularly lacked transformational resilience, thus limiting adaptation implementation after the earthquake disaster. Thus, this community typifies community three in Figure 5, due to the experience of high levels of vulnerability and low levels of resilience. Incorporating both vulnerability and resilience concepts in an adaptation framework allowed for an exploration of these complexities, providing a deeper understanding of the impacts of the hazard; this could similarly be applied within the context of climate change and an exploration of the capacities available to resist, respond, adapt and transform.

Through an integration of vulnerability and resilience as an adaptation framework, using the term vulnerability itself has come under critique. Handmer (2003) suggests that using the term vulnerability is unnecessarily negative, and proposes using resilience as a comprehensive term describing the overall vulnerabilities and resilience levels of individuals and groups. This point could be expanded upon to argue that adaptation could be used as a comprehensive term describing overall vulnerability reduction and resilience building strategies. This viewpoint presents a positive approach to adaptation and transformation, which may be more appealing to decision-makers and stakeholders:

We are all vulnerable, but we are also all resilient, and we all have adaptive capacity. Building resilience and capacity is politically appealing and a practical policy response to communities in difficulties – labeling or stigmatizing communities as particularly vulnerable or incapable is not usually politically appealing and is often strongly opposed by the communities involved. (Handmer, 2003, p. 60)

Mixed reactions to identifying unequal vulnerability were found to exist among municipal practitioners during a study conducted by the authors in Metro Vancouver, Canada (Oulahen, Mortsch, Tang, & Harford, in press). In seeking local practitioner views on measuring social vulnerability to flood hazards, some practitioners expressed resistance to the idea of identifying parts of their community as highly vulnerable, while other practitioners embraced the approach. Those who resisted the label did so because they feared negative identification could stigmatize neighborhoods found to be vulnerable. Practitioners who embraced the findings of differential vulnerability thought that neighborhoods identified as vulnerable could receive increased attention from decision-makers which could serve to reduce their vulnerability. In one municipality, the perceived stigmatization and negative implications of being identified 'vulnerable' in part led practitioners to not participate in the study. If a more positive framing was used, such as identifying resilience or adaptive capacity, perhaps the dialogue and engagement with practitioners from that municipality would have been more successful.

While Handmer (2003) highlighted the importance of a more positive approach, he also acknowledged the political and administrative usefulness of making distinctions between vulnerable and resilient groups. The downfall of using empowering language has also been noted through the use of this language as a pretext for limiting government and societal responsibility. Thus, there is still a need to fully develop an adaptation framework that is cognizant of the vulnerabilities and resiliencies in the community, although we still recognize the value of framing the issues in positive language that may be more likely to be embraced by locals.

6. Moving adaptation thinking forward

In order to move this adaptation framework forward, the researchers involved in the Coastal Cities at Risk (CCaR) project are working on an adaptation framework that incorporates concepts of vulnerability and resilience that will use Vancouver, Canada, as a case study. CCaR is a five-year multi-disciplinary project examining the vulnerabilities and resiliencies associated with climate-related hazards in major coastal cities. The project is particularly focused on furthering understanding of how we think about the adaptation and resilience of communities impacted by climate change hazards and using this information to further develop methods for measuring resilience and developing adaptation options. To this end, the project is using the work of the Multidisciplinary Center for Earthquake Engineering Research (MCEER) as a basis for further developing a resilience index. The MCEER research group developed a resilience framework which focuses on defining attributes and determinants of resilience through a focus on four key factors: robustness, redundancy, resourcefulness and rapidity (Bruneau et al., 2003; Tierney & Bruneau, 2007). These four components or four R's of resilience offer a method for understanding the pre-existing vulnerabilities that are likely to result in higher impacts as well as affect the ability of households and communities to resist the consequences of a changing climate. The four R's also provide a method for understanding how households and communities can recover quickly and efficiently following a hazardous event. Particularly, our research team is focused on applying the concept to social aspects of resilience, defining appropriate measurements for each of the four R's of resilience, and developing methods of understanding adaptability and transformational capacity. The four R's of resilience will be applied on both a short timescale for examining climate hazards as well as a long timescale for examining climate change stressors.

7. Conclusion

Incorporating vulnerability and resilience concepts into an integrative framework can advance thinking, planning and implementation of climate change adaptation. This framework can make such a contribution only if the strengths of each concept are applied, particularly if the knowledge gained from vulnerability research can be properly integrated into the concept of resilience. Adaptation is not a neutral process; if it is to be a process for achieving social change and transformation, then resilience, like vulnerability, must be understood as a construction dependent on socio-economic structural mechanisms. If not, the concept of resilience could serve only to entrench existing social structures that produce unequal risk. As planning for adaptation continues to gain traction among governments, organizations and stakeholders, the process will be well served to be guided by a framework that integrates the powerful concepts of vulnerability and resilience.

Disclosure statement

No potential conflict of interest was reported by the authors.

Funding

This collaboration was undertaken as part of work that is supported by the Coastal Cities at Risk: Building Capacity for Managing Climate Change in Coastal Megacities (CCaR) project. This project is supported by the Social Sciences and Humanities Research Council of Canada, the Natural Sciences and Engineering Research Council of Canada, the Canadian Institutes for Health Research and the International Development Research Council under the International Research Initiative on Adaptation to Climate Change (IRIACC).

References

Adger, W. (2000). Social and ecological resilience: Are they related? *Progress in Human Geography, 24*(3), 347–364.

Adger, W. (2006). Vulnerability. *Global Environmental Change, 16*(3), 268–281.

Agard, J., Schipper, L., Birkmann, J., Campos, M., Dubeux, C., Nojiri, Y., ... Bilir, E. (2014). *WGII AR5 glossary. IPCC 5th assessment report*. Retrieved from http://ipcc-wg2.gov/AR5/images/uploads/WGIIAR5-Glossary_FGD.pdf

Alesch, D. J. (2004, January). *Complex urban systems and extreme events: Towards a theory of disaster recovery*. 1st international conference of Urban Disaster Reduction, Kobe, Japan.

Anderson, M., & Woodrow, P. (1998). *Rising from the ashes: Development strategies in times of disaster*. London: Lynne Rienner Publishers.

Arctic Council. (2013). *Arctic resilience interim report 2013*. Stockholm: Stockholm Environment Institute and Stockholm Resilience Centre.

Barnett, J., Lambert, S., & Fry, I. (2008). The hazards of indicators: Insights from the environmental vulnerability index. *Annals of the Association of American Geographers, 98*(1), 102–119.

Berkes, F. (2007). Understanding uncertainty and reducing vulnerability: Lessons from resilience thinking. *Natural Hazards, 41*, 283–295.

Birkmann, J. (2006). Measuring vulnerability to promote disaster-resilient societies: Conceptual frameworks and definitions. In J. Birkmann (Ed.), *Measuring vulnerability to natural hazards: Towards disaster resilient society* (pp. 9–54). Tokyo: United Nations University Press.

Birkmann, J. (2007). Risk and vulnerability indicators at different scales: Applicability, usefulness and policy implications. *Environmental Hazards, 7*(1), 20–31.

Birkmann, J., & Wisner, B. (2006). *Measuring the un-measurable: The challenge of vulnerability*. Bonn: United Nations University, Institute for Environment and Human Security.

Blaikie, P., Cannon, T., Davis, I., & Wisner, B. (1994). *At risk: Natural hazards, people's vulnerability and disasters*. London: Routledge.

Bruneau, M., Chang, S. E., Eguchi, R. T., Lee, G. C., O'Rourke, T. D., Reinhorn, A. M., ... von Winterfeldt, D. (2003). A framework to quantitatively assess and enhance the seismic resilience of communities. *Earthquake Spectra, 19*(4), 733–752.

Buckle, P. (2001/2002). Managing community vulnerability in a wide area disaster. *The Australian Journal of Emergency Management, 16*(4), 13–18.

Buckle, P., Mars, G., & Smale, S. (2000). New approaches to assessing vulnerability and resilience. *Australian Journal of Emergency Management, 15*(2), 8–14.

Burby, R. J. (2001). Flood insurance and floodplain management: The US experience. *Environmental Hazards, 3*, 111–122.

Cannon, T. (2000). Vulnerability analysis and disasters. In D. Parker (Ed.), *Floods* (pp. 43–55). Oxford: Routledge.

Cannon, T., & Müller-Mahn, D. (2010). Vulnerability, resilience and development discourses in context of climate change. *Natural Hazards, 55*(3), 621–635.

Cannon, T., Twigg, J., & Rowell, J. (2003). *Social vulnerability, sustainable livelihoods and disasters* (Report for DFID's Conflict and Humanitarian Assistance Department and Sustainable Livelihood Office). London: DFID. Retrieved from ipcc-wg2.gov/njlite_download.php?id=6377

Cardona, O. D., van Aalst, M. K., Birkmann, J., Fordham, M., McGregor, G., Perez, R., ... Sinh, B. T. (2012). Determinants of risk: Exposure and vulnerability. In C. B. Field, V. Barros, T. F. Stocker, D. Qin, D. J. Dokken, K. L. Ebi, ... P. M. Midgley (Eds.), *Managing the risks of extreme events and disasters to advance climate change adaptation*. A special report of working groups I and II of the intergovernmental panel on climate change (IPCC) (pp. 65–108). Cambridge: Cambridge University Press.

Christmann, G. B., Balgar, K., & Mahlkow, N. (2014). Local constructions of vulnerability and resilience in the context of climate change. *A comparison of Lübeck and Rostock. Social Sciences, 3*, 142–159.

Cutter, S. L. (1996). Vulnerability to environmental hazards. *Progress in Human Geography, 20*(4), 529–539.

Cutter, S., Barnes, L., Berry, M., Burton, C., Evans, E., Tate, E., & Webb, J. (2008). A place-based model for understanding community resilience to natural disasters. *Global Environmental Change, 18*(4), 598–606.

Downing, T. E. (2012). Views of the frontiers in climate change adaptation economics. *WIREs Climate Change, 3*, 161–170.

Eakin, H., & Luers, A. (2006). Assessing the vulnerability of social-environmental systems. *Annual Review of Environmental Resources, 31*, 365–394.

Ferrier, N. (2008). *Fundamentals of emergency management: Preparedness*. Toronto, ON: Edmond Montgomery Publications.

Folke, C. (2006). Resilience: The emergence of a perspective for social-ecological systems analysis. *Global Environmental Change, 16*(3), 253–267.

Folke, C., Carpenter, S. R., Walker, B., Scheffer, M., Chapin, T., & Rockström, J. (2010). Resilience thinking: Integrating resilience, adaptability and transformability. *Ecology and Society, 15*(4), 20–28.

Foster, H. (1995). Disaster mitigation: The role of resilience. In D. Etkin (Ed.), *Proceedings of a tri-lateral workshop on natural hazards* (pp. 93–108). Toronto: Environmental Adaptation Research Group (EARG), Environment Canada.

Füssel, H. (2007). Vulnerability: A generally applicable conceptual framework for climate change research. *Global Environmental Change, 17*, 155–167.

Füssel, H., & Klein, R. (2006). Climate change vulnerability assessments: An evolution of conceptual thinking. *Climatic Change, 75*, 301–329.

Gunderson, L. (2010). Ecological and human community resilience in response to natural disasters. *Ecology and Society, 15*(2), 18–18.

Handmer, J. (2003). We are all vulnerable. *Australian Journal of Emergency Management, 18*(3), 55–60.

Haque, C., & Etkin, D. (2007). People and community as constituent parts of hazards: The significance of societal dimensions in hazards analysis. *Natural Hazards, 41*(2), 271–282.

Hewitt, K. (1983). The idea of calamity in a technocratic age. In K. Hewitt (Ed.), *Interpretations of calamity from the viewpoint of human ecology* (pp. 3–32). Boston, MA: Allen & Unwin.

Hewitt, K. (1997). *Regions of risk: A geographical introduction to disaster*. Essex: Addison Wesley Longman Limited.

Hewitson, B., Janetos, A. C., Carter, T. R., Giorgi, F., Jones, R. G., Kwon, W. T., ... van Aalst, M. K. (2014). Chapter 21, regional context. In C. B. Field, V. R. Barros, D. J. Dokken, K. J. Mach, M. D. Mastrandrea, T. E. Bilir, ... L. L. White (Eds.), *Climate change 2014: Impacts, adaptation, and vulnerability. Part A:*

Global and sectoral aspects. Contribution of working group II to the fifth assessment report of the inter-governmental panel on climate change (pp. x–xx). Cambridge: Cambridge University Press.

Holling, C. (1973). Resilience and stability of ecological systems. *Annual Review of Ecology and Systematics, 4*, 1–23.

IFRC. (1996). *Vulnerability and capacity assessment: Toolbox*. Geneva: International Federation of Red Cross and Red Crescent Societies.

IPCC. (2007). *Climate change 2007: Impacts, adaptation and vulnerability*. Contribution of working group II to the fourth assessment report of the intergovernmental panel on climate change (M. L. Parry, O. F. Canziani, J. P. Palutikof, P. J. van der Linden, & C. E. Hanson, Eds.). Cambridge: Cambridge University Press.

IPCC. (2014). Summary for policymakers. In C. B. Field, V. R. Barros, D. J. Dokken, K. J. Mach, M. D. Mastrandrea, T. E. Bilir, … L. L. White (Eds.), *Climate change 2014: Impacts, adaptation, and vulnerability. Part A: Global and sectoral aspects*. Contribution of working group II to the fifth assessment report of the intergovernmental panel on climate change (pp. 1–32). Cambridge: Cambridge University Press.

Janssen, M., Schoon, M. L., Ke, W., & Borner, K. (2006). Scholarly networks on resilience, vulnerability and adaptation within the human dimensions of global environmental change. *Global Environmental Change, 16*, 240–252.

Joakim, E. (2008). *Assessing the 'Hazards of Place' model of vulnerability: A case study of waterloo region* (Unpublished master's dissertation). Waterloo: Wilfrid Laurier University.

Joakim, E. (2013). *Resilient disaster recovery: A critical assessment of the 2006 Yogyakarta, Indonesia earthquake using a vulnerability, resilience and sustainable livelihoods framework* (Doctoral thesis). Waterloo: University of Waterloo. Retrieved from https://uwspace.uwaterloo.ca/bitstream/handle/10012/7315/Joakim_Erin.pdf?sequence=1

Kasperson, J., & Kasperson, R. (2001). *Global environmental risk*. London: United Nations University Press/Earthscan.

Kelly, P., & Adger, W. (2000). Theory and practice in assessing vulnerability to climate change and facilitating adaptation. *Climatic Change, 47*, 325–352.

Kenny, S., Clarke, M., Fanany, I., & Kingsbury, D. (2010). Deconstructing Aceh's reconstruction. In M. Clarke, I. Fanany, & S. Kenny (Eds.), *Post-disaster reconstruction: Lessons from Aceh* (pp. 3–28). London: Earthscan Ltd.

King, D., & MacGregor, C. (2000). Using social indicators to measure community vulnerability to natural hazards. *Australian Journal of Emergency Management, 15*(3), 52–57.

Klein, R., Nicholls, R., & Thomalla, F. (2003). Resilience to natural hazards: How useful is this concept? *Environmental Hazards, 5*, 35–45.

Kuhlicke, C. (2010). Resilience: A capacity and a myth: Findings from an in-depth case study in disaster management research. *Natural Hazards, 67*(1), 61–76.

Kulig, J. (2000). Community resilience: The potential for community health nursing theory development. *Public Health Nursing, 17*(5), 374–385.

Kumpfer, K. (1999). Factors and processes contributing to resilience. In M. Glantz & J. Johnson (Eds.), *Resilience and development: Positive life adaptations* (pp. 179–224). New York, NY: Kluwer Academic Publishers.

Lavell, A., Oppenheimer, M., Diop, C., Hess, J., Lempert, R., Li, J., … Myeong, S. (2012). Climate change: New dimensions in disaster risk, exposure, vulnerability, and resilience. In C. B. Field, V. Barros, T. F. Stocker, D. Qin, D. J. Dokken, K. L. Ebi, … P. M. Midgley (Eds.), *Managing the risks of extreme events and disasters to advance climate change adaptation*. A special report of working groups I and II of the intergovernmental panel on climate change (IPCC) (pp. 25–64). Cambridge: Cambridge University Press.

Levina, E., & Tirpak, D. (2006). *Key adaptation concepts and terms*. OECD/IEA project for the annex 1 expert group on the UNFCCC. Retrieved from http://www.oecd.org/environment/cc/36278739.pdf

MacClune, K., & Optiz-Stapleton, S. (2012). *Building urban resilience to climate change: What works where, and why*. Boulder, CO: Institute for Social and Environmental Transition-International.

Magis, K. (2010). Community resilience: An indicator of social sustainability. *Society and Natural Resources, 23*(5), 401–416.

Maguire, B., & Cartwright, S. (2008). *Assessing a community's capacity to manage change: A resilience approach to social assessment*. Canberra: Australian Government: Bureau of Rural Sciences.

Maguire, B., & Hagan, P. (2007). Disasters and communities: understanding social resilience. *Australian Journal of Emergency Management, 22*, 16–20.

Mayunga, J. (2007, July 22–28). *Understanding and applying the concept of community disaster resilience: A capital-based approach*. Draft of summary paper for the Summer Academy for Social Vulnerability and Resilience Building, Munich, Germany. Retrieved from http://www.ehs.unu.edu/file.php?id=296

Miller, F., Osbahr, H., Boyd, E., Thomalla, F., Bharwani, S., Ziervogel, G., ... Nelson, D. (2010). Resilience and vulnerability: Complementary or conflicting concepts? *Ecology and Society*, *15*(3), Article no. 11.

Munich Re. (2007). *Natural catastrophes 2006: Analyses, assessments, positions*. Munich: Munchener Ruck.

National Research Council. (2011). *National earthquake resilience: Research, implementation, and outreach*. Washington, DC: National Academy Press.

Nelson, D. R., Adger, N., & Brown, K. (2007). Adaptation to environmental change: Contributions of a resilience framework. *The Annual Review of Environment and Resources*, *32*, 395–419.

O'Brien, K., Eriksen, S., Nygaard, L., & Schjolden, A. (2007). Why different interpretations of vulnerability matter in climate change discourses. *Climate Policy*, *7*, 73–88.

Oulahen, G., Mortsch, L., Tang, K., & Harford, D. (In press). Unequal vulnerability to flood hazards: "Ground truthing" a social vulnerability index of five municipalities in Metro Vancouver, Canada. *Annals of the Association of American Geographers*.

Paton, D. (2006). Disaster resilience: Building capacity to co-exist with natural hazards. In D. Paton & D. Johnston (Eds.), *Disaster resilience: An integrated approach* (pp. 3–10). Springfield, IL: Charles C. Thomas Publisher Ltd.

Peet, R., Robbins, P., & Watts, M. (2011). *Global political ecology*. New York, NY: Routledge.

Pelling, M. (2003). *The vulnerability of cities: Natural disasters and social resilience*. London: Earthscan Publications Ltd.

Pelling, M. (2011). *Adaptation to climate change: From resilience to transformation*. London: Routledge.

Ride, A., & Bretherton, D. (2011). *Community resilience in natural disasters*. New York, NY: Palgrave Macmillan.

Ronan, K. R., & Johnston, D. M. (2005). *Promoting community resilience in disasters: The role for schools, youth and families*. New York, NY: Springer Science+Business Media.

Rose, A. (2007). Economic resilience to natural and man-made disasters: Multidisciplinary origins and contextual dimensions. *Environmental Hazards*, *7*(4), 383–398.

Rose, A. (2011). Resilience and sustainability in the face of disasters. *Environmental Innovation and Societal Transitions*, *1*, 96–100.

Rose, A., & Krausmann, E. (2013). An economic framework for the development of a resilience index for business recovery. *International Journal of Disaster Risk Reduction*, *5*, 73–83.

Sapountzaki, K. (2012). Vulnerability management by means of resilience. *Natural Hazards*, *60*, 1267–1285.

Sen, A. (1981). *Poverty and famines: An essay on entitlement and deprivation*. Oxford: Clarendon Press.

Smit, B., & Wandel, J. (2006). Adaptation, adaptive capacity and vulnerability. *Global Environmental Change*, *16*(3), 282–292.

Tierney, K. (2014). *The social roots of risk: Producing disasters, promoting resilience*. Stanford, CA: Stanford University Press.

Tierney, K., & Bruneau, M. (2007). Conceptualizing and measuring resilience: A key to disaster loss reduction. *TR News*, *250*, 14–17.

Turner, B. L., Kasperson, R. E., Matson, P. A., McCarthy, J. J., Christensen, L., Eckley, N., ... Schiller, A. (2003). A framework for vulnerability analysis in sustainability science. *Proceedings of the National Academy of Science*, *100*(14), 8074–8079.

UNISDR. (2009). *UNISDR terminology on disaster risk reduction*. Geneva: Author. Retrieved from http://www.unisdr.org/files/7817_UNISDRTerminologyEnglish.pdf

Villágran, J. C. (2006). *Vulnerability: A conceptual and methodological review*. Bonn: United Nations University: Institute for Environment and Human Security.

Wisner, B., Blaikie, P., Cannon, T., & Davis, I. (2004). *At risk: Natural hazards, people's vulnerability and disasters* (2nd ed.). London: Routledge.

Yohe, G., & Tol, R. S. (2002). Indicators for social and economic coping capacity – Moving toward a working definition of adaptive capacity. *Global Environmental Change*, *12*, 25–40.

Foundations of community disaster resilience: well-being, identity, services, and capitals

Scott B. Miles

If community disaster resilience is to mature into a robust and lasting area of research, methodologically facilitated dialogue between empirical observations and theory is necessary. However, methodological and empirical research has outpaced community disaster resilience theory. To address this gap, a theoretical framework called WISC is presented. WISC is named after four constructs of the framework: well-being, identity, services, and capitals. WISC relates the two concepts of community and infrastructure, broadly defined, to the four constructs it is named after. The 4 constructs are respectively defined by 29 variables. The broadest interpretation of WISC is that infrastructure supports and facilitates components of community within human settlements. Infrastructure is represented as combinations of capitals and services; community is represented by connections of identity and well-being. Ultimately, well-being of a community is dependent on that community's collective capital. But these two constructs are mediated by the intervening constructs of identity and services. WISC goes beyond existing frameworks by addressing essential elements of theory building that have been overlooked in the literature, while synthesizing other frameworks and areas of knowledge. WISC provides a powerful foundation for posing and evaluating hypotheses, improving data collection efforts, and, most importantly, enabling critical theory building.

1. Introduction

In 1986, Henry Quarantelli – pioneer of the social science of disasters and founder of the Disaster Research Center – stated during his presidential address to the International Research Committee on Disasters that the field of disaster research would only advance if scholars prioritize the construction of theoretical frameworks. Twenty years later, he published this sentiment while lamenting that theory building in disaster research had still barely begun (Quarantelli, 2005). Quarantelli stresses that the disaster research agenda for the twenty-first century 'need[s] more theory and abstract thinking and less mucking around in practical matters and concrete details. The heart of any scientific activity is basic knowledge and curiosity driven, and is not concerned with immediate outcomes or products' (2005, p. 329). Perry (2005) observes that there is a vast backlog of disaster research that requires integration in theoretically meaningful ways. 'Progress [in disaster research] will be measured when many [theoretical frameworks] exist – some complimentary some competing – and when researchers use them' (p. 324). Lindell (2013) cautions that the field of disaster research will move forward only when there is both inductive (building theory from data) and deductive (making projections about data based on theory) inquiry. Of course, both inductive and deductive research of disasters requires the existence of complimentary and competing theoretical frameworks.

Community disaster resilience is a relatively new focus of disaster researchers that is rapidly growing in popularity. Depending on the reference database searched, as many or more articles on the subject have been published in the past five years than all previous years. The popularity suggests that many disaster researchers believe that a deeper understanding of the resilience of communities to disasters will eventually benefit society. Considering the above imperatives for disaster research, construction of theoretical frameworks about community disaster resilience is needed if the subject is to be a robust and long-lasting subfield of disaster research that makes meaningful contributions to knowledge, as well as practice.

Unfortunately, to date there is little theoretical research on community resilience to disasters in the literature. The National Research Council report 'Disaster Resilience: A National Imperative' makes no mention of existing theory or the need for theoretical research in their recommendations to meet this national imperative (NRC, 2012). Instead, there is an urgency to develop indicators, metrics, and capacity-building interventions to help communities become more resilient (Gilbert, 2011; Jordan & Javernick-Will, 2012; Manyena, 2006). Clearly, methodological research and, to a lesser extent, empirical research have outpaced theory-building efforts (Brown & Westaway, 2011; Jordan & Javernick-Will, 2012; Kulig, Edge, Townshend, Lightfoot, & Reimer, 2013). Few published works about community disaster resilience explicitly discuss or develop theory (Liao, 2012; Norris, Stevens, Pfefferbaum, Wyche, & Pfefferbaum, 2008; Wilson, 2012). There does not appear to be any deductive research studies of the subject, where theoretically based hypotheses are empirically analyzed. Very few peer-reviewed studies even suggest the need for posing and evaluating hypotheses (Norris et al., 2008). This is true for closely related research areas that could be drawn upon (mitigation and recovery, in particular), which are also under-theorized and in need of detailed, testable theoretical frameworks that link knowledge about the social, economic, political, and built environments (Chang & Rose, 2012; Tierney, 1989). This situation is not necessarily surprising given that disaster 'researchers have often been more concerned with solving problems that are important for governmental institutions and practitioners than with advancing theory' (Tierney, 2007, p. 516).

The research literature does include many alternative definitions and conceptual models of community resilience to disasters. However, these do not constitute mature theoretical contributions. There are meta-studies that review the various definitions and frameworks, but few attempt to comprehensively and meaningfully synthesize relevant components into a theoretically grounded proposal. Most, such as NRC (2012), conclude with the selection a single alternative or propose a new one. More critically, Kulig et al. (2013) review five popular conceptual models of community disaster resilience and conclude 'that several of them are incongruent, needing conceptual clarification and empirical testing or a determination of their usefulness for assessment, and are not suitable for theoretical development of the concept' (p. 772). Indeed, this might be because researchers feel that resilience is not worthy of theoretical development (Anderies, Walker, & Kinzig, 2006; Wilson, 2012). Norris et al. (2008), who developed a highly cited conceptual framework, suggest that resilience is just a metaphor – not a theory – whose utility is measured by its tendency to inspire effective interventions and policies.

It is not true that community disaster resilience is unworthy of theory building and by implication, based on Quarantelli's and others' standards, that it is not an area for scientific disaster research. It is quite the opposite: in line with Alexander (2013), resilience is one of the most promising threads in the literature for understanding the relationships between static and dynamic components of communities and their evolution between pre- and post-event contexts. Within disaster research, definitions 'delineate an area of study and in so doing set the stage for knowledge accumulation and theory construction' (Perry, 2007, p. 3). Given the vigorous discussion of definitions and conceptual models, it is appropriate and constructive to begin the difficult work of theory building as part of a collective research agenda on community disaster resilience. Antecedently, there must be a shift in attitudes of researchers that community disaster resilience can be elevated and evolved toward theory.

It is critical to not get distracted by debates regarding the origins and applicability of discipline-specific definitions of resilience (Cutter, Ash, & Emrich, 2014). Instead, it is better to present and debate definitions that explicitly delineate community disaster resilience from discipline-specific or unrelated conceptions of resilience. More important is to build and integrate knowledge that is specific to the needs of describing human settlements and their many communities after hazard events (Cutter et al., 2014). Collective and concerted construction of complimentary and competing theories will unlock the possibility of conducting both inductive and deductive research on the resilience of communities to disasters, allowing researchers to have dialogues between data and theory. While the practical benefits may not be immediate, this dialogue will eventually lead to significantly more impactful means of monitoring, simulating, and promoting the resilience of communities.

Furthering the state-of-the-art of community disaster resilience scholarship and practice requires synthesizing a coherence of knowledge for the sole purpose of better understanding the subject. This theoretical study attempts to affirm, challenge, extend, and interweave disaster resilience-focused literature, while cohering this knowledge with relevant non-resilience literature. This is opposed to the more common attempts to unify or debate disciplinary perspectives of resilience that were not originally developed for the study of community disaster resilience. For instance, definitions or frameworks of engineering resilience for understanding the response of buildings to hazards are not enough to facilitate investigating the question of whether a built environment engineered to be completely resilient to hazards is best for promoting social, cultural, or economic identity and well-being. Alternatively, psychosocial resilience frameworks do not include constructs for understanding the impact on people's sense of pride or attachment to place associated with rebuilding a 'new' or 'better' built environment.

This study also attempts to ground relevant components of existing community resilience frameworks with essential elements of theory. Whetten (1989) states that there are four essential elements needed to make a theoretical contribution to knowledge: (1) the elements (concepts, constructs, and variables) that must be considered to explain the theory, (2) how and why these elements relate, (3) the assumptions underlying the theory, and (4) the limitations for generalizing hypotheses generated from the theory. Based on Rowlands (2005), these four elements are adopted and addressed. The goal of this is to add precision to community disaster resilience and avoid it being a catch-all disaster metaphor that is symbolic of everything and thus nothing.

The theoretical framework presented here is called WISC. WISC is named after the four constructs of the framework – well-being, identity, services, and capitals. WISC represents foundational (static) components of resilience to disasters for describing the state of communities at a given point in time. The dynamic components for understanding changes of communities across space and time are beyond the scope of this paper. WISC provides additional utility to existing frameworks by unpacking and synthesizing their respective components and theoretically linking these components with knowledge from outside of the disaster literature. WISC adds to the few frameworks that integrate well-being as a significant component of disaster resilience. WISC is one of the few comprehensive frameworks that include community identity as a core component. Finally, it is the only instance in the community disaster resilience literature where infrastructure is defined explicitly as the connection between the services and capitals accessible to communities.

2. Community disaster resilience frameworks

A brief review of existing frameworks is useful in identifying the purposes of these frameworks and how components of resilience can be organized. This helps to distinguish the purpose of WISC (theory-building and researching hypotheses), as well as its current scope. Additionally, the review is useful in illustrating some elements that are missing in the literature that are needed to make robust theoretical contributions. The selection and critique of frameworks below are not done to suggest that the frameworks are somehow wrong or not useful. Many of the frameworks could very well form the basis of theory and have informed the development of WISC. The purpose of the discussion is to help situate the endeavor of constructing theoretical frameworks and, specifically, WISC with more familiar frameworks.

A number of frameworks have emerged to help qualitatively describe disaster resilience (Berkes, 2007; Cutter et al., 2008; Norris et al., 2008; Paton & Johnston, 2001). Norris et al. (2008) is one of the most influential and cited of these frameworks (Kulig et al., 2013). Norris et al. (2008) argue that resilience requires four fundamental capacities for adapting to hazards and other disturbances: economic development, social capital, information/communication, and community competence. Their framework explicitly and implicitly integrates constructs or variables from other frameworks, such as the four capacities of Berkes (2007): ability to deal with change and uncertainty, presence of diversity, capacity to synthesize disparate knowledge and information, and opportunities for self-organization. It similarly integrates the components from Paton and Johnston (2001): sense of community, coping style, self-efficacy, and social support.

Kulig et al. (2013) argue that the framework of Norris et al. (2008) falls short of a theoretical contribution because the boundaries and limitations of components in the framework are 'unclear' and 'conflated' (p. 763). These elements are essential for evolving concept to theory (Grover, Lyytinen, & Srinivasan, 2008; Rowlands, 2005; Whetten, 1989). Without these elements, Kulig et al. (2013) argue that there is no way to pose and evaluate hypotheses using Norris et al. (2008). As noted above, the intention of Norris et al. (2008) was not to develop theory to foster hypothesis-driven research but to elucidate a means for promoting practical interventions and policies. A reading of the above-cited qualitative frameworks suggests that an overt statement of assumptions, theoretical boundaries, and limits to generalizability are collectively missing. These qualitative frameworks are presented more as metaphor or a 'collection of ideas about how to interpret complex systems' (Anderies et al., 2006).

In addition to qualitative frameworks, there are a growing number of quantitative frameworks for calculating indices of community resilience (Cutter, Burton, & Emrich, 2010; Joerin & Shaw, 2011; Sherrieb, Norris, & Galea, 2010; Verrucci, Rossetto, Twigg, & Adams, 2012). These frameworks often include a conceptual model that graphically depicts what indicators are used to compute the respective index and in some cases how the indicators are combined. These conceptual models are typically a means to present input data requirements, rather than to represent theoretical knowledge or facilitate the posing of hypotheses (Kulig et al., 2013). However, just as in the case of the qualitative frameworks in the literature, these conceptual models can be viewed as proposals for theoretical consideration.

Kulig et al. (2013) take the example of the quantitative framework by Sherrieb et al. (2010) for computing a community resilience index. This framework employs an approach similar to the other quantitative frameworks cited above. Kulig et al. (2013) disagree with the claim of Sherrieb et al. (2010) that their framework is theoretically grounded because assumptions and limits are not stated. This results in a 'wish list' of indicators that cannot be used to pose and research hypotheses (Kulig et al., 2013, p. 763). Another critique by Kulig et al. (2013) of the Sherrieb et al. (2010) framework and similar ones is that indicators of resilience and variables for understanding resilience are often conflated, resulting in tautological untestable relationships. Kulig et al. (2013) conclude their review to say that

> [i]n most cases [framework] indicators are more properly conceived as antecedents or consequences of resilience and should therefore be posited as independent from resilience, so that we can conduct the empirical investigations to determine the most important conditions for its enhancement. (p. 772)

Or as Norris et al. (2008) put it, 'there is no variable called "relativity" in the Theory of Relativity' (p. 146). Community resilience cannot be a variable in a theoretical framework if the framework is used to guide scientific research because resilience is the subject under study.

Alexander (2013) argues that the concept of resilience holds most promise in enabling researchers to bridge the static and dynamic components of disasters across pre- and post-event contexts. Thus, a useful way of organizing components of community disaster resilience is to develop separate static and dynamic conceptual models (NRC, 2012; Rose, 2007). The value of having explicit and distinct static and dynamic models of theoretical knowledge is cited within several disciplines, including sociology, ecology, economics, urban planning, and information science (Abbink, Braber, & Cohen, 1995; Batty, 2009; Blaha & Premerlani, 1991; Carroll, Phillips, Schumaker, & Smith, 2003; Perry-Smith & Shalley, 2003). For example, Walker and Salt (2006) presents a popular framework where ecological resilience is represented as a static landscape upon which a ball might dynamically move, depending on attributes of the metaphorical landscape and a given disturbance event.

A static model of community disaster resilience must identify and organize critical variables and relationships between those variables that are sufficient for describing conditions of a community before and after a disaster. A static model presents the ontology of community disaster resilience. It provides the vocabulary to explain the state of a community at any point in time or space relative to some event. NRC (2012) states that 'effective community resilience is similar to a healthy human body' (p. 17). Adopting this metaphor, one can view a static model as a skeleton that shows what components are meaningful and how those components fit together. Conversely, a dynamic model represents how and why the variables of a static model change across time and space. It helps explain the happenings of community resilience and provides the vocabulary to describe the evolution of states. A dynamic model can be thought of as the muscles and metabolic system that propel the metaphorical skeleton.

While it is important to make the distinction between static and dynamic representations of resilience, researchers often do not explicitly do so or acknowledge the need (Irajifar, Alizadeh, & Sipe, 2013; Kulig et al., 2013). Kulig et al. (2013) observe that authors of one popular framework misclassified their own framework as dynamic, when only static components are included. (The distinction is often clear depending on whether time (dynamics) is considered.) Promisingly, there are a growing number of research studies that propose and link static and dynamic conceptual models, though not all make the distinction explicit (Bruneau et al., 2003; Cutter et al., 2008; Miles & Chang, 2006, 2011; Simonovic & Peck, 2013; Wilson, 2014; Zhou, Wang, Wan, & Jia, 2010).

The present paper only describes static components for understanding community resilience to disasters. This is not because dynamic components are unnecessary, but because of the length needed to describe the static components of WISC. In the future, WISC can be evolved to incorporate dynamic components, as advocated elsewhere (Miles, 2014a, 2014b).

3.   The WISC framework and conceptual model

Before introducing WISC and its conceptual model, it is useful to note that concepts, constructs, and variables are hierarchical abstractions of increasing specificity (Whetten, 1989). Concepts, constructs, and variables are the components that define the 'what' in a theoretical framework. These components are the necessary first step in any theory-building process (Whetten, 1989). The conceptual model of a theoretical framework is simply a graphical depiction of the components to help illustrate how these components relate (Corley & Gioia, 2011).

The parsimony of WISC comes from the small number of explicitly defined constructs that frame wide-ranging knowledge and insights about the subject of community resilience to disasters. The small number of familiar constructs is intended to inspire and facilitate creation of hypotheses specific to the subject. WISC includes a large number of variables because in the early stage of any theory-building process – the current stage for community resilience theory building – it is key to err on the side of too many components (Whetten, 1989). Additionally, a variety of variables ensure internal validity when operationalizing a framework because of the availability of multiple proxies for representing the framework constructs. After a brief introduction below, the components of WISC are discussed in the following two sections. In the remainder of this section the assumptions, theoretical boundaries, and limits to generalizability of WISC are discussed.

Figure 1 presents the conceptual model of WISC; it summarizes and relates two concepts (community and infrastructure), 4 constructs (well-being, identity, services, and capitals), and 29 variables with respect to a theoretical boundary (human settlement). The broadest interpretation of Figure 1 is that components of infrastructure support and facilitate components of community (UN-HABITAT, 1987). Communities 'sit atop' infrastructure. Ultimately, well-being of a community (top of Figure 1) is dependent on its capital (bottom of Figure 1) (Costanza, 2000). While community well-being is dependent on capitals, these constructs are mediated by the two intervening constructs of identity and services. In the context of resilience, 'disasters are forcing us to recognize the interrelatedness of ... different capitals and their valuable contributions to sustainable ... well-being' (Costanza & Farley, 2007, p. 252).

The theoretical spatial and conceptual boundary of WISC is a particular human settlement. A human settlement contains multiple communities and types of infrastructure. The United Nations defines human settlements as 'the spatial dimension as well as the physical expression of economic and social activity ... no matter how small or physically or economically isolated' (UN-HABITAT, 1987, p. 3). For WISC, a human settlement is not necessarily synonymous with any particular jurisdiction; instead human settlements have overlapping and fragmented

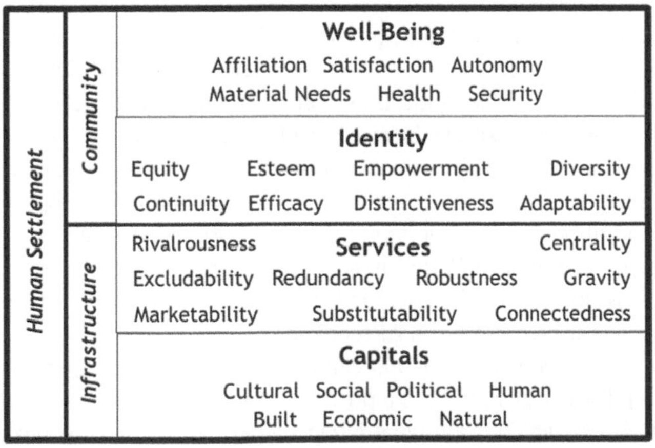

Figure 1.   Conceptual model of static community resilience for the theoretical framework WISC, showing relationships between the concepts of community and infrastructure, constructs of well-being, identity, services, capitals, and 29 collective variables for the 4 constructs.

governance structures, both formal and informal (Amin, 2007). The theoretical representations of WISC apply to communities within a human settlement, but not to the human settlement itself (or beyond). WISC is for studying *community* resilience to disasters, not human settlement resilience to disasters. Within WISC, human settlements are assumed to be infinitely resilient and so do not collapse due to any hazard event. Empirically, this is true much more often than not (Page, 2005; Vale & Campanella, 2005). WISC instead focuses on communities with the assumption that their resilience is variable and heterogeneous. Practically, these assumptions only mean that care and precision have to be put into explicitly defining the boundary of a human settlement under study, as well as the criteria for identifying the communities of interest within it.

Human settlements can be conceived as a community of communities (Hempel, 1999; Keller, 1988; Pavlich, 2001). A community is defined for WISC as any social group that refer to themselves as 'members' or 'us' (Dalby & Mackenzie, 1997; Haslam, Jetten, Postmes, & Haslam, 2009; Lewis & Kelman, 2010). Communities are more than a 'list of the socio-demographic groups that can be used to classify individuals (e.g. gender, age, ethnicity, and religion)' (Haslam et al., 2009, p. 6). Communities might manifest as 'family and friends, work and sports teams, community and religious groups, regional and national entities' (p. 6). There has been debate whether community should be physically defined or socially defined (Hidalgo & Hernandez, 2001; Puddifoot, 1995; Scannell & Gifford, 2010). Puddifoot (1995) asserts that community is constructed by both physical place and social relationships (via common interests, values, culture, etc.). This assertion is adopted for WISC to maximize applicability of the framework. Members of the same community can reside in multiple different human settlements and jurisdictions; they can also be members of multiple other communities (Sonn & Fisher, 1999).

Defining the theoretical temporal boundary of WISC requires adoption of definitions for disaster and vulnerability – concepts that facilitate linkages to components of time and dynamism. These definitions for WISC are drawn from social vulnerability theory (Wisner, Blaikie, Cannon, & Davis, 2003) because of the central role of time and access to resources (infrastructure). Vulnerability manifests in the present as specific unsafe conditions but is socially constructed over time by more distant forces (spatially and temporally), such as national policies, historical decisions, cultural norms, or institutionalized power dynamics. These forces influence what resources people or groups can access or the capacity they have to manage them. A hazard event at a given point in time may breach community vulnerabilities and result in a disaster. Livelihoods and community functioning are disrupted beyond a community's capacity to cope using its own resources (United Nations, 2004).

To be consistent with the adopted definitions for vulnerability and disaster, the temporal boundary for WISC is one just large enough to represent the social construction of the vulnerability of the communities under investigation, as well as to represent meaningful changes to some WISC variables caused by a single hazard event. This is likely to be on the order of years to decades. For example, in appropriate applications of WISC, orogenisis would be assumed fixed and climate change impacts should be decomposed into separate, sequential hazard events. The temporal boundary is defined to avoid a normative statement about when a community has recovered. The goal of WISC is to help understand components of community resilience to disasters; it is not necessarily intended to predict the potential for recovery, which would require an operationalizable definition of recovery.

WISC is applicable only to communities that can experience or have experienced disaster. If there is no (potential for) disaster there will be a positive conclusion (complete resilience) from a negative premise (no disaster), which is a logical fallacy. There will always be hazard events that communities have the capacity to deal with. However, communities can never be resilient to all hazard events. If a community enters a new state that is invulnerable to a previously harmful hazard event, WISC is no longer useful for studying the community for that particular vulnerability–hazard combination. This means that hypotheses based on WISC can be empirically evaluated only using communities where a hazard event has changed the state of some WISC variables (i.e. impacted those communities). In comparing two or more communities, all communities must have been impacted to make meaningful cross-case conclusions. This constraint may seem unnecessary but it distinguishes vulnerability and resilience, which are often confused or erroneously seen as opposites (Cutter et al., 2014; Kennedy, Ashmore, Babister, & Kelman, 2008; Lewis & Kelman, 2010; Zhou et al., 2010). In WISC vulnerability and resilience are treated as potentially independent. This allows for the possibility that high vulnerability does not lead to negative assessments of a community's resilience. The constraint ensures that vulnerability is a premise in arguments or hypotheses about community resilience to disasters. If vulnerability is treated as the opposite of resilience, it is part of the conclusion of any hypothesis about resilience. This makes the relationship between vulnerability and resilience untestable.

4.   Infrastructure: the relationship between capitals and services

For WISC, infrastructure refers to any combination of a given service and the capital used to derive that service. Services are the link between capitals and their benefits to communities (Ash et al., 2010; Costanza, 2000; Costanza & Farley, 2007; Robeyns, 2005). The definition of infrastructure as any combination of capitals and services goes beyond the typical understanding of infrastructure as physical lifelines, such as roads, utilities, and other horizontal elements of the built environment. Infrastructure can be inanimate or animate, material or non-material. An example is entrepreneurial social infrastructure, defined as collective action derived from social capital (Flora & Flora, 1993).

People and communities view capitals and services as inextricably linked and typically as a 'black box' (Graham & Marvin, 2001; Monstadt, 2009). Services give capitals meaning because capitals are primarily experienced via the services derived from them. For example, employees need transportation infrastructure in order to go to work. Employees might have a negative experience when the service of mobility is disrupted, regardless of the specific capitals the service depends on, such as a freeway or bike path. A service can be derived from a single capital but more often depends on multiple capitals, as well as other intermediate services.

In the post-disaster context, temporary capital can provide the same service as damaged 'permanent' capital. This is illustrated by the provision of water before and after the 2011 Great East Japan earthquake (Kuwata & Ohnishi, 2012). Prior to the earthquake, the capital that provided residents of Miyagi Prefecture drinking water was a modern water network. Drinking water service was disrupted because of damage from the earthquake. However, the service was partially restored using the alternative capital of water trucks until repairs could be made to the pipes and pumps of the water network. Of course, the temporary infrastructure may have negatively impacted communities because of issues such as delivery times, access to smaller volumes of water, or the inability for water trucks to reach remote locations.

The incorporation of capital and infrastructure is common within the community resilience literature, as described below. In general, the distinction of infrastructure as capitals and services is uncommon in the disaster literature (Davis, 2013; Kameda, 2000). Specific to community resilience, it appears that no other framework explicitly links capitals and services as infrastructure.

4.1.   *Capitals*

Originally an economic construct, capital refers to a stock of assets used to create or obtain additional assets or derive services. The construct has been broadened to the idea of community capital, which refers to any asset, whether corporeal, material, or non-material, that is utilized as part of the metabolic flows supporting human settlements (Costanza, 2000; Emery & Flora, 2006; Putnam, 2001). There is no accepted set of community capital variables. There is also little justification in the literature for preferring one set of capitals to another. In some cases, the choice depends on the context and scale of application. For example, Costanza (2000) uses four constructs – built, natural, human, and social – to define economic systems. Inclusion of economic capital as a variable in that framework would be circular. In the case of a framework for community resilience to disasters the inclusion of economic capital is not circular. As another example, the triple bottom line framework (social, natural, and economic capital) was developed specifically to raise understanding of the social and environmental impacts of economic growth (Elkington, 1998), so purposely does not include other capitals.

The capitals construct is widespread, either explicitly or implicitly, within the community resilience literature (Gilbert, 2011). While the construct is popular, there is no agreement on what specific variables are minimally sufficient. This seems to be the case for non-disaster community capital frameworks, as well. Aldrich (2012) links just social capital to community resilience. Norris et al. (2008) incorporate social and economic capital. Bruneau et al. (2003) propose technical, organizational, social, and economic variables of capital. The framework of Berkes (2007) is an example where the term resources is used instead of capital; access to political, social, and ecological resources are core components of the framework. Paton and Johnston (2001) focuses on just social and personal resources, but argues that these influence people's ability to access physical and economic resources. Cutter et al. (2014) propose the capitals of social, economic, housing and infrastructure, institutional, community, and environmental. Gilbert (2011) includes human, built, economic, government, social, and natural capital. The similar framework of Cimellaro and Arcidiacomo (2013) has two variables of capital in place of social capital: 'lifestyle and community competence' and socio-cultural capital. The prominence of the community capitals concept in the disaster research literature has notably influenced practice. For example, the Canterbury Earthquake Recovery Authority has adopted a recovery strategy that explicitly considers economic, social, cultural, natural, and built capital (CERA, 2012).

For the WISC framework, seven variables of community capital are adopted: *social, political, cultural, human, economic, built, and natural*. This list is taken from Emery and Flora (2006). The choice should be treated as an assumption that is intended to capture the wide range of propositions in the literature. Perhaps not all variables of capital will influence all other components of WISC; perhaps some variables, as defined, will have no influence on other components. The validity of this list is subject to future deductive research. Certainly there will be some relationships between the variables of capitals. For example, it is known that built capital mediates a community's access to natural capital (Costanza, 2000; Gunderson, 2010; Monstadt, 2009).

4.2. *Services*

Services are typically defined in economic terms as measurable flows, such as perishable goods, that are provided and consumed by communities. Suggesting that this definition is too narrow, Costanza (2008) defines services as the benefits that communities derive from capital, where the 'end or goal [of services] is sustainable human well-being' (p. 351). Unlike the economic definition, which assumes that people explicitly perceive and exchange services, Constanza's (2008) definition recognizes that many services, such as carbon sequestration, go largely unnoticed and provide benefits without market-based exchange.

The definition adopted for the WISC framework recognizes that different services require different variables of capital (Costanza, 2000; Olewiler, 2006). Services have variables; the specific values of these variables make each variable unique and potentially beneficial or detrimental (Davis, 2013). The services construct consists of nine variables (see Figure 1 for the list). All of the services variables are relevant to all six of the capitals variables. A particular value or instance of a variable not only characterizes the particular service, but also the capitals that it is connected to. In this way, some variables that might typically be associated with a type of capital are associated with services. The values of the variables are all a matter of degree and should not be considered binary or necessarily mutually exclusive. For example, a service can be partially rivalrous, not absolutely rivalrous or non-rivalrous.

Rivalrous means that consumption by community members results in capital depletion and thus less opportunity for consumption by others; non-rivalrous means the opposite (Fisher, Turner, & Morling, 2009; Frischmann, 2005). *Excludable* means that consumption of a service can be limited by physical and financial means, as well as explicit or institutionalized access qualifications – for example, through institutionalized racism (Fisher et al., 2009; Frischmann, 2005; Green, Bates, & Smyth, 2007). Non-excludable means that access cannot be realistically limited, such as access to oxygen to breath. *Marketability* refers to the degree to which a particular infrastructure service is managed by private markets, or, conversely, whether it is managed as a public commons (Frischmann, 2005). Together, the variables of rivalrousness, excludability, and marketability define the relative opportunity for community members to access and consume particular infrastructure services (Costanza, 2008).

Redundancy identifies whether some element of infrastructure can be omitted without significant overall loss of meaning or function. This contrasts with the definition of Bruneau et al. (2003), which conflates the potential for substitution (i.e. *substitutability*) with that of omission. For example, *redundancy* describes a case in which a high voltage transformer goes offline, but power customers still receive electricity via one or more additional transmission lines serving the area. In an example of substitutability, the redundant transmission lines do not exist, but the customer has or can obtain a generator (substitute) to meet her service needs.

Robustness means the strength or ability to resist degradation or loss, as defined in Bruneau et al. (2003). It also refers to the brittleness of a service and its underlying capitals and whether failure will occur suddenly or gradually (Barrett, Eubank, Kumar, & Marathe, 2004; Kahan, Allen, & George, 2009). Contrasting examples are the gradual decline of storm buffering as mangroves die off versus the lights suddenly going out if an electrical transformer explodes on a distribution pole.

Centrality refers to the degree to which infrastructure is centralized. For example, the political infrastructure that exists between the informal political coalitions of the precarious settlement residents in Guatemala City is less centralized than that of the formal government of the Guatemala Metropolitan Region (Miles, Green, & Svekla, 2011). *Gravity* refers to the level of importance of the infrastructure with respect to different communities at various scales (Kahan et al., 2009). For instance, crop-yield decline resulting from climate change may impact migrant workers in one way, local economic development in another way, and national economic development in yet another way. Finally, the *connectedness* variable represents the relative degree of an infrastructure's network effect. A network effect is when the benefit to one user of an infrastructure network increases if additional users join the network (Frischmann, 2005). Social media infrastructure, such as Facebook and Twitter, is an example of high connectedness. In the case of electric infrastructure, on the other hand, the addition of new customers to a grid does not significantly benefit existing customers; there is little network effect and so connectedness is low.

5. Community: the relationship between well-being and identity

Above, the theoretical boundary of community was given as a social group with members who distinguish themselves as 'us' because of both geographical location and social relationships. This boundary makes a distinction between human settlements, jurisdictions, and communities.

More importantly, it helps to develop criteria for identifying communities to study and compare. However, it is not specifically intended to help construct arguments or hypotheses about community resilience.

Instead, within WISC the community concept comprises the constructs well-being and identity. This is because the challenges of community are reflected in identity and well-being (Sonn, 2002). Having a primary community 'reinforces ... identities and ... provide[s] structures and social support systems that are crucial to ... well-being' (Sonn & Fisher, 1999, p. 715). Not identifying with a community can have negative effects on well-being and, in turn, resilience (Jetten, Haslam, Haslam, & Alexander, 2012; Sonn & Fisher, 1998). Haslam et al. (2009) summarize the connection of community, identity, and well-being, introducing some variables of WISC in the process: '[Community identities] make us feel distinctive and special, efficacious and successful. They enhance our self-esteem and sense of worth. These effects can buffer well-being when it is threatened, and can also help people cope ... ' (p. 3). As an alternative to the two constructs of WISC, McMillan and Chavis (1986) set out four constructs for understanding the concept of community: membership, influence, shared emotional connection, and integration and fulfillment of needs. Within WISC, the first two constructs of McMillan and Chavis (1986) are considered components of identity, while the latter two are part of well-being.

5.1. *Well-being*

In most of the literature the goal of community resilience is a functioning system, agent, or community. The goal of resilience, however, should go beyond safety or functioning and ultimately ensure the well-being of communities and their members (Adler, 2006; Costanza & Farley, 2007). Norris et al. (2008) observe that the construct of well-being sets a higher bar than is typical in the literature and is an appropriate and necessary standard for community resilience. An increasing number of studies in the disaster literature have focused on the construct of well-being as central to community resilience (Brown & Westaway, 2011; Colliard & Baggio, 2007; Kellezi, Reicher, & Cassidy, 2009; Kirmayer, Sehdev, Whitley, Dandeneau, & Isaac, 2009; Norris et al., 2008; Nyamwanza & Nyamwanza, 2012). WISC builds upon these studies and incorporates insights from outside disaster research.

Costanza et al. (1997) argue that well-being must be conceived using both non-market-based and market-based variables. Norris et al. (2008) define well-being 'as high and non-disparate levels of mental and behavioral health, role functioning [at home, school, and/or work], and quality of life' (p. 133). In relating ecosystem services to well-being, Ash et al. (2010) propose five variables of well-being achievements: material needs, security safety and predictability, mental/physical health, social relations, affiliation, role functioning, mutual respect, ability to help, freedom of choice, autonomy, and control. Nussbaum (2003) sets out 10 variables of well-being capacities intended to help evaluate and design policies for social change. The variables of Nussbaum (2003) that are not redundant of Ash et al. (2010) can be related to achievements of pleasure and satisfaction (i.e. capability for imagination, emotions, reason, play, and interaction with other species).

Synthesizing and simplifying insights from Ash et al. (2010), Norris et al. (2008), and Nussbaum (2003), WISC incorporates six variables of well-being: *material needs, security, health, affiliation, autonomy,* and *satisfaction*. As Robeyns (2005) and Gasper (2004) observe, it is important to distinguish well-being as either capabilities (Nussbaum, 2003) or achievements (Ash et al., 2010). Two people given equal capabilities may freely choose different achievements that meet their culturally rooted idea of a good life (Nyamwanza & Nyamwanza, 2012). There is a meaningful difference between someone who has the capability (access) to eat healthily and does so versus someone with similar capability that 'freely' chooses not to. In employing WISC, one must choose to treat the six constructs as either 'opportunities for' or 'achievements of'. This decision is dependent on the particular study objectives. For consistency with social vulnerability theory (Wisner et al., 2003), which focuses on access to resources, it is more important to focus on capabilities (Nussbaum, 2003; Robeyns, 2005). Alternatively, an achievement perspective makes quantitative measurement easier (Gasper, 2004).

5.2. *Identity*

Variables of the construct community identity have been empirically linked to variables of well-being (Adger, Barnett, Chapin, & Ellemor, 2011; David & Bar-Tal, 2009; Haslam et al., 2009; Kellezi et al., 2009; Zottarelli, 2008). Puddifoot (1995) observes that community identity is the link between services and quality of life. A strong case has been made within the disaster and community resilience literature for inclusion of the construct of identity or, more precisely, particular variables of identity (Berkes, 2007; Cutter et al., 2008; Kulig et al., 2013; Norris et al., 2008; Paton, Millar, & Johnston, 2001). Adger et al. (2011) define disaster risk as the risk of harmful changes to community identity and, in turn, well-being. Community identity provides a strong connection to social vulnerability theory because it 'constitutes a foundation for a variety of social effects, from humans' ability to feel, act, and think as members of a social group to intergroup behaviors, such as discrimination, confrontation, and cooperation'. (David & Bar-Tal, 2009) Consideration of identity is important for applications of community disaster resilience because it has been shown to play a large role in place-based improvements and planning (Manzo & Perkins, 2006). Tobin (1999) argues that it is doubtful that successful community resilience planning 'can be accomplished without due consideration of the contextual issues of place' (p. 23).

The WISC construct of community identity is built up from the work of Breakwell (1992) and Twigger-Ross and Uzzell (1996). These studies empirically show that community identity is influenced by four variables: *efficacy, esteem, distinctiveness, continuity, and empowerment*. To these, the WISC framework adds *equity, empowerment, diversity*, and *adaptability* (Berkes & Ross, 2012).

The variables of *esteem* and *efficacy* are included in the framework of Norris et al. (2008). Efficacy and esteem have been functionally related to indicators of competence, community attachment, sense of place, social support, and coping (Adger et al., 2011; Haslam et al., 2009; Kellezi et al., 2009), which are commonly included within resilience frameworks. Holding in high esteem the community a person most identifies with has been shown to improve availability of social support and reduce the importance of individual coping strategies after disasters (Kennedy et al., 2008). This was observed after Hurricane Katrina within the Vietnamese community of east New Orleans who maintained their collective sense of worth after the hurricane (Campanella, 2006). As a result, community members formed new neighborhood groups to rebuild and decontaminate homes, as well as provide tetanus shots and acupuncture to fellow community members.

An important role of community identity is promoting a feeling and understanding of *distinctiveness* (Haslam et al., 2009). The variable of distinctiveness is strongly linked to people's sense of place and attachment (Twigger-Ross & Uzzell, 1996). Hazards and vulnerability play a role in the construction of a community's identity (Dalby & Mackenzie, 1997). For example, residents of San Francisco likely feel distinctive from residents of Miami in part because of their identification with earthquakes and not hurricanes. The meaning of the impacts of a hazard event for a particular community is tied to what makes that community feel distinctive from other communities (Kellezi et al., 2009). A rural community may place greater meaning on the loss of a few small businesses. Whereas some communities in the Seattle area may be more distressed by impacts to Boeing, Starbucks, or Microsoft, if they do not feel that small businesses make their community unique.

Closely related to distinctiveness is *diversity*, which is a central variable of the framework of Berkes (2007). Diversity refers to the relative variety of infrastructure constructs (not just built infrastructure) within or connected to a human settlement (e.g. through trade) (Berkes, 2007; Norris et al., 2008). Gunderson (2010) notes that numerical diversity is less important than functional diversity. Further, the diversity of infrastructure that communities have access to is more important than the overall diversity of infrastructure access across an entire human settlement (Wisner et al., 2003). This is because diversity of access promotes the livelihoods that communities identify with and are identified by (Berkes & Ross, 2012).

Community identity requires maintenance of some acceptable level of *continuity* in the face of threats to infrastructure or access to it (Gillson, 2009; Pendall, Foster, & Cowell, 2007). If continuity of identity is disrupted – for example, because of an extreme event (Adger et al., 2011; Kellezi et al., 2009) – or if membership significantly changes – for example, because of disaster migration (David & Bar-Tal, 2009; Haslam et al., 2009) – there tends to be negative consequences for the ability to cope, public health, and well-being (Iyer & Jetten, 2011). *Adaptability*, which is incorporated into many definitions of resilience (Manyena, 2006), is also included in the WISC framework. Community adaptive capacity to deal with hazard events relies on cultural infrastructure, feedback loops of experimentation and learning, and the synthesis of knowledge from different sources, collective experiences, and value systems (Berkes & Ross, 2012; Gunderson, 2010). Adaptability is considered here to be largely synonymous with absorptive capacity, flexibility, and creativity. If given too much priority or considered in isolation, however, adaptability can have a negative impact on communities (Lewis & Kelman, 2010). This is because a community might effectively absorb a hazard event, while retaining its pre-event vulnerability.

Understanding of identity requires consideration of the variable *equity* (Norris et al., 2008). Here equity is applied to communities and not necessarily individuals. Communities that have poor access to infrastructure have high potential to suffer inequitable impacts from hazard events (Cutter, 2003; Wisner et al., 2003). This differential vulnerability produces heterogeneous patterns of recovery with respect to different demographics (Miles & Chang, 2011; Tierney, 2006). Population recovery after the 1906 San Francisco earthquake provides a useful example. By 1910, population in most neighborhoods returned to pre-event levels. However, multiple neighborhoods, such as Outer Mission and Chinatown, inequitably experienced significant increases in unemployment (Davies, 2011). In contrast to equity, the variable of *empowerment* relates identity to whether agents or communities can access the infrastructure they need or want (Adger et al., 2011; Kellezi et al., 2009). Empowerment also relates to whether they can improve their ability to avoid loss and achieve recovery, which can positively influence esteem before and after a disaster (Muldoon, Schmid, & Downes, 2009). The acceptance and ultimate success of post-disaster decisions and outcomes requires that affected communities are empowered to participate in the reconstruction and recovery of their collective identity (Berkes & Ross, 2012; Kamani-Fard, 2012).

6. Conclusion

This paper proposes a foundation for enabling future theoretical research toward maturing community disaster resilience knowledge and creating greater balance with the methodological and empirical work conducted to date. The constructs of WISC are offered as minimally sufficient to characterize community resilience. Using WISC, existing studies can be extended, deepened, and rooted in more formal knowledge particular to community disaster resilience. WISC is developed under the supposition that a theoretical framework built specifically for understanding community resilience to disasters will alleviate the need for debating disciplinary specific definitions of resilience and avoid community disaster resilience becoming a catch-all. This will quicken the pace of establishing knowledge about the subject through methodologically facilitated dialogues between data and theory.

WISC enables researchers and practitioners to generalize community resilience across past disasters, forecast it for future disasters, and decide what data are important to collect (and how), given limited opportunities and resources for data collection. The framework benefits efforts to develop and evaluate associated computational models, geovisualization interfaces, and decision support systems by theoretically justifying what elements and relationships are represented.

The constructs of well-being, identity, services, and capitals provide a simple but powerful foundation for making arguments and testing hypotheses to investigate community disaster resilience. A pressing set of arguments that can be studied with WISC is whether recovery should be defined as 'back to normal' or 'to a new normal' and whether recovery should be rapid or deliberate (Olshansky, Johnson, Horne, & Nee, 2008). WISC can help to reveal nuance in these dichotomous arguments, potentially allowing for all assertions to be true based on selected components of community resilience and their hypothesized relationships. For example, assuming a community's well-being is desirable, members of the community may not welcome changes to capitals, services, and identity that result in a new normal of reduced well-being. It is likely that most community members will want their electricity service back as fast as possible to power their lights as normal. Some may welcome the chance to replace the capital of a damaged nuclear power plant with solar power generation, while the value systems of others may color this change as undesirable. A community that identifies strongly with their natural capital may choose to strengthen their identity in the event that built capital is damaged by liquefaction. Such a community may be eager to abandon that built capital in favor of services provided by additional parkland. Another community may not feel their parks make them distinctive and so not view a similar future as better.

Many other arguments for inquiry are possible. WISC can frame research to understand how much damage or loss can occur to different community capital before a community's identity changes enough to negatively impact their well-being. In what instances can critical services be maintained after an event using temporary strategies and alternative capitals to ensure continuity of positive well-being? It is useful to question how relevant financial loss due to direct damage of built capital is to constructs of community well-being, identity, and resilience. How far beyond life safety does the vulnerability of distinctive built capital have to be reduced in order to protect a community's identity? Resources may be more efficiently used to maintain services associated with built capital that is not strongly tied to a community's identity, rather than the undistinguished built capital itself.

WISC is intended to shift attitudes about the value of community disaster resilience as theory and stimulate research that strongly incorporates theory. The WISC constructs are universally relevant and justifiably related. However, the variables of WISC are necessarily provisional and so only loosely related at this point. The components and relationships of WISC need to be debated, evaluated, strengthened, and revised. Future research studies can be designed to propose and investigate hypotheses between WISC variables to gain insights into which are relevant, which can be discarded without losing explanatory or predictive power, and which should be added. In parallel, WISC should be extended in order to represent dynamic aspects of community resilience (Miles, 2014a, 2014b).

Acknowledgement

The author expresses gratitude to the responsible editor and reviewers for providing encouragement and constructive comments, as well as to the many colleagues who gave feedback on conference presentations associated with this paper.

Disclosure statement

No potential conflict of interest was reported by the author.

Funding

This work was supported by the National Science Foundation under Civil, Mechanical, and Manufacturing Innovation Program [grant number 0927356].

References

Abbink, G. A., Braber, M. C., & Cohen, S. I. (1995). A SAM-CGE demonstration model for Indonesia: Static and dynamic specifications and experiments. *International Economic Journal, 9*(3), 15–33.

Adger, W. N., Barnett, J., Chapin, F. S.III, & Ellemor, H. (2011). This must be the place: Underrepresentation of identity and meaning in climate change decision-making. *Global Environmental Politics, 11*(2), 1–25.

Adler, M. (2006). Policy analysis for natural hazards: Some cautionary lessons from environmental policy analysis. *Duke Law Journal, 56*(1), 1–50.

Aldrich, D. P. (2012). *Building resilience*. Chicago, IL: University of Chicago Press.

Alexander, D. E. (2013). Resilience and disaster risk reduction: An etymological journey. *Natural Hazards and Earth System Sciences Discussions, 1*(2), 1257–1284.

Amin, A. (2007). Re-thinking the urban social. *City, 11*(1), 100–114.

Anderies, J. M., Walker, B. H., & Kinzig, A. P. (2006). Fifteen weddings and a funeral: Case studies and resilience-based management. *Ecology and Society, 11*(1). Retrieved from http://www.ecologyandsociety.org/vol11/iss1/art21/

Ash, N., Blanco, H., Brown, C., Vira, B., Zurek, M., Garcia, K., & Tomich, T. (2010). *Ecosystems and human well-being*. Washington, DC: Island Press.

Barrett, C., Eubank, S., Kumar, V., & Marathe, M. (2004). Understanding large scale social and infrastructure networks: A simulation based approach. *SIAM News, 37*(4), 1–5.

Batty, M. (2009). Urban modeling. In N. Thrift & R. Kitchin (Eds.), *International encyclopedia of human geography* (pp. 51–58). Springer: Oxford.

Berkes, F. (2007). Understanding uncertainty and reducing vulnerability: Lessons from resilience thinking. *Natural Hazards, 41*(2), 283–295.

Berkes, F., & Ross, H. (2012). Community resilience: Toward an integrated approach. *Society & Natural Resources: An International Journal, 26*(1), 5–20.

Blaha, M., & Premerlani, W. (1991). *Object oriented modeling and design*. Upper Saddle River, NJ: Prentice Hall.

Breakwell, G. M. (1992). *Social psychology of identity and the self concept*. Waltham, MA: Academic Press.

Brown, K., & Westaway, E. (2011). Agency, capacity, and resilience to environmental change: Lessons from human development, well-being, and disasters. *Annual Review of Environment and Resources, 36*, 321–342.

Bruneau, M., Chang, S. E., Eguchi, R. T., Lee, G. C., O'Rourke, T. D., Reinhorn, A. M., … von Winterfeldt, D. (2003). A framework to quantitatively assess and enhance the seismic resilience of communities. *Earthquake Spectra, 19*(4), 733–752.

Campanella, T. J. (2006). Urban resilience and the recovery of New Orleans. *Journal of the American Planning Association, 72*(2), 141–146.

Carroll, C., Phillips, M. K., Schumaker, N. H., & Smith, D. W. (2003). Impacts of landscape change on wolf restoration success: Planning a reintroduction program based on static and dynamic spatial models. *Conservation Biology, 17*(2), 536–548.

CERA. (2012). *Recovery strategy for Greater Christchurch – Mahere Haumanutanga o Waitaha* (p. 48). Christchurch: Canterbury Earthquake Recovery Authority.

Chang, S. E., & Rose, A. (2012). Towards a theory of economic recovery from disasters. *International Journal of Mass Emergencies and Disasters, 30*(2), 171–181.

Cimellaro, G. P., & Arcidiacomo, V. (2013). Resilience-based design for urban cities. In D. Serre, B. Barroca, & R. Laganier (Eds.), *Resilience and urban risk management* (pp. 127–141). Leiden: CRC Press.

Colliard, C., & Baggio, S. (2007). *Well-being and resilience after the Tsunami* (p. 143). Terres des hommes.

Corley, K. G., & Gioia, D. A. (2011). Building theory about theory building: What constitutes a theoretical contribution? *Academy of Management Review, 36*(1), 13–32.

Costanza, R. (2000). Social goals and the valuation of ecosystem services. *Ecosystems, 3*(1), 4–10.

Costanza, R. (2008). Ecosystem services: Multiple classification systems are needed. *Biological Conservation, 141*, 350–352.

Costanza, R., d'Arge, R., de Groot, R., Farber, S., Grasso, M., Hannon, B., … van den Belt, M. (1997). The value of the world's ecosystem services and natural capital. *Nature, 387*(6630), 253–260.

Costanza, R., & Farley, J. (2007). Ecological economics of coastal disasters: Introduction to the special issue. *Ecological Economics, 63*(2–3), 249–253.

Cutter, S. L. (2003). The vulnerability of science and the science of vulnerability. *Annals of the Association of American Geographers, 93*(1), 1–12.

Cutter, S. L., Ash, K. D., & Emrich, C. T. (2014). The geographies of community disaster resilience. *Global Environmental Change, 29*, 65–77.

Cutter, S. L., Barnes, L., Berry, M., Burton, C., Evans, E., Tate, E., & Webb, J. (2008). A place-based model for understanding community resilience to natural disasters. *Global Environmental Change, 18*(4), 598–606.

Cutter, S. L., Burton, C. G., & Emrich, C. T. (2010). Disaster resilience indicators for benchmarking baseline conditions. *Journal of Homeland Security and Emergency Management, 7*(1), 1–24.

Dalby, S., & Mackenzie, F. (1997). Reconceptualising local community: Environment, identity and threat. *Area, 29*(2), 99–108.

David, O., & Bar-Tal, D. (2009). A sociopsychological conception of collective identity: The case of national identity as an example. *Personality and Social Psychology Review, 13*(4), 354–379.

Davies, A. R. (2011). *Saving San Francisco.* Philadelphia, PA: Temple University Press.

Davis, C. A. (2013). *Quantifying post-earthquake water system functionality.* Presented at the Sixth China-Japan-US Trilateral Symposium on Lifeline Earthquake Engineering, Chengdu, China.

Elkington, J. (1998). *Cannibals with Forks: The triple bottom line of 21st century business [reprint].* Gabriola Island, BC: New Society Publishers.

Emery, M., & Flora, C. (2006). Spiraling-up: Mapping community transformation with community capitals framework. *Community Development, 37*(1), 19–35.

Fisher, B., Turner, R., & Morling, P. (2009). Defining and classifying ecosystem services for decision making. *Ecological Economics, 68*(3), 643–653.

Flora, C. B., & Flora, J. L. (1993). Entrepreneurial social infrastructure: A necessary ingredient. *Annals of the American Academy of Political and Social Science, 529*, 48–58.

Frischmann, B. (2005). An economic theory of infrastructure and sustainable infrastructure commons. *Minnesota Law Review, 89*(4), 917–1030.

Gasper, D. (2004). *Human well-being: Concepts and conceptualizations* (No. 2004/06). WIDER Discussion Papers, World Institute for Development Economics (UNU-WIDER).

Gilbert, S. W. (2011). *Disaster resilience: A guide to the literature* (No. NIST Special Publication 1117) (p. 125). U.S. Department of Commerce National Institute of Standards and Technology.

Gillson, L. (2009). Landscapes in time and space. *Landscape Ecology, 24*(2), 149–155.

Graham, S., & Marvin, S. (2001). *Splintering urbanism: Networked infrastructures, technological mobilities and the urban condition.* New York, NY: Routledge.

Green, R., Bates, L. K., & Smyth, A. (2007). Impediments to recovery in New Orleans' upper and lower ninth ward: One year after Hurricane Katrina. *Disasters, 31*(4), 311–335.

Grover, V., Lyytinen, K., & Srinivasan, A. (2008). Contributing to rigorous and forward thinking explanatory theory. *Journal of the Association for Information Systems, 9*(2), 40–47.

Gunderson, L. (2010). Ecological and human community resilience in response to natural disasters. *Ecology and Society, 15*(2). Retrieved from http://www.ecologyandsociety.org/vol15/iss2/art18/

Haslam, S. A., Jetten, J., Postmes, T., & Haslam, C. (2009). Social identity, health and well-being: An emerging agenda for applied psychology. *Applied Psychology, 58*(1), 1–23.

Hempel, L. C. (1999). Conceptual and analytical challenges in building sustainable communities. In D. A. Mazmanian & M. E. Kraft (Eds.), *Toward sustainable communities* (pp. 43–74). Cambridge, MA: MIT Press.

Hidalgo, M. C., & Hernandez, B. (2001). Place attachment: Conceptual and empirical questions. *Journal of Environmental Psychology, 21*(3), 273–281.

Irajifar, L., Alizadeh, T., & Sipe, N. (2013). Disaster resiliency measurement frameworks: State of the art. In S. Kajewski, K. Manley, & K. Hampson. Presented at the World Building Congress, Brisbane, Australia.

Iyer, A., & Jetten, J. (2011). What's left behind: Identity continuity moderates the effect of nostalgia on well-being and life choices. *Journal of Personality and Social Psychology, 101*(1), 94–108.

Jetten, J., Haslam, C., Haslam, A. S., & Alexander, S. H. (2012). *The social cure*. East Sussex, UK: Psychology Press.

Joerin, J., & Shaw, R. (2011). Chapter 3 mapping climate and disaster resilience in cities. In R. Shaw & A. Sharma (Eds.), *Community, environment and disaster risk management* (Vol. 6, pp. 47–61). Bingley: Emerald Group.

Jordan, E., & Javernick-Will, A. (2012). Measuring community resilience and recovery: A content analysis of indicators. Presented at the Construction Research Congress, West Lafayette. (pp. 2190–2199).

Kahan, J. H., Allen, A. C., & George, J. K. (2009). An operational framework for resilience. *Journal of Homeland Security and Emergency Management*, *6*(1), 1–51.

Kamani-Fard, A. (2012). The sense of place in the new homes of post-Bam earthquake reconstruction. *International Journal of Disaster Resilience in the Built Environment*, *3*(3), 220–236.

Kameda, H. (2000). Engineering management of lifelines systems under earthquake risk. Presented at the World Conference on Earthquake Engineering, Auckland, New Zealand.

Keller, S. (1988). The American dream of community: An unfinished agenda. *Sociological Forum*, *3*(2), 167–183.

Kellezi, B., Reicher, S., & Cassidy, C. (2009). Surviving the Kosovo conflict: A study of social identity, appraisal of extreme events, and mental well-being. *Applied Psychology*, *58*(1), 59–83.

Kennedy, J., Ashmore, J., Babister, E., & Kelman, I. (2008). The meaning of "Build back better": Evidence from post-tsunami Aceh and Sri Lanka. *Journal of Contingencies and Crisis Management*, *16*(1), 24–36.

Kirmayer, L. J., Sehdev, M., Whitley, R., Dandeneau, S. F., & Isaac, C. (2009). Community resilience: Models, metaphors and measures. *Journal of Aboriginal Health*, *5*(1), 62–117.

Kulig, J. C., Edge, D. S., Townshend, I., Lightfoot, N., & Reimer, W. (2013). Community resiliency: Emerging theoretical insights. *Journal of Community Psychology*, *41*(6), 758–775.

Kuwata, Y., & Ohnishi, Y. (2012). Emergency-response capacity of lifelines after wide-area earthquake disasters (pp. 1475–1486). Presented at the Proceedings of the International Symposium on Engineering Lessons Learned from the Great East Japan Earthquake, Tokyo, Japan.

Lewis, J., & Kelman, I. (2010). Places, people and perpetuity: Community capacities in ecologies of catastrophe. *ACME: an International E-Journal for Critical Geographies*, *9*(2), 191–220.

Liao, K.-H. (2012). A theory on urban resilience to floods – a basis for alternative planning practices. *Ecology and Society*, *17*(4). Retrieved from http://www.ecologyandsociety.org/vol17/iss4/art48/

Lindell, M. K. (2013). Disaster studies. *Current Sociology*, *61*(5–6), 797–825.

Manyena, S. (2006). The concept of resilience revisited. *Disasters*, *30*(4), 434–450.

Manzo, L. C., & Perkins, D. D. (2006). Finding common ground: The importance of place attachment to community participation and planning. *Journal of Planning Literature*, *20*(4), 335–350.

McMillan, D. W., & Chavis, D. M. (1986). Sense of community: A definition and theory. *Journal of Community Psychology*, *14*, 6–23.

Miles, S. B. (2014a). Theorizing community resilience to earthquakes. Presented at the Tenth National Conference on Earthquake Engineering, Anchorage, AK.

Miles, S. B. (2014b). Theorizing community resilience to improve computational modeling. Presented at the International Conference on Vulnerability and Risk Analysis Management, Liverpool, UK.

Miles, S. B., & Chang, S. E. (2006). Modeling community recovery from earthquakes. *Earthquake Spectra*, *22*(2), 439–458.

Miles, S. B., & Chang, S. E. (2011). ResilUS: A community based disaster resilience model. *Cartography and Geographic Information Science*, *38*(1), 36–51.

Miles, S. B., Green, R. A., & Svekla, W. (2011). Disaster risk reduction capacity assessment for precarious settlements in Guatemala City. *Disasters*, *36*(3), 365–381.

Monstadt, J. (2009). Conceptualizing the political ecology of urban infrastructures: Insights from technology and urban studies. *Environment and Planning A*, *41*(8), 1924–1942.

Muldoon, O. T., Schmid, K., & Downes, C. (2009). Political violence and psychological well-being: The role of social identity. *Applied Psychology*, *58*(1), 129–145.

Norris, F. H., Stevens, S. P., Pfefferbaum, B., Wyche, K. F., & Pfefferbaum, R. L. (2008). Community resilience as a metaphor, theory, set of capacities, and strategy for disaster readiness. *American Journal of Community Psychology*, *41*(1–2), 127–150.

NRC. (2012). *Disaster resilience: A national imperative*. Washington, DC: National Academy Press.

Nussbaum, M. (2003). Capabilities as fundamental entitlements: Sen and social justice. *Feminist Economics*, *9*(2–3), 33–59.

Nyamwanza, A. M., & Nyamwanza, A. M. (2012). Livelihood resilience and adaptive capacity: A critical conceptual review. *Jàmbá: Journal of Disaster Risk Studies*, *4*(1), 1–6.

Olewiler, N. (2006). Environmental sustainability for urban areas: The role of natural capital indicators. *Cities, 23*(3), 184–195.

Olshansky, R., Johnson, L., Horne, J., & Nee, B. (2008). Longer view: Planning for the rebuilding of New Orleans. *Journal of the American Planning Association, 74*(3), 273–287.

Page, S. E. (2005). Are we collapsing? A review of Jared Diamond's Collapse: How societies choose to fail or succeed. *Journal of Economic Literature, XLIII*, 1049–1062.

Paton, D., & Johnston, D. (2001). Disasters and communities: Vulnerability, resilience and preparedness. *Disaster Prevention and Management, 10*(4), 270–277.

Paton, D., Millar, M., & Johnston, D. (2001). Community resilience to volcanic hazard consequences. *Natural Hazards, 24*, 157–169.

Pavlich, G. (2001). The force of community. In H. Strang & J. Braithwaite (Eds.), *Restorative justice and civil society* (p. 250). Cambridge, UK: Cambridge University Press.

Pendall, R., Foster, K., & Cowell, M. (2007). *Resilience and regions: Building understanding of the Metaphor* (Working Paper 2012-7). Berkeley Institute for Urban and Regional Development.

Perry, R. W. (2005). Disasters, definitions and theory construction. In R. W. Perry & E. L. Quarantelli (Eds.), *What is a disaster?* (pp. 311–324). Bloomington, IN: Psychology Press.

Perry, R. W. (2007). What is a disaster? In H. Rodriguez, E. L. Quarantelli & R. Dynes (Eds.), *Handbook of disaster research* (pp. 1–15). New York, NY: Springer.

Perry-Smith, J. E., & Shalley, C. E. (2003). The social side of creativity: A static and dynamic social network perspective. *Academy of Management Review, 28*(1), 89–106.

Puddifoot, J. E. (1995). Dimensions of community identity. *Journal of Community & Applied Social Psychology, 5*, 357–370.

Putnam, R. D. (2001). *Bowling alone*. New York, NY: Simon & Schuster.

Quarantelli, E. L. (2005). A social science research agenda for the disasters of the 21st century. In R. W. Perry & E. L. Quarantelli (Eds.), *What is a disaster?* (pp. 325–396). Bloomington, IN: Xlibris.

Robeyns, I. (2005). The capability approach: A theoretical survey. *Journal of Human Development, 6*(1), 93–117.

Rose, A. (2007). Economic resilience to natural and man-made disasters: Multidisciplinary origins and contextual dimensions. *Environmental Hazards, 7*(4), 383–398.

Rowlands, B. (2005). Grounded in practice: Using interpretive research to build theory. *The Electronic Journal of Business Research Methodology, 3*(1), 81–92.

Scannell, L., & Gifford, R. (2010). Defining place attachment: A tripartite organizing framework. *Journal of Environmental Psychology, 30*(1), 1–10.

Sherrieb, K., Norris, F. H., & Galea, S. (2010). Measuring capacities for community resilience. *Social Indicators Research, 99*(2), 227–247.

Simonovic, S. P., & Peck, A. (2013). Dynamic resilience to climate change caused natural disasters in coastal megacities quantification framework. *British Journal of Environment and Climate Change, 3*(3), 378–401.

Sonn, C. C. (2002). Immigrant adaptation. In A. T. Fisher, C. C. Sonn & B. J. Bishop, *Psychological sense of community* (pp. 205–222). Boston, MA: Springer US.

Sonn, C. C., & Fisher, A. T. (1998). Sense of community: Community resilient responses to oppression and change. *Journal of Community Psychology, 26*(5), 457–472.

Sonn, C. C., & Fisher, A. T. (1999). Aspiration to community: Community responses to rejection. *Journal of Community Psychology, 27*(6), 715–726.

Tierney, K. (2006). Social inequality, hazards, and disasters. In R. J. Daniels, D. F. Kettl, & H. Kunreuther (Eds.), *On risk and disaster: Lessons from Hurricane Katrina* (pp. 109–137). Philadelphia, PA: University of Pennsylvania Press.

Tierney, K. J. (1989). Improving theory and research on hazard mitigation: Political economy and organizational perspectives. *Annual Review of Sociology, 7*(3), 367–396.

Tierney, K. J. (2007). From the margins to the mainstream? Disaster research at the crossroads. *Annual Review of Sociology, 33*, 503–525.

Tobin, G. A. (1999). Sustainability and community resilience: The holy grail of hazards planning? *Global Environmental Change Part B: Environmental Hazards, 1*(1), 13–25.

Twigger-Ross, C. L., & Uzzell, D. L. (1996). Place and identity processes. *Journal of Environmental Psychology, 16*(3), 205–220.

UN-HABITAT. (1987). *A new agenda for human settlements* (p. 32). Nairobi, Kenya: United Nations Commission on Human Settlements.

United Nations. (2004). *Living with risk – A global review of disaster reduction initiatives*. Geneva, Switzerland: Author.

Vale, L. J., & Campanella, T. J. (2005). *The resilient city*. New York, NY: Oxford University Press.

Verrucci, E., Rossetto, T., Twigg, J., & Adams, B. J. (2012). Multi-disciplinary Indicators for evaluating the Seismic Resilience of Urban Areas. Presented at the 14th Annual World Conference on Earthquake Engineering, Lisboa, Portugal.

Walker, B., & Salt, D. (2006). *Resilience thinking*. Washington, DC: Island Press.

Whetten, D. A. (1989). What constitutes a theoretical contribution? *Academy of Management Review*, *14*(4), 490–495.

Wilson, G. A. (2012). *Community resilience and environmental transitions*. New York, NY: Earthscan.

Wilson, G. A. (2014). Community resilience: Path dependency, lock-in effects and transitional ruptures. *Journal of Environmental Planning and Management*, *57*(1), 1–26.

Wisner, B., Blaikie, P., Cannon, T., & Davis, I. (2003). *At risk: Natural hazards, people's vulnerability and disasters (Second.)*. New York, NY: Routledge.

Zhou, H., Wang, J., Wan, J., & Jia, H. (2010). Resilience to natural hazards: A geographic perspective. *Natural Hazards*, *53*(1), 21–41.

Zottarelli, L. (2008). Post-hurricane Katrina employment recovery: The interaction of race and place. *Social Science Quarterly*, *89*(3), 592–607.

Temporal and spatial change in disaster resilience in US counties, 2010–2015

Susan L. Cutter ⓘ and Sahar Derakhshan

ABSTRACT

The allure of disaster resilience studies continues to garner interest by policy makers and academics alike. While there are advances in assessing communities' resilience to natural hazards at different scales, monitoring changes in resilience lags behind. This paper updates the 2010 Baseline Resilience Index for Communities (BRIC) using the six different domains of disaster resilience. The purpose is to test for significant spatial and temporal change in county index values by providing a comparative assessment of increased or decreased resilience over a five-year period across the U.S. The significance of monitoring change is to empirically demonstrate the dynamic nature of resilience and the causal mechanisms that lead to increasing or decreasing resilience in places. Such evidence sets the stage for implementing intervention policies or programs designed to enhance disaster resilience. The national distribution of BRIC index values in 2015 is generally similar to the 2010 BRIC distribution, but there are some notable regional differences. For example, there is a decrease in resilience in the South, the Great Lakes states, and the Central U.S., with improved resilience in the west and Pacific Coast states. The individual domains of institutional resilience and community capital, have the highest and lowest level of variation, respectively.

1. Introduction

Increasing policy and practitioner interest in resilience have facilitated efforts to develop measurement schemes to assess disaster resilience at community to regional scales. Such efforts range from the Rockefeller Foundation's 100 Resilient Cities (Rockefeller Foundation, 2017) to Robert Wood Johnson Foundation's Culture of Health program and its articulation with community resilience principals (Chandra et al., 2011; Plough et al., 2013) to FEMA's community resilience indicators (FEMA, 2016).

The landscape of disaster resilience measurement tools and approaches is quite messy as many researchers have noted (Beccari, 2016; Cutter, 2016a). Some of the approaches are place-specific such as urban (Rockefeller Foundation, 2017; UNISDR, 2017a) or rural areas (Cutter, Ash, & Emrich, 2016), while others take a broader spatial perspective but narrow the measurement to one type of disaster causal agent such as flooding (Szoenyi et al., 2016) or one type of affected environment such as coastal (Sempier, Swann, Emmer, Sempier, & Schneider, 2010) or one specific sector such as critical infrastructure (Petit et al., 2013).

As the number of measurement schemes and tools proliferate, so too has the number of critiques of community resilience measurement (Linkov et al., 2013; Matyas & Pelling, 2015; Ostadtaghizadeh, Ardalan, Paton, Jabbari, & Khankeh, 2015; Sharifi, 2016). At present, the measurement critiques of resilience fall into five thematic areas:

(1) Basic operationalization of the concept with consistent and discrete variables (Sharifi, 2016);
(2) Prediction and validation of outcomes (Linkov et al., 2013)
(3) Scales and units of analysis including the spatial mismatch between local actions, responsibilities, and the processes that shape resilience at regional to global scales (MacKinnon & Derickson, 2013).
(4) Utilization of reductionist approaches to address resilience to what and resilience for whom (Cutter, 2016b; Weichselgartner & Kelman, 2015).
(5) Lack of dynamic approaches to measure changes over time and across space (Cutter, 2016b; Sharifi, 2016).

This paper addresses the last critique of resilience assessments by providing a time series of changes in resilience patterns. Employing a replication of the baseline resilience indicators for communities (BRIC) over two different time periods (2010 and 2015), the dynamic nature of community resilience is illustrated through an examination of spatial and temporal changes in the baseline and the drivers of increasing or decreasing resilience across U.S. counties.

2. Modeling resilience

There are many definitional concerns regarding the term resilience, which are quite variable depending on disciplinary perspective (humanities, social sciences, natural sciences, health sciences, or engineering), methodological approach (qualitative to quantitative), and theoretical orientation (e.g. positivism, realism, pragmatism, post-structuralism) (Alexander, 2013; Weichselgartner & Kelman, 2015). Even when narrowing down the term to focus on community resilience to disasters, it still remains a somewhat amorphous concept, although increasingly there are common elements that define what it means for a community to become resilient (Cutter, 2016a; Patel, Rogers, Amlot, & Rubin, 2017).

For this article, we use the definition put forward in the US National Academies report (USNAS, 2012, p. 1) which defines disaster resilience as 'the ability to prepare and plan for, absorb, recover from, and more successfully adapt to adverse events.' This policy-relevant definition is consistent with the one used by the UNISDR (2017b) as it develops indicators for the implementation of the Sendai Framework:

> "The ability of a system, community, or society exposed to hazards to resist, absorb, accommodate, adapt to, transform and recover from the effects of a hazard in a timely and efficient manner, including through the preservation and restoration of its essential basic structures and functions through risk management."

2.1. BRIC – The Baseline Resilience Index for Communities

The BRIC measurement follows the theoretical framework of a place-based model for understanding community resilience to natural hazards, called Disaster Resilience of Place (DROP) model (Cutter et al., 2008). DROP conveys the relationship between the inherent resilience (e.g. pre-existing capabilities and assets that allow a community to function during non-crisis time periods) and the adaptive resilience (capacities and flexibility that allow communities to adjust and develop creative solutions to post-event problems) in places (Rose, 2007; Tierney & Bruneau, 2007). Both of these forms of resilience affect the ability of a community to recover from the event. BRIC, however, only measures the inherent resilience (pre-event) within communities. It does not measure the processes or strategies within communities for coping with, undertaking rapid change, or adapting to some adverse event or disturbance in both short and longer term contexts.

BRIC employs a capitals approach to understanding community disaster resilience. A capitals approach suggests that communities are integrated systems, made up of differing and intersecting subsystems (or capitals) – economic, social, natural, and so forth – all of which contribute to its functioning and well-being (USNAS, 2012). While the specific types of capitals may vary depending on the study, there is consistency in the overall types of capitals that influence disaster recovery – natural (environmental), built environment (physical), economic (financial), human, social, political, and cultural (Miles, 2015; NIST, 2016; Ritchie & Gill, 2011). There are six different capitals used as sub-indices in the construction of BRIC and 49 individual variables used to represent the capitals. The variables initially derived from and justified via the extant literature and were then thoroughly tested for their applicability for measuring each of the capitals (Cutter, Ash, & Emrich, 2014, 2016). For example, communities with relatively stable populations are more resilient institutionally, than those that have rapid population shifts which stress provision of government services such as water/sewage, public safety, and building inspections in rapidly growing communities, or significantly reduce such local government services in declining populations. Both situations (rapid growth or decline) stress a community and influence its capacity to respond to and recover from a disaster (Sherrieb, Norris, & Galea, 2010; USNAS, 2012). The variable (population change over previous five-year period) was used to capture such pre-existing conditions. In addition to their known importance in influencing disaster resilience, we used variables found in national databases and open data sources to insure consistency across study units and over time, ease of data collection, and replication. The variables and sub-index structure are as follows (Table 1): social (10 variables), economic (8 variables), community capital (7 variables), institutional (10 variables), housing /infrastructural (9 variables), and environmental (5 variables).

2.2. Data and methods

Data were gathered the same sources as the 2010 construction (Tables 1 and 2) with updates to 2015 or as close to that year as possible from the source data. As was the case for the 2010 BRIC, the primary data source for the 2015 BRIC update was the US Census Bureau's database and American Community Survey five-year estimates from 2010 to 2014. Updated data from American Red Cross and National Flood Insurance

Table 1. Indicator sets and variable descriptions (Cutter et al., 2014).

Resilience concept	Variable description
Social resilience	
Educational attainment equality	Absolute difference between % population over 25 with college education and % population over 25 with less than high school education (Inverted: more equality is more resilient)
Pre-retirement age	% Population below 65 years of age
Transportation Access	% Households with at least one vehicle
Communication capacity	% Households with telephone service available
English language competency	% Population proficient English speakers
Non-special needs	% Population without sensory, physical, or mental disability
Health insurance	% Population under age 65 with health insurance
Mental health support	Psychosocial support facilities per 10,000 persons
Food provisioning capacity	Food security rate
Physician access	Physicians per 10,000 persons
Economic resilience	
Homeownership	% Owner-occupied housing units
Employment rate	% Labor force employed
Race/ethnicity income equality	Gini coefficient (Inverted; more equality is more resilient)
Non-dependence on primary/tourism sectors	% Employees not in farming, fishing, forestry, extractive industry, or tourism
Gender income equality	Absolute difference between male and female median income (Inverted; more equality is more resilient)
Business size	Ratio of large to small businesses
Large retail-regional/national geographic distribution	Large retail stores per 10,000 persons
Federal employment	% Labor force employed by federal government
Community Capital resilience	
Place attachment-not recent immigrants	% Population not foreign-born persons who came to US within previous 5 years
Place attachment-native born residents	% Population born in state of current residence
Political engagement	% Voting age population participating in recent election
Social capital-religious organizations	# affiliated with a religious organization per 10,000 persons
Social capital-civic organizations	# civic organizations per 10,000 persons
Social capital-disaster volunteerism	# Red Cross volunteers per 10,000 persons
Citizen disaster preparedness and response skills	# Red Cross training workshop participants per 10,000 persons
Institutional resilience	
Mitigation spending	Ten year average per capita spending for mitigation projects
Flood insurance coverage	% Housing units covered by National Flood Insurance Program
Performance regimes-state capital	Distance from county seat to state capital (Inverted; closer is more resilient)
Performance regimes-nearest metro area	Distance from county seat to nearest county seat within a Metropolitan Statistical Area (Inverted; closer is more resilient)
Political & jurisdictional fragmentation	# governments and special districts per 10,000 persons (Inverted; less fragmented is more resilient)
Disaster aid experience	# Presidential Disaster Declarations divided by # of loss-causing hazard events for ten year period
Local disaster training	% Population in communities covered by Citizen Corps programs
Population stability	Population change over previous five-year period (Inverted; less change is more resilient)
Nuclear plant accident planning	% Population within 10 miles of nuclear power plant
Crop insurance coverage	Crop insurance policies per square mile
Housing/ Infrastructural resilience	
Sturdier housing types	% housing units not mobile homes
Temporary housing availability	% vacant housing units that are for rent
Medical care capacity	# hospital beds per 10,000 persons
Evacuation routes	Major road egress points per 10,000 persons
Housing stock construction quality	% housing units built prior to 1970 or after 2000
Temporary shelter availability	# hotels/motels per 10,000 persons
School restoration potential	# public schools per 10,000 persons
Industrial re-supply potential	Rail miles per square mile

(Continued)

Table 1. Continued.

Resilience concept	Variable description
High speed internet infrastructure	% Population with access to broadband internet service
Environmental resilience	
Local food suppliers	Farms marketing products through Community Supported Agriculture per 10,000 persons
Natural flood buffers	% Land in wetlands
Efficient energy use	Megawatt hours per energy consumer (Inverted; more efficient is more resilient)
Pervious surfaces	Average percent perviousness
Efficient water use	Water Supply Stress Index (Inverted; more efficient is more resilient)

Table 2. Data sources and period of coverage.

Dataset	Data Provider	Time period	
		2015 BRIC	2010 BRIC
United States Federal Government			
USA Counties Database	Census Bureau	2015	2007
Small Area Health Insurance Estimates		2014	2010
County Business Patterns		2014	2009–2010
Tiger/Line		2015	2010
Current Population Estimate		2010, 2015	2005, 2012
American Community Survey 5Year Estimates		2010–2014	2006–2010
Hazard Mitigation Grant Program	Federal Emergency Management Agency	2006–2015	2000–2009
Presidential Disaster Declarations Database		2006–2015	2000–2009
Citizen Corps Councils		2010	2010
National Flood Insurance Program		2015	2010
National Land Cover Dataset	US Geological Survey	2011	2006
National Atlas		2014	2010
Quarterly Census of Employment and Wages	Bureau of Labor Statistics	2015	2010
Census of Agriculture	Department of Agriculture	2012	2007
National Center for Education Statistics (NCES)	Department of Education	2014	2009–2010
Electricity Consumption	Energy Information Administration	2015	2010
Broadband Internet Access	Federal Communications Commission	2016	2010
Water Supply Stress Index	Forest Service	2015	2005
Nuclear Power Plants Database	Nuclear Regulatory Commission	2012	2010
Railroad Network	Oak Ridge National Laboratory	2014	2010
Academic/Non Profit			
Spatial Hazard Events and Losses Database for the US (SHELDUS)	Univ. South Carolina Hazards and Vulnerability Research Institute	2006–2015	2000–2009
Religious Congregations and Membership Study	Association of Religion Data Archives	2010	2010
Farm Subsidies	Environmental Working Group	2014	2010
Map the Meal Gap	Feeding America	2014	2010
2016 Presidential Election	Politico	2016	2012
Volunteers and Preparedness Training	American Red Cross	2016	2013
County Health Rankings and Roadmaps	Robert Wood Johnson Foundation & Univ. Wisconsin	2016	N/A
Million Dollar Database	Dun and Bradstreet	N/A	2010

Program were provided by personal contacts – the same as the previous version. The Hazard Events and Losses Database (SHELDUS) accessed through the Hazards and Vulnerability Research Institute (HVRI) at the University of South Carolina provided estimates of disaster aid experience. The Citizen Corps Councils database is no longer available online, hence the latest available data are from 2010. Overall, there are no major differences in data sources and format between 2010 and 2015 with the exception of the County

Health Ranking and Roadmaps database used in lieu of Dun and Bradstreet's Million Dollar Database (the original source data for the 2010 BRIC formulation).

The 2015 BRIC database includes 3,142 counties including Alaska and Hawaii. These states were not included in 2010 BRIC (which only contained the 3,108 counties in the conterminous U.S. because of limited data availability). Only one county unit (Bedford City, Virginia) appeared in 2010 BRIC but not in the 2015 version as it rejoined Bedford County in 2013.

Given the utilization of national sources, the data update for all counties is consistent in terms of temporal and spatial quality, with the geographic exceptions noted in the previous paragraph. While there was one substitution for a data source between 2010 and 2015, it is the authors opinion that this was minor and did not have any specific implications for the findings given the nature of the overall construction of the index (see below). The variables included in the time-series analysis represent the best and most current data available from national sources.

2.3. Data processing

The first step for processing the gathered data is to normalize and convert the raw count variables into percentages, averages, rates, or differences (Cutter et al., 2014). In order to normalize the variables, we use a min–max scaling of 0 to 1 for each indicator. Larger values (closer to 1) represent greater resilience according to the BRIC construction framework, so in some instances indicator values needed to be inverted to reflect this cardinality. The second step is to compute the value for each of the six resilience sub-indices by calculating the mean value in each sub-index. The overall 2015 BRIC score sums the six resilience sub-indices to create the final score. Resilience scores theoretically vary from zero to six indicating lower to higher resilience levels, respectively. Therefore, the final BRIC score provides a relative measure not an absolute measurement of resilience among places.

As was done for 2010 BRIC, we used Cronbach's alpha to test the internal consistency of the 2015 resilience index construction, given the addition of Alaska and Hawaii to the dataset. The alpha value for all the 49 indicators is 0.623, which is a moderate level of interrelatedness and close to the alpha value for 2010 index (0.65) (Cutter et al., 2014). As expected, there is little inter-item correlation among the sub-indices (Table 3) implying general independence from one another. Lastly, there was no significant inter-item covariance between indicators. Social and housing/infrastructural resilience variables are the

Table 3. Cronbach's alpha and inter-item correlation mean for each resilience category's indicators.

Resilience category	Number of indicators	Cronbach's alpha 2015	Inter-item correlation (mean) 2015	Cronbach's alpha 2010 BRIC
Social	10	0.470	0.077***	0.533
Economic	8	0.123	−0.005***	0.242
Community Capital	7	0.344	0.059***	0.317
Institutional	10	−0.117	0.003***	0.074
Housing/ Infrastructural	9	0.428	0.095***	0.411
Environmental	5	−0.126	−0.015***	−0.028
BRIC Total	49	0.623	0.023**	0.650

Based on the ANOVA test: $*p < .05$, $**p < .01$, and $***p < .001$.

most internally consistent, and the environmental, economic and institutional variables are the least internally consistent, again as was found in the earlier construction of 2010 BRIC.

3. Disaster resilience in 2015

For the entire U.S. including Alaska and Hawaii, 2015 BRIC scores ranged from a minimum of 2.059 to the maximum value is 3.234, with a mean of 2.73 and a standard deviation of 0.147. The spatial distribution of BRIC scores is mapped using standard deviations (Figures 1 and 2). For visualization purposes, maps for Alaska and Hawaii appear separately (Figure 2).

The spatial patterns in 2015 are similar to 2010 BRIC for the continental US. Regionally, higher resilience values extend from the upper Midwest to western Ohio. Southern Louisiana also has very high resilience scores, a seemingly counter-intuitive finding explained below. The lowest resilience scores continue in the West, Southwest, Texas borderlands, and Appalachian counties. Pockets of low resilience also appear in Arkansas, southwestern Florida, eastern Texas, and in eastern Alabama. Alaskan counties have among the lowest resilience scores in the nation.

Resilience scores in each six categories vary statistically and spatially (Figures 2 and 3). Social resilience has a higher average compared to other categories (mean = 0.66), while housing/infrastructural resilience has the lowest average (mean = 0.26) and the most variation in scores (standard deviation of 0.059). The environmental resilience index has the least variation (standard deviation of 0.036).

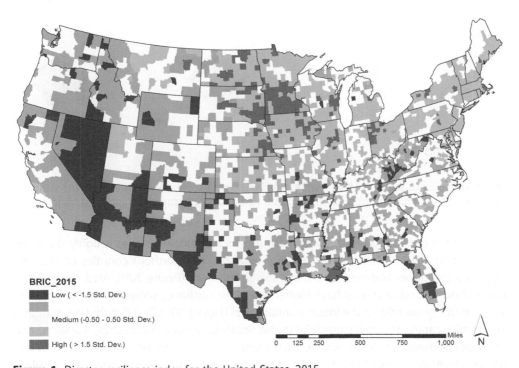

Figure 1. Disaster resilience index for the United States, 2015.

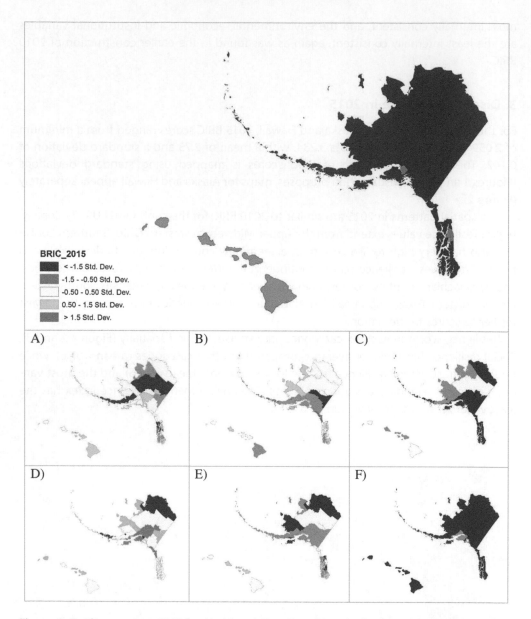

Figure 2. Resilience scores 2015 for Alaska and Hawaii, with scores for six categories: (A) social, (B) economic, (C) community capital, (D) institutional, (E) housing/infrastructural, and (F) environmental.

The spatial distribution of the drivers for the six sub-indexes shows roughly the same pattern as in 2010 (Cutter et al., 2014). For example, the northern counties of U.S. and especially the upper Midwest have more social resilience (Figure 3(A)). Also, the areas of upper Midwest and Northeast have higher economic resilience scores, while the opposite is true in many counties in the intermountain west (Figure 3(B)). The distribution of higher community capital scores (Figure 3(C)) in the South (Louisiana, Mississippi, Alabama) and the upper Midwest (Iowa, Minnesota, Wisconsin) is explained by large percentages of religious adherents as well as low mobility among residents (indicating strong levels of place

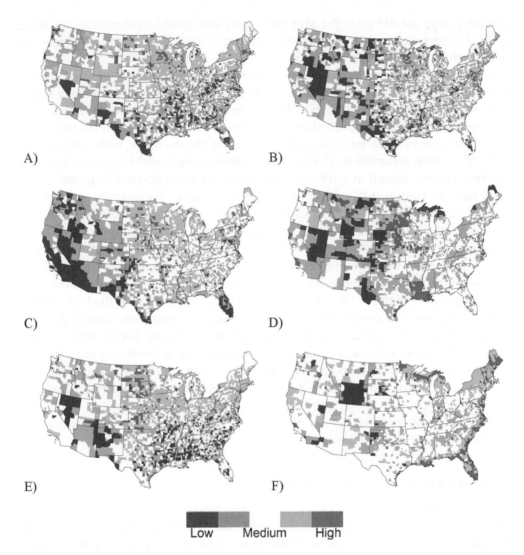

Figure 3. Resilience scores 2015 for the continental U.S. for six categories: (A) social, (B) economic, (C) community capital, (D) institutional, (E) housing/infrastructural, and (F) environmental. Data classified into low, medium, and high using the standard deviation method.

attachment). The western U.S. and Florida have the lowest level of community capital resilience, reflective of less civic engagement, volunteerism, and place attachment. The highest institutional resilience scores (Figure 3(D)) occur in the Midwest and in Louisiana, the latter explained by high mitigation spending as a direct consequence of large and frequent Presidential disaster declarations. In addition to mitigation spending, southern Louisiana also has a large percentage of housing covered by the National Flood Insurance Program (NFIP), another indicator of institutional resilience. The spatial pattern for housing/infrastructural resilience (Figure 3(E)) is similar to social resilience. Lastly, the environmental resilience is higher along the eastern coast and northeast (Figure 3(F)), while Wyoming scores lowest on environmental resilience due to lower levels of efficient energy use. Although the 2015 BRIC shows little spatial variation from the broad regional patterns in the 2010 version, there has been considerable change in individual counties that warrant further discussion.

Higher scores in the economic and institutional capital along with community capital characterize the most resilient regions. Lower scores on social and community capital appear in less resilient regions, but these differences are not statistically significant between the more resilient and less resilient places. When examining the ten most resilient counties, for example, institutional resilience is the primary driver, with eight counties rating in the 90th percentile on this sub-index (Table 4). The overall top-ranked Louisiana counties rank the highest in institutional resilience and environmental (especially in the percentage of land in wetlands). In addition, six counties of these top counties also appear in the top 90th percentile for community capital and economic capital. In contrast, the primary driver for the least resilient counties is the lack of community capital; all of the counties fall in the lowest 10th percentile on the sub-index. Similarly, eight of the least resilient counties are in the lowest percentile on social resilience. The ten least resilient counties rank in the lowest 10th percentile on at least three of the six sub-indices (Table 4). An interesting aspect of this is the case of Aleutians East, Alaska which is the least resilient county in the nation scoring the lowest in the country on community capital, yet it has the second highest score for economic resilience. This is largely due to higher median household incomes, racial and ethnic income equality, and a diversified economy (e.g. manufacturing including canning and processing of fish and fish products, management/administration, and natural resource extraction). The higher ranking in economic resilience appears as an outlier given the very low placement of the county in the other capitals – notably community and institutional – and is insufficient to improve the overall score given the additive nature of BRIC.

4. The geography of temporal change

The difference between 2010 and 2015 BRIC values range from –0.247 to 0.392 (mean= 0.084, standard deviation = 0.061), with Alaska and Hawaii deleted from the spatial change analysis due to the lack of 2010 BRIC scores. A one-way ANOVA indicates a statistically significant difference between 2010 and 2015 BRIC means (F = 6.5, p < 0.05). The BRIC scores for each 2010 and 2015 dataset were placed into three classes based on standard deviations from the yearly mean – low (<–0.5 Std. Dev.), medium (–0.5–0.5 Std. Dev.), and high (>0.5 Std. Dev.) and then assigned a numerical value of –1 for low, 0 for medium, and + 1 for high. The difference between values for 2010 and 2015 were calculated (values range from –2 to +2) and then mapped into a five class category based on the resilience change score. The majority of counties (79.2%) had no change in their resilience index score. For those that registered a change in overall score, approximately 11% of counties are more resilient and 9.5% less resilient. The computed changes show a decrease in resilience in the South, in the Great Lakes states, and in the Great Plains states, and improved resilience in the mountain west, Pacific Coast states, and portions of the South (Figure 4).

The Moran's I spatial autocorrelation test for spatial clustering was used to examine the spatial changes in overall resilience. The test showed a significant positive spatial correlation (Moran's I = 0.22, z= 20.64, p < 0.05) with a slight clustering of similar values. Clusters of counties with improved resilience were in the mountain west from Montana in the

Table 4. Counties with highest and lowest resilience scores, with scores and ranks for each category (ranks are in parenthesis).

Rank	County, State	2015 BRIC score	Social	Economic	Housing/Infrastructural	Community Capital	Institutional	Environmental
Most resilient								
1	St. Charles, LA	3.234	0.74 (53)	0.50 (254)	0.27 (1360)	0.41 (423)	0.64 (1)	0.66 (51)
2	St. Bernard, LA	3.150	0.72 (260)	0.46 (1224)	0.30 (646)	0.37 (1267)	0.56 (5)	0.70 (11)
3	St. John the Baptist, LA	3.139	0.68 (1346)	0.48 (597)	0.28 (1097)	0.39 (852)	0.63 (2)	0.66 (48)
4	Brown, MN	3.113	0.73 (95)	0.50 (202)	0.30 (688)	0.47 (19)	0.50 (52)	0.58 (944)
5	Putnam, OH	3.111	0.77 (5)	0.53 (13)	0.27 (1179)	0.48 (11)	0.44 (425)	0.58 (1195)
6	Red Lake, MN	3.097	0.71 (339)	0.52 (50)	0.27 (1263)	0.43 (158)	0.50 (72)	0.64 (102)
7	Nicollet, MN	3.095	0.73 (77)	0.53 (20)	0.27 (1199)	0.41 (370)	0.53 (17)	0.59 (840)
8	Eddy, ND	3.086	0.70 (642)	0.44 (2031)	0.35 (110)	0.55 (2)	0.46 (226)	0.55 (2524)
9	Fillmore, NE	3.084	0.70 (721)	0.47 (754)	0.33 (190)	0.57 (1)	0.43 (647)	0.55 (2749)
10	Waseca, MN	3.084	0.71 (337)	0.51 (116)	0.31 (430)	0.43 (157)	0.52 (25)	0.57 (1454)
Least resilient								
3142	Aleutians East, AK	2.059	0.53 (3121)	0.59 (2)	0.18 (2783)	0.04 (3142)	0.30 (3132)	0.39 (3128)
3141	Kalawao, HI	2.105	0.58 (2977)	0.34 (3113)	0.25 (1828)	0.13 (3140)	0.39 (1749)	0.38 (3131)
3140	North Slope, AK	2.143	0.65 (1874)	0.44 (1948)	0.13 (3070)	0.24 (3080)	0.32 (3107)	0.33 (3142)
3139	Denali, AK	2.145	0.68 (1269)	0.32 (3134)	0.21 (2450)	0.17 (3135)	0.37 (2772)	0.36 (3141)
3138	La Paz, AZ	2.156	0.55 (3079)	0.40 (2817)	0.13 (3076)	0.19 (3130)	0.37 (2775)	0.48 (3097)
3137	Presidio, TX	2.162	0.49 (3137)	0.40 (2792)	0.14 (3063)	0.21 (3119)	0.32 (3108)	0.57 (1564)
3136	Hudspeth, TX	2.193	0.48 (3139)	0.41 (2677)	0.16 (2960)	0.19 (3129)	0.35 (2977)	0.57 (1535)
3135	Esmeralda, NV	2.201	0.57 (3024)	0.30 (3140)	0.24 (2001)	0.20 (3128)	0.31 (3122)	0.56 (2505)
3134	Nye, NV	2.204	0.58 (2942)	0.37 (3047)	0.08 (3138)	0.24 (3077)	0.34 (3058)	0.56 (2376)
3133	Stewart, GA	2.248	0.49 (3138)	0.40 (2891)	0.14 (3037)	0.25 (3065)	0.39 (2034)	0.56 (2238)

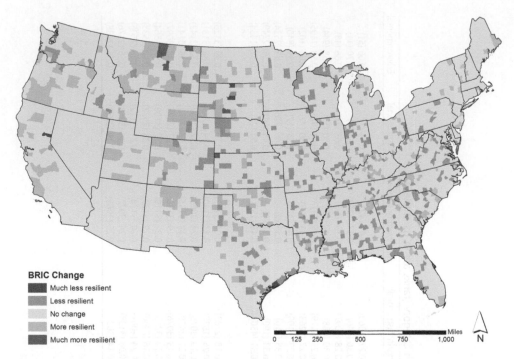

Figure 4. Temporal change in BRIC disaster resilience index for the United States, 2010–2015.

north to Arizona and New Mexico in the south. Counties with decreasing resilience clustered in the Deep South region stretching from Louisiana to the Florida panhandle. Overall, about 88.9% of counties do not have a significantly higher or lower value compared to their neighboring counties. However, there are 53 outliers (22 change from high–low, 31 change low–high), which represents about 1.7% of all counties. The outliers depicted in Figure 5 indicate counties where their resilience index has changed significantly more than their neighboring counties. Explanations for these changes in resilience appear in the next section.

5. Drivers of spatiotemporal change

To more fully explain the regional outliers and clusters of BRIC values from 2010 to 2015, the variability in each sub-index was calculated and mapped for the continental U.S. only (Figure 6). The greatest variation in the two time periods is seen in institutional resilience (Std. Dev = 0.60), and the least variation is in community capital resilience (Std. Dev = 0.49). The average normalized score for community capital resilience has the greater positive change (more resilient), and institutional resilience has the greater negative change (less resilient). On average, environmental and infrastructural sub-indices are more resilient in 2015 compared to 2010, and social and economic sub-indices are less resilient. Several individual variables influenced the temporal change in sub-indices. For example, the difference in internet access (infrastructural resilience) has the lowest average and greatest variation (Std. Dev= 0.29), reflecting more relative change of internet access among communities. The variation in this variable reflects differences in communities

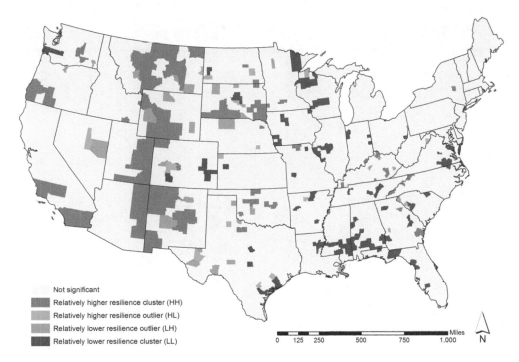

Figure 5. Spatial change in BRIC using spatial clusters and spatial outliers of BRIC from Moran's I, 2010–2015.

that met or failed to meet the advanced threshold for service set by the Federal Communications Commission with an existing speed benchmark is 25 Mbps/3 Mbps for fixed services (it was set at 3 Mbps/768 kbps fixed broadband in 2010) (FCC, 2016).

Another example is disaster aid experience (an input into institutional resilience), showing more Presidential Disaster Declarations per loss-causing events since 2010 in some counties. The average positive change in community capital is due to more political engagement and increases in number of Red Cross volunteers. Improvement in social resilience is partly a function of more people affiliating with religious organizations. Improvements in economic resilience are a function of more race/ethnicity income equality (economic resilience) in selected counties.

In order to compare the spatial distribution of clusters and outliers in each sub-index, the Moran's I spatial autocorrelation test was done for each resilience category (Table 5). The BRIC change data has a statistically significant positive spatial correlation and a

Table 5. Moran's I spatial autocorrelation test for the BRIC change data in each sub-index.

Resilience category	Moran's I	z-Score	p-Value
Social	0.28	26.20	<.05
Economic	0.07	6.79	<.05
Community Capital	0.17	15.57	<.05
Institutional	0.54	50.22	<.05
Housing/ Infrastructural	0.28	25.92	<.05
Environmental	0.44	41.39	<.05
2015 BRIC	0.22	20.64	<.05

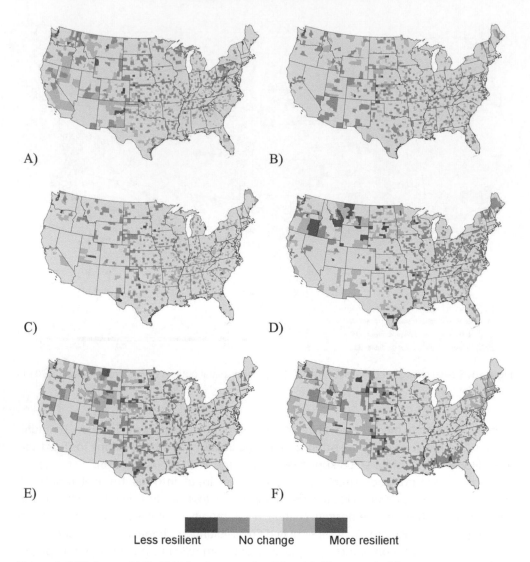

Figure 6. BRIC change 2010–2015 for six categories: (A) social, (B) economic, (C) community capital, (D) institutional, (E) housing/infrastructural, and (F) environmental. Data classified into low, medium, and high using the standard deviation method.

clustering of similar values in all sub-indices. However, the institutional resilience change (Moran's I = 0.54) is more spatially clustered than other sub-indices. On the other hand, the economic resilience change has the lowest level of spatially clustered values among sub-indices (Moran's I = 0.07) and there are no distinct patterns of clusters or outliers in this category. However, the number of outliers (high–low and low–high) in the economic resilience change includes 74 counties (2.4% of all counties), which is more than other categories and illustrates communities with drastic change from 2010 to 2015 in their economic resilience.

The counties exhibiting the most change in their BRIC scores (change values of −2 much less resilient; +2 much more resilient) are shown in Table 6 (and Figure 4). Counties that

Table 6. Counties with highest variation, with differences between 2010 and 2015 scores in each category and their five level class in parenthesis (–2 to +2).

County, state	BRIC difference	Social	Economic	Housing/ infrastructural	Community capital	Institutional	Environmental
Much less resilient (Score class –2)							
Sully County, SD	–0.192	0.022 (0)	0.042 (0)	0.017 (1)	0.011 (0)	–0.10 (–2)	–0.18 (–2)
Campbell County, SD	–0.169	–0.02 (–1)	–0.02 (0)	0.023 (0)	0.005 (0)	–0.09 (–2)	–0.05 (–1)
Richmond County, VA	–0.164	–0.10 (–2)	0.035 (0)	–0.06 (0)	0.001 (0)	–0.006 (–1)	–0.02 (–1)
Matagorda County, TX	–0.131	0.018 (0)	0.071 (0)	–0.05 (–1)	–0.14 (–1)	0.027 (0)	–0.04 (–1)
Sierra County, CA	–0.080	–0.02 (–1)	0.025 (–1)	–0.10 (–1)	–0.01 (0)	0.016 (0)	0.019 (1)
Much more resilient (Score class + 2)							
Mellette County, SD	0.357	–0.01 (0)	0.086 (1)	0.120 (2)	0.017 (0)	0.082 (2)	0.064 (1)
Petroleum County, MT	0.353	0.107 (2)	0.098 (0)	0.072 (1)	–0.01 (–1)	0.102 (1)	–0.01 (0)
Daniels County, MT	0.346	0.057 (0)	0.077 (1)	0.111 (0)	–0.003 (0)	0.147 (0)	–0.04 (–1)
Lexington city, VA	0.308	0.134 (2)	0.068 (0)	0.010 (0)	0.050 (1)	0.017 (0)	0.026 (0)
Blaine County, MT	0.286	0.015 (0)	0.069 (1)	0.090 (2)	–0.01 (0)	0.105 (2)	0.019 (1)
San Juan County, WA	0.273	0.016 (0)	0.045 (0)	0.020 (0)	0.130 (1)	0.041 (0)	0.018 (0)
Cheyenne County, KS	0.265	0.013 (0)	0.125 (2)	–0.02 (–1)	–0.01 (–1)	0.048 (0)	0.117 (1)
Winchester city, VA	0.248	0.057 (1)	0.069 (1)	–0.01 (0)	0.082 (1)	0.017 (0)	0.033 (0)

became less resilient over the five-year period did so because of lower environmental, social and institutional scores. On the other hand, those counties that became more resilient improved in their economic, social, and institutional resilience compared to other counties. The 2015 resilience score for Sully County (2015 population 1,426) and Campbell County (2015 population 1,397) in South Dakota, for example, is lower than their score in 2010 due to reductions in the environmental and institutional resilience sub-indexes. Specifically, the loss of resilience is a reflection of less efficient energy use, fewer local food suppliers, less mitigation spending, and less crop insurance coverage in these two counties. Richmond County, Virginia (population 8,774) also became less resilient due to a reduction in its social and environmental resilience, specifically less equality in educational attainment and less efficient energy use. The reduction in resilience in Sierra County, California (population 2,947), on the other hand, was driven by less health insurance coverage and access to physicians, as well as decreased employment and homeownership rates.

The resilience score of Mellette County in South Dakota has the greatest positive change in all counties since 2010, based on changes in institutional and infrastructural resilience – a function of more internet access, disaster aid experience, and population stability (population 2,102). The improved score of Petroleum County (population 489) in Montana is reflective of greater social, infrastructural and institutional resilience in the five-year period. The county improved its disaster resilience due to improved health insurance coverage, more mental health support facilities, more disaster aid experience, and sturdier housing types. Greater economic resilience is the main factor for the improved score in Daniels County, Montana (population 1,755), primarily due to higher employment rates and more income equality since 2010.

Overall, the significant improvement in resilience among those counties with the greatest change results from increases in economic resilience and institutional resilience from 2010–2015. However, when examined individually as mentioned above, the capitals driving the change are quite variable for individual places as expected. The same is true for those counties experiencing the greatest decline in resilience.

6. Discussion

In this study of 2015 BRIC, the regional concentration of highest and lowest resilience index values is similar to the 2010 BRIC distribution for the continental US. While, the distribution of changes in the BRIC index from 2010 to 2015 indicates a decrease in resilience in the South, in the Great Lakes states, and in the Central US states, improved resilience occurred in the mountain west and Pacific Coast states. The drivers of change in improving resilience were institutional variables, particularly more political engagement as measured by the percentage of the voting age population participating in the 2016 election. Improvements in internet access also account for greater resilience among counties. Among the six sub-indexes, the most stable element in resilience change is community capital which shows minimal variability over the 5-year time period. On the other hand, institutional resilience appears to be the most volatile with the highest variability among all the sub-indexes. Specifically, differences in the five-year average of mitigation spending, flood and crop insurance coverage, population stability, and disaster-aid experience offer concrete explanations as to why some counties increase or decrease in their resilience. Clearly, we need more detailed analyses at the local scale to completely understand the landscape of resilience for local places.

The temporal analysis suggests a measurable improvement in overall in BRIC scores during the five-years. The community capital and housing/infrastructure capital remain static during the time-frame, while environmental resilience has declined slightly. The policy implications of this finding are insightful. For example, the community capital appears to be an ascribed characteristic of communities that is not easily changed because it relates to the basic core culture, essence, and ethos of communities. If this capital is not present within a community, there is little that national or state-level policies can do to improve this. Another capital that remains somewhat constant is housing/infrastructure resilience, although it was slightly lower in 2015 than in 2010. Interestingly, the housing and infrastructure in many communities is built out, that is, there is little opportunity to change many of the variables within the sub-index such as rail miles per square mile, or the per capita number of schools, medical care facilities, or hotels/motels in such a short time frame (5 years). The one variable that did change within the sub-index was access to broadband internet as noted previously.

More opportunities for change appear in the social, economic, and institutional capital. Often predicated on external (national) policies implemented at state levels, improvements in educational equality (compulsory education), provision of affordable health insurance, disaster insurance coverage (crop and flood), and mitigation spending are all dependent on federal resources sent to states and counties. The receipt of federal disaster assistance (post-event response and longer term recovery) improves institutional resilience as seen by the relatively high overall resilience in scores like southern Louisiana. Such financial disaster assistance also may help stimulate local economies post-event that are otherwise ailing by improving employment rates, incomes, and changes in employment sectors. There is considerable speculation in the literature on the role of disasters (and the resources that flow from Presidential Disaster Declarations) as economic engines, although this is has not been systematically addressed (Strobl, 2011; Xiao, 2011).

Environmental resilience shows a decline nationally based on the last five years, but has both regional and local significance because of local land use and development regulations. The decrease in pervious surfaces is often due to development pressures and local decision-making about growth and prosperity. Locally-driven policies to restrict development especially in sensitive environments may be the most efficient way to improve environmental resilience, but such policies are controlled by municipal and/or unincorporated county jurisdictions, who may be averse to such restrictions, because of other priorities.

The field of disaster resilience measurement, especially in the US is still in the development stage with very few empirically-based studies at the national scale. This partially explains the descriptive nature of this paper in examining temporal and spatial changes in county-level disaster resilience. The lack of validation is an ongoing issue of concern, not only for this research but also for the disaster resilience field in general. For example, there are few studies that attempt validation of resilience metrics (Burton, 2015; Sherrieb et al., 2010) using external measures. Using Mississippi counties, Sherrieb et al. validated their capital measures – economic development, social – using survey data on collective efficacy (as a measure of capacities) and the Social Vulnerability Index (SoVI) in an effort to examine the county's ability to bounce back and reduce the physical and mental health problems after disasters. The other study, also focused on Mississippi examined the role of community resilience in predicting disparities in the temporal and spatial patterning of disaster recovery, finding that social and economic resilience were the primary drivers of recovery, five-years after Hurricane Katrina destroyed much of the Gulf Coast (Burton, 2015). Lastly, there have been no formal studies examining the validity of inherent resilience in moderating the impact of and recovery from disasters, although both the Sherrieb et al. (2010) and Burton (2015) studies suggest that the pre-existing resilience within communities does partially influence post-disaster outcomes in terms of recovery trajectories as well as reductions in health outcomes.

7. Conclusion

We cannot manage what we do not measure and this statement belies the policy relevance of resilience measurement. Yet, there is the fundamental issue of perspective – should we measure **what is** in order to establish some starting point by which to assess how various policies or interventions work to improve disaster resilience? Or should we measure **what should be** in order to predict a desired outcome (e.g. less deaths, fewer economic losses, faster recovery)? The latter brings up a number of ongoing and interesting issues starting with the very nature of the outcome – Resilience for whom? Resilience to what? These questions posit a differential in the assessment of the outcome that could have dramatically dissimilar results depending on for whom and to what. Such questions may or may not be amenable to statistical prediction. Moreover, focusing on predicted quantitative outcomes ignores the process role of resilience as a capacity building endeavor, its articulation best captured by qualitative methods.

The Baseline Resilience Index for Communities (BRIC) is a first approximation for measuring **what is** – documenting a baseline for estimating and comparing the communities' resilience to natural hazards, spatially and temporally. Community leaders and policy makers can monitor progress over time for each county, assess how a county compares to neighboring counties, identify the capitals contributing to higher or lower levels of resilience, and *de facto* show where improvements are possible. Such information provides a basis for longer term spatial (or development planning) in addition to emergency preparedness. BRIC provides a suitable descriptive and dynamic monitoring tool to identify larger-scale patterns and drivers of changes in disaster resilience. Whether at the individual county level or a larger regional scale, BRIC captures and describes the pre-existing inherent resilience within places. At a broad policy level BRIC is currently in use as part of the US Federal Emergency Management Agency's National Risk Index (USFEMA, 2017), an interactive tool designed for state and local officials to provide a more holistic picture of 'at risk communities.' The National Risk Index combines hazard likelihood, social vulnerability, built environment exposure, and community resilience into a baseline multi-hazard product at both the county and census-tract levels. The purpose is to provide the risk-based knowledge for inclusion in local, state, and federal mitigation plans. BRIC is the community resilience component in the index.

But at this juncture, the BRIC measurement does not address the discrepancies between local actions, roles and responsibilities of key stakeholders, and governance structures and capacities which influence the underlying processes of enhancing resilience that ultimately produce differential resilience patterns on the landscape that are evident today. More detailed and contextually specific analyses of local places are needed to address this concern and possible policy remedies for improvement. The findings presented here demonstrate the variability in disaster resilience patterns and that they do change over relatively small time periods. Many of the counties with the greatest change in scores – improving resilience or declining resilience – are sparsely populated with low densities. This could account for higher percentages of change given the low numbers. This aspect of the empirical measurement of change certainly warrants further attention.

The truism, all disasters are local implies that differences in exposure and vulnerability within a community means that the impacts are also variant within the affected region. It makes sense that disaster resilience is also locally variable and while national policies may set the stage or context for improvement, ultimately it is left to the community to articulate and implement resilience goals and ways to measure its progress in achieving them. The variations in and drivers of disaster resilience aptly illustrates that one size fits all, top-down (national to local) policy interventions to improve resilience become problematic at best, and ineffective at worst as it ignores the uniqueness of places and the willingness of communities to enhance their resilience to the next disaster coming their way.

Disclosure statement

No potential conflict of interest was reported by the authors.

Funding

The authors received no direct funding for this work, but it is an outgrowth of previous research supported by the U.S. National Science Foundation, Award #1132755.

ORCID

Susan L. Cutter ⓘ http://orcid.org/0000-0002-7005-8596

References

Alexander, D. E. (2013). Resilience and disaster risk reduction: An etymological journey. *Natural Hazards and Earth System Sciences, 13*, 2707–2716. doi:10.5194/nhess-13-2707-2013

Beccari, B. (2016). A comparative analysis of disaster risk, vulnerability and resilience composite indicators. *PLoS Currents.* doi:10.1371/currents.dis.453df025e34b682e9737f95070f9b970

Burton, C. G. (2015). A validation of metrics for community resilience to natural hazards and disasters using the recovery from hurricane katrina as a case study. *Annals of the Association of American Geographers, 105*(1), 67–86. doi:10.1080/00045608.2014.960039

Chandra, A., Acosta, J., Stern, S., Uscher-Pines, L., Williams, M. V., Yeung, D., … Meredith, L. S. (2011). *Building community resilience to disasters.* Santa Monica, CA: Rand.

Cutter, S. L. (2016a). The landscape of disaster resilience indicators in the USA. *Natural Hazards, 80*, 741–758. doi:10.1007/s11069-015-1993-2

Cutter, S. L. (2016b). Resilience to what? Resilience for whom? *The Geographical Journal, 182*(2), 110–113. doi:10.1111/geoj.12174

Cutter, S. L., Ash, K. D., & Emrich, C. T. (2014). The geographies of community disaster resilience. *Global Environmental Change, 29*, 65–77. doi:10.1016/j.gloenvcha.2014.08.005

Cutter, S. L., Ash, K. D., & Emrich, C. T. (2016). Urban-rural differences in disaster resilience. *Annals of the Am Assoc Geographers, 106*(6), 1236–1252.

Cutter, S. L., Barnes, L., Berry, M., Burton, C., Evans, E., Tate, E., & Webb, J. (2008). A place-based model for understanding community resilience to natural disasters. *Global Environmental Change, 18*(4), 598–606. doi:10.1016/j.gloenvcha.2008.07.013

Federal Communications Commission. (2016). *2016 broadband progress report.* Washington, DC: Federal Communications Commission. Retrieved from https://www.fcc.gov/document/fcc-releases-2016-broadband-progress-report

FEMA Mitigation Framework Leadership Group. (2016). *Draft interagency concept for community resilience indicators and national-level measures.* Washington, DC: FEMA. Retrieved from https://www.fema.gov/media-library-data/1466085676217-a14e229a461adfa574a5d03041a6297c/FEMA-CRI-Draft-Concept-Paper-508_Jun_2016.pdf

Linkov, I., Eisenberg, D. A., Bates, M. E., Chang, D., Convertino, M., Allen, J. H., … Seager, T. P. (2013). Measurable resilience for actionable policy. *Environmental Science and Technology, 47*(18), 10108–10110.

MacKinnon, D., & Derickson, K. D. (2013). From resilience to resourcefulness: A critique of resilience policy and activism. *Progress in Human Geography, 37*(2), 253–270. doi:10.1177/0309132512454775

Matyas, D., & Pelling, M. (2015). Positioning resilience for 2015: The role of resistance, incremental adjustment and transformation in disaster risk management policy. *Disasters, 39*(Suppl. 1), S1–S18. doi:10.1111/disa.12107

Miles, S. (2015). Foundations of community disaster resilience: Well-being, identity, services, and capitals. *Environmental Hazards, 14*(2), 103–121. doi:10.1080/17477891.2014.999018

National Institute of Standards and Technology Special Publication. (2016). *Community resilience planning guide for buildings and infrastructure systems* (Vols. 1 and 2, pp. 1190–1191). Washington, DC: Department of Commerce. Retrieved from https://nvlpubs.nist.gov/nistpubs/SpecialPublications/NIST.SP.1190v1.pdf1

Ostadtaghizadeh, A., Ardalan, A., Paton, D., Jabbari, H., & Khankeh, H. R. (2015, April 8). Community disaster resilience: A systematic review on assessment models and tools. *PLOS Currents Disasters.* doi:10.1371/currents.dis.f224ef8efbdfcf1d508dd0de4d8210ed

Patel, S. S., Rogers, M. B., Amlot, R., & Rubin, G. J. (2017, February 1). What do we mean by 'community resilience'?: A systematic literature review of how it is defined in the literature. *PLOS Currents Disasters.* doi:10.1371/currents.dis.db775aff25efc5ac4f0660ad9c9f7db2

Petit, F. D., Bassett, G. W., Black, R., Buehring, W. A., Collins, M. J., Dickinson, D. C., … Peereenboom, J. P. (2013). *Resilience measurement index: An indicator of critical infrastructure resilience.* Argonne, IL: Argonne National Laboratory ANL/DIS-13-01. Retrieved from http://www.ipd.anl.gov/anlpubs/2013/07/76797.pdf

Plough, A., Fielding, J. E., Chandra, A., Wiliams, M., Eisenman, D., Wells, K. B., ... Magaña, A. (2013). Building community disaster resilience: Perspectives from a large urban county department of public health. *American Journal of Public Health, 103*(7), 1190–1197. doi:10.2105/AJPH.2013.301268

Ritchie, L. A. and Gill, D. A. (2011). Considering community capitals in disaster recovery and resilience, PERI Scope, Public Entity Risk Institute, 14(2). https://www.riskinstitute.org/peri/component/option,com_deeppockets/task,catContShow/cat,86/id,1086/Itemid,84/

Rockefeller Foundation. (2017). *100 resilient cities*. Retrieved from http://www.100resilientcities.org

Rose, A. (2007). Economic resilience to natural and man-made hazards: Multidisciplinary origins and contextual dimensions. *Environmental Hazards, 7*, 383–398.

Sempier, T. T., Swann, D. L., Emmer, R., Sempier, S. H., & Schneider, M. (2010). *Coastal Community Resilience Index: A Community Self-Assessment. NOAA Mississippi-Alabama Sea Grant MASGP-08-14*. Retrieved from http://masgc.org/assets/uploads/publications/662/coastal_community_resilience_index.pdf

Sharifi, A. (2016). A critical review of selected tools for assessing community resilience. *Ecological Indicators, 69*, 629–647. doi:10.1016/j.ecolind.2016.05.023

Sherrieb, K., Norris, F. H., & Galea, S. (2010). Measuring capacities for community resilience. *Social Indicators Research, 99*(2), 227–247. doi:10.1007/s11205-010-9576-9

Strobl, E. (2011). The economic growth impact of hurricanes: Evidence from U.S. Coastal counties. *Review of Economics and Statistics, 93*(2), 575–589. doi:10.1162/REST_a_00082

Szoenyi, M., Nash, D., Bürer, M., Keating, A., McQuistan, C., & Campbell, K. (2016). *Risk nexus: Measuring flood resilience-our approach*. Zurich: Zurich Insurance. Retrieved from https://www.zurich.com/_/media/dbe/corporate/docs/corporate-responsibility/zurich-flood-resilience-measurement-paper-feb-2016.pdf?la=en&hash=6F2D3D582431B4C10A134997F450DAA19C7F1E67

Tierney, K., & Bruneau, M. (2007, May–June). Conceptualizing and measuring resilience: A key to disaster loss reduction. *TR-News, 250*, 14–17.

UN Office for Disaster Risk Reduction. (2017a). *Disaster resilience scorecard for cities*. Geneva: UNISDR. Retrieved from http://www.unisdr.org/campaign/resilientcities/assets/documents/guidelines/03%20Preliminary%20Assessment_Disaster%20resilience%20scorecard%20for%20cities_UNISDR.pdf

UN Office for Disaster Risk Reduction. (2017b). *Terminology*. Retrieved from https://www.unisdr.org/we/inform/terminology#letter-r

US Federal Emergency Management Agency. (2017). *National Risk Index*. https://data.femadata.com/FIMA/NHRAP/NationalRiskIndex/National_Risk_Index_Summary.pdf

US National Academy of Science. (2012). *Disaster resilience: A national imperative*. Washington, DC: The National Academies Press.

Weichselgartner, J., & Kelman, I. (2015). Geographies of resilience: Challenges and opportunities of a descriptive concept. *Progress in Human Geography, 39*(3), 249–267. doi:10.1177/0309132513518834

Xiao, Y. (2011). Local economic impacts of natural disasters. *Journal of Regional Science, 51*(4), 804–820. doi:10.1111/j.1467-9787.2011.00717.x

Assessing community resilience: mapping the community rating system (CRS) against the 6C-4R frameworks

Ajita Atreya and Howard Kunreuther

ABSTRACT

This paper introduces an holistic approach to assessing community resilience in the United States with respect to hazards by inventorying a community's strengths: Financial, Human, Natural, Physical, Political and Social, as sources of capital (6 Capitals, or 6Cs) and characterizing four properties of its resilience (4R) (robustness, resourcefulness, redundancy and rapidity). We link the 6C-4R framework to the National Flood Insurance Program's (NFIP) Community Rating System (CRS). There is a positive correlation between the 6C-4R framework and the CRS, demonstrating the extent to which that system might therefore be used to measure resilience holistically in an effective and efficient manner. We also provide illustrative examples of resilience strategies linked to the 6C-4R framework that were adopted by Ottawa, Illinois, Birmingham, Alabama and Cedar Rapids, Iowa, USA, the last being a community that joined the CRS in 2010 following a severe flood in 2008. The CRS does not cover all the aspects of a community's status and activities so in order to make informed decisions and prioritize the implementation of resilience-improving activities, community-wide cost–benefit analyses of CRS activities would be useful in the future as inputs for further developing a strategy for reducing future flood losses.

1. Introduction

Community resilience has become an important concern due to the increasing scale and frequency of natural and technological disasters. The term 'resilience' has been defined in the literature in a variety of ways. A report by the Community and Regional Resilience Institute (CARRI, 2013) presented forty-five of the most widely cited definitions of resilience published between 1973 and 2009, twenty-five of which defined community resilience.[1] The report concluded that it is difficult to select one definition of resilience as 'the best' since each offers a positive contribution to the concept within its domain.

The Zurich Flood Resilience Alliance defines resilience as: *The ability of a system, community, or society to pursue its social, ecological, and economic development and growth objectives, while managing its disaster risk over time, in a mutually reinforcing way* (Keating et al., 2017). This definition underscores that resilience is not only the ability to manage disaster risk but also the ability to grow and thrive in the face of the hazards faced by the system, community or society. We use Keating et al.'s definition as it closely aligns with the framework that we discuss in the paper.

Resilience is a desirable property of natural and human systems given the wide variety of threats that communities, regions and countries face today (UN/ISDR, 2002). There is general agreement that the term implies a proactive and positive expression of community engagement in reducing losses from natural hazards (Cutter et al., 2008). There is also consensus that resilience is multifaceted and can be achieved via improvements in several sectors of a community such as social, economic, institutional, infrastructural and natural/ecological (Adger, 2000; Bruneau et al., 2003; Burton, 2015; Cutter, 2016; Cutter, Burton, & Emrich, 2010; Gunderson & Holling, 2001; Norris, Stevens, Pfefferbaum, Wyche, & Pfefferbaum, 2008; NRC, 2012; Tierney & Bruneau, 2007; Tobin, 1999).

Several tools have been developed by federal agencies, national and international organizations, communities and cities to measure resilience at different levels (local to global), pre-disaster to post-disaster, top down to bottom up, hazard specific to general proneness to multi-hazards. For example, the San Francisco Planning and Urban Research (SPUR) framework measures resilience by determining the community's ability to recover from earthquakes with a focus on building and infrastructure.[2] Baseline Resilience Indicators for Communities (BRIC) measure overall pre-existing community resilience by evaluating the community's economic, social, institutional, ecosystem, and infrastructure capacities.[3] Resilience United States (ResilUS) measures recovery over time of critical infrastructure.[4] The NOAA Coastal Resilience Index assists communities by specifying key indicators that provide a preliminary assessment of a community's disaster resilience.[5] This index can be used to determine whether the community can function well after a disaster in the areas of critical infrastructure, transportation, community plans, mitigation measures, social systems and business plans.

Other tools such as THRIVE (Toolkit for Health and Resilience in Vulnerable Environments) developed by the Prevention Institute are focused on health resilience (see https://www.preventioninstitute.org/tools/thrive-tool-health-resilience-vulnerable-environments). CART (Communities Advancing Resilience Toolkit) proposed by Pfefferbaum, Pfefferbaum, and Van Horn (2011) is designed to enhance community resilience by bringing stakeholders together to address community issues through a process that includes assessment, feedback, planning, and action. This publicly available toolkit includes a field-tested community resilience survey and other assessment and analytical tools (Pfefferbaum et al., 2013). Other measurement tools include community conversations, neighborhood infrastructure maps, key informant interviews, capacity and vulnerability assessment, etc.

In this paper, we aim to demonstrate that a holistic framework with its associated matrices can be used to assess community resilience by systematically inventorying a community's strengths across several sources of capital. Focusing on flood resilience, we tie a multi-dimensional resilience framework to the National Flood Insurance Program's (NFIP) Community Rating System (CRS) and demonstrate how the CRS could play a role in gauging and thereby enhancing that community resilience. The CRS is a voluntary incentive program that encourages communities to go beyond the NFIP's basic requirements to reduce the flood risk. It is a long-standing program that measures and rewards the extent of efforts undertaken to improve flood resilience of a community. If it can be shown to map well to the multi-dimensional resilience framework, which we believe our analysis achieves, then this serves the dual purpose of easily conveying via CRS scores the extent of resilience and justifying the insurance discounts provided to community residents in the floodplain.

In the next section we introduce our framework – the 6 capitals (the 6Cs) measures of resilience – financial, human, natural, physical, political, social, and the properties of resilience (the 4Rs). Section 3 describes the NFIP's CRS program and shows how it can be tied to the 6C-4R framework. In Section 4, we discuss the strategies that communities have adopted to become more resilient. A case study of Cedar Rapids, Iowa (Section 5) examines that city in the context of the 6C-4Rs and proposes strategies to further enhance resilience in the community. The concluding section summarizes the key findings and proposes future research directions.

2. The 6C (Capitals) and 4R (Resilience Properties) framework

2.1. The 6 Capital (6C) framework

The Sustainable Livelihood (SL) framework (Chambers & Conway, 1992) included five types of capital: *financial, human, natural, physical and social* that characterize the wellbeing of a community holistically while at the same time suggesting ways to enhance its resiliency.[6] Another important element for building resilience in a community is the ability of those in power to support change: We thus introduce a sixth C, *political capital* that measures the ability of the community to influence decisions, engage state and federal agencies in projects, and discover new funding sources to enhance community resiliency. It is important to note that there are interdependencies across these six capitals so that enhancing one form of capital may lead to resilience or vulnerability in some or all of the other five. For example, spending more on infrastructure (physical capital) may tax communities' wherewithal (i.e. reducing its financial capital).[7]

2.1.1. Financial capital
Financial capital denotes the financial resources at the household and community levels that can support the community becoming more resilient to future hazards. Financial resources at the household level consist of savings and investments, access to credit, wealth (property) and regular inflow of income via wages and salaries, pensions and other retirement benefits and remittances. At the community level, financial resources to reduce or recoup losses may include insurance and funds to invest in mitigation measures. Among other factors, financial capital can be measured by evaluating household incomes, property values and investments in a community (Peacock et al., 2010). Higher levels of financial capital increase the abilities of households and communities to absorb disaster impacts and speed the recovery process.

2.1.2. Human capital
Human capital refers to skills, knowledge, health and access to labor that enable people to cope with and recover from the impacts of hazards. Human capital can be measured by evaluating employment records, formal and informal educational attainments and infant mortality rates (Cumming et al., 2005; Cutter, Boruff, & Shirley, 2003). Knowledge

about hazard exposure and hazard protection, and awareness of resources available for disaster management are important aspects of human capital to foster community resilience. For example, knowledge about the flood risk helps individuals make informed decisions on where to locate their homes.

2.1.3. Natural capital

There is wide variation in the resources that make up natural capital, ranging from intangible public goods such as the atmosphere and biodiversity, to divisible assets used directly for production (trees, land, etc.). Natural capital can be measured using the proportion of wetlands, undeveloped lands such as parks, and forests in a community (Cutter et al., 2008; Mayunga, 2007). Wetlands, in particular, soak up excess flood water and reduce the impacts of floods and therefore, can be considered a very important defense system against flooding.

2.1.4. Physical capital

Physical capital refers to infrastructure, such as electricity, water, and transportation lifelines, and the built environment of a community such as residential, commercial and public buildings. The ability of the community infrastructure to withstand the shocks of natural disaster is a proxy for the strength of its physical capital. Physical capital can be measured in a number of ways, such as the percentage of buildings constructed to code as a proxy for the strength of the structures in the community; the number of such as schools and other buldings that can be used as temporary shelters in case of evacuation; and the percentage of non-mobile homes, representing the vulnerability of the population (Cutter et al., 2010; John Heinz III Center, 2002). Strengthening physical capital is expensive so that the economic health of a community largely dictates the resilience of its physical capital.

2.1.5. Political capital

Political capital characterizes how decisions are made in the community and how outside resources are obtained and utilized to foster resiliency.[8] Individuals and special interest groups or persons possess political capital if they have the voice, power, and ability to influence the distribution of resources (Jacobs, 2007). In some communities, elected officials have voice and power, while in others, an unelected community leader may have the reputation of being the key decision maker. Following Jacobs (2007), we define political capital as the ability to influence decisions, engage state and federal agencies in the projects, discover new funding sources and possess the leverage to accomplish things. Communities that are strongly integrated with higher levels of government are more likely to receive the resources they need to cope and recover from a disaster (Berke, Kartez, & Wenger, 1993; Lindell & Prater, 2003).

2.1.6. Social capital

Social capital reflects networks such as political or civic bodies that increase people's trust and ability to work together and expand their access to wider institutions. Social capital can be measured through variables such as voter participation, the number and roles played by non-profit organizations, voluntary associations and religious organizations (Cutter et al., 2010; Mayunga, 2007). These measurements reflect the availability of networks to foster connectedness among the people living in the community should a disaster strike. The greater the membership in community organizations and associations, the higher is the collective action whereby individuals join together with other residents of the community to solve a common problem (Yip et al., 2007).

2.2. The 4 Resilience properties (4Rs)

Bruneau et al. (2003) and Tierney and Bruneau (2007) developed the 4R framework by characterizing four properties of resilient infrastructure: robustness, rapidity, redundancy and resourcefulness.

2.2.1. Robustness
Ability of a system to withstand given levels of stress without suffering degradation or loss of functionality.

2.2.2. Rapidity
Capacity to meet priorities and achieve goals in a timely manner to contain losses, recover functionality and avoid future disruption.

2.2.3. Redundancy
Extent to which alternative elements or other measures are substitutable, that is, capable of satisfying functional requirement in the event of disruption.

2.2.4. Resourcefulness
Capacity to identify problems, establish priorities, and mobilize resources when existing conditions threaten to disrupt the system.

Table 1 illustrates how the 6Cs and 4Rs intersect in their characterization of community resilience. For example, robust physical capital can be measured by the *ability of critical infrastructure and housing to withstand shock* that enhances resilience in a community. Similarly, the rapidity of the recovery process after a disaster is ensured by financial capital that is via *insurance availability*.

3. The NFIP's community rating system (CRS) program

The National Flood Insurance Program in the United States, established in 1968, provides flood insurance to homeowners residing in flood-prone communities. Those having a mortgage from a federally-insured lender are required to purchase flood insurance if they live in special flood hazard areas (SFHAs), defined as having a likelihood of annual flooding that is greater than 1 in 100.

In 1990, the NFIP introduced the Community Rating System (CRS), a voluntary incentive program that recognizes and encourages community floodplain management activities that exceed the minimum NFIP requirements. The CRS is a federal program that directly engages local governments in integrating flood risk reduction in their plans to enhance community flood resilience (Sadiq & Noonan, 2015). Currently, over 1200 communities participate in the CRS and earn points across 19 creditable activities in the following

Table 1. Indicators of 6Cs and their associated 4Rs.

Capital (C)/ Resilience (R)	Financial	Human	Natural	Physical	Political	Social
Robustness	Income/Wealth	Family emergency plans	Floodplain conservation	Ability of critical infrastructure and housing to withstand shock	Ability to manage a disaster efficiently	Proportion of socially vulnerable population
Rapidity	Insurance Availability			Multiple communication sources	Respond immediately to the disaster	Interval to restore life line services
Redundancy	Alternative sources of income	Skills for alternative income		Availability of alternate routes	Generate reports at local, state and federal level	Availability of housing options for victims
Resourcefulness	Funds set aside for emergency	Ability to maintain income sources	Regulations to limit the use of natural resources	Availability of materials for restoration	Creating funding for disaster management	Capacity to address the needs of victims

Table 2. CRS credits, associated classes, and the NFIP premium discounts.

CRS Class	Credit points	Premium reduction for residences in SFHA (%)	Premium reduction for residences outside SFHA (%)
1	>4500	45	10
2	4000–4499	40	10
3	3500–3999	35	10
4	3000–3499	30	10
5	2500–2999	25	10
6	2000–2499	20	10
7	1500–1999	15	5
8	1000–1499	10	5
9	500–999	5	5
10	0–499	None	None

four major categories: public information, mapping and regulations, flood damage reduction, warning and responses.

Based on credit points earned, communities achieve class levels that determine the amount of residents' flood insurance premium reductions. Premiums are discounted in increments of 5% for insured properties in the high risk SFHAs, and between 5% and 10% for those outside of the SFHAs. Table 2 shows the CRS classes, required credit points and associated premium reductions for SFHA and non-SFHA properties in communities that participate in the CRS program.

All the activities under the four major categories are listed in Table 3. The numbers in parenthesis denote the maximum points or credits that a community can earn for that

Table 3. Activities included in the community rating system activity series.

Public Information (Series 300)
310 Elevation Certificates (116)
320 Map Information Service (90)
330 Outreach Projects (350)
340 Hazard Disclosure (80)
350 Flood Protection Information (125)
360 Flood Protection Assistance (110)
370 Flood Insurance Promotion (110)
Mapping and Regulations (Series 400)
410 Floodplain Mapping (802)
420 Open Space Preservation (2020)
430 Higher Regulatory Standards (2042)
440 Flood Data Maintenance (222)
450 Storm water Management (755)
Flood Damage Reduction (Series 500)
510 Floodplain Management Planning (622)
520 Acquisition and Relocation (2250)
530 Flood Protection (600)
540 Drainage System Maintenance (570)
Warning and Response (Series 600)
610 Flood Warning and Response (395)
620 Levees (235)
630 Dams (160)

------------➤ **310 Elevation Certificates**
(1) Credit for elevation certificate since CRS application (38 pts)
(2) Credit for post-FIRM elevation certificates (48 pts)
(3) Credit for pre-FIRM elevation certificates (30 pts)

Note: The maximum possible points are as of 2014.

particular activity. For example, a community can earn a maximum of 116 points for the activity 'Elevation Certificates.' Within each activity there are one or more discrete elements and each receives a certain number of credit points. 'Elevation Certificates' for instance, encompasses three elements: (1) credits for maintaining elevation certificates on all buildings in the SFHA since CRS application, (2) credits for maintaining elevation certificates before the CRS application but after the initial date of Flood Insurance Rate Maps (FIRM) and (3) credits for maintaining elevation certificates on buildings built pre-FIRM.

As seen in Table 3, series 300 (Public Information) credits programs that advise people about the flood hazard, encourage the purchase of flood insurance, and provide information about ways to reduce flood damage. Series 400 (Mapping and Regulations) credits programs that provide increased protection to new development. Series 500 (Flood Damage Reduction) credits programs that provide increased protection for areas in which existing development is at risk. Series 600 (Warning and Response) credits measures that protect life and property during a flood, through flood warnings and response programs.

Credit is also given for maintaining levees and dams and for programs that prepare for their potential failure. Communities can also propose alternative approaches to reducing the risk of flooding. FEMA evaluates the alternative approaches and provides appropriate credit for those activities. For example, communities that are prohibited by state law from adopting and enforcing building codes may submit comprehensive building construction regulations, administration and inspection procedures for review to determine the equivalent Building Code Effectiveness Grading Schedule (BCEGS) classification. Such regulations must be enforced throughout the community, not just in the floodplain.

The distribution of CRS points across the activities captures the community's preparedness with respect to flood damage. In this paper, we map CRS activities into the 6Cs and 4Rs with respect to their contribution to improving resiliency. Certain activities, such as open space preservation, storm water management, and buyouts cut across several capitals and impact several of the 4Rs. While CRS activities are not an exhaustive list that enhances community resilience, we believe they reflect a wide range of measures for tracking the progress of communities in preparing for and recovering from floods.

As shown in Table 4, each of the 19 CRS creditable activities can be related to at least one of the 6Cs which in turn links to one or more of the 4Rs. Most of these links are through the social (10 links), physical (9) and human capitals (7), with natural, financial and political capitals only showing three links in total between them.

4. Examples of strategies for resilient communities

In this section we examine particular strategies across each of the 6Cs adopted by exemplary communities that have appeared to improve their resilience to future flooding and have yielded benefits to residents via lower flood insurance premiums from involvement in the CRS program.

4.1. Financial capital

Property owners are likely to recover more quickly from a disaster if they have flood coverage than if they are uninsured. Insurance speeds recovery by making available funds for rebuilding in the immediate aftermath of a flood (Kousky & Shabman, 2012). A report

Table 4. Our interpretation of how CRS Activities Link to the 6C-4R framework.

Series	CRS Activities	Capital	Resilience Properties
Series 300	**Public Information Activities**		
	Elevation Certificates	Physical /Human	Robustness
	Map Information Services	Human	Resourcefulness
	Outreach Projects	Social/Human	Rapidity, Resourcefulness
	Hazard Disclosure	Social	Resourcefulness
	Flood Protection Information	Social	Resourcefulness
	Flood Protection Assistance	Human	Resourcefulness
	Flood Insurance Promotion	Financial	Rapidity/Redundancy
Series 400	**Mapping and Regulation**		
	Floodplain Mapping	Human/Social	Resourcefulness
	Open Space Preservation	Natural/Social	Robustness
	Higher Regulatory Standards	Physical/Social	Resourcefulness Robustness
	Flood Data Maintenance	Human	Resourcefulness
	Storm Water Management	Physical/Social	Robustness
Series 500	**Flood Damage Reduction**		
	Floodplain Management Planning	Social	Rapidity, Resourcefulness
	Acquisition and Relocation	Physical/Political	Robustness, Rapidity
	Flood Protection	Physical	Robustness
	Drainage System Maintenance	Physical /Social	Robustness
Series 600	**Warning and Response**		
	Flood Warning and Response	Physical / Social/Human	Resourcefulness
	Levees	Physical	Robustness
	Dams	Physical	Robustness

that examined the aftermath of Hurricane Katrina found that residences that had been insured before the storm were 37% more likely to have been rebuilt (Turnham et al., 2011).

In its 2016 annual report, the Technical Mapping Advisory Council (TMAC) recommended that

> FEMA should communicate to the property owner and the relevant interested parties on the cost of risk-rated insurance today and over time for new and existing structures to make the risk transparent. The data should include the benefits and cost that mitigation measures will have on these premiums.[9]

Risk-based insurance premiums provide accurate information to property owners on the nature of the flood hazard. Those who invest in mitigation measures to reduce their flood risk should receive a premium discount to reflect lower claims payments that would be expected in the future. In 2013, the CRS added *flood insurance promotion* under the **public information activities** (Table 3) for communities to play a more active role in encouraging households to purchase and maintain adequate flood insurance coverage.

4.2. Human capital

Public education plays a vital role in making people aware of the consequences of a disaster and ways to cope with the event. For example, public education about floods can encourage preparedness in the home, in schools, in the work place, and at healthcare facilities; communications can also raise public awareness of safe water sources, evacuation routes, flood zones, and community response plans (Keim, 2008).

Ottawa, Illinois located at the confluence of the Illinois River and the Fox River provides an example of how public education can lead to flood mitigation and flood control efforts so as to make the city resilient. The City held public outreach meetings in conjunction with

FEMA and formed a Flood Commission composed of staff and residents. It also developed a website with flood information that provides information on early warnings concerning the Fox and Illinois Rivers, current river stages and how properties throughout the city will be impacted by each flood stage and what residents should do in the event of a flood for flood safety.[10] (FEMA, 2011)

As a result of this initiative, the residents of Ottawa voted to buy out a school in the district that was heavily damaged when Hurricane Ike flooded the Illinois River and the Fox River in 2008. The Ottawa city council acquired the school and the surrounding property from the school district for $375,000 and built a new school well outside of the 100-year floodplain using FEMA funds combined with special state allocations and local resources. (FEMA, 2011; News Tribune, 2012). This activity also highlights the importance of *financial capital* and *political capital* to build resilience in a community. The decision to move the school to higher ground was justified in 2013 when flooding exceeded the 2008 record stage by 1.5 feet and would have seriously damaged the school if it had not been moved to safer ground.

After Ottawa's past and ongoing floodplain management activities the city qualified for a Class 5 CRS rating – making it among the top 6% of CRS communities. Ottawa earns CRS credit for public outreach, higher standards for mapping and regulations, preserving open space, effective stormwater management, natural hazard planning, acquisitions (buyouts), and drainage maintenance. As a result, Ottawa's residents receive a 25% reduction in their flood insurance premiums.

4.3. Physical capital

Well-enforced building codes are shown to reduce the damages caused by natural disasters as characterized by *avoided losses* and *reduction in insurance claims*. Whilst not explicitly related to flooding, jurisdictions with effective and well-enforced building codes reduce hail damage on the order of 10% to 20% compared to those without these codes based on Missouri hail claim insurance data from 2008 to 2010 (Czajkowski & Simmons, 2014). Hurricane Charley in 2004 demonstrated that homes built to the 1996 wind resistant standard in Florida had a claim frequency that was 60% less than those built prior to that year (Kunreuther & Michel-Kerjan, 2011). In Florida, homes built during that period sustained excessive damage from Hurricane Andrew in 1992 due to the lack of enforcement of building codes in the period between 1960 and 1992. If these building codes had been enforced, the damage from the hurricane would been reduced by at least 33% (Fronstin & Holtmann, 1994).

The CRS program incentivizes communities with higher standards (under series 430), one of which is adoption of stringent building codes. In addition to reducing the claims and losses, communities can earn up to 100 points in CRS credit by adopting International Series of building codes (or their equivalent) and hence reduce their insurance premium.

4.4. Natural capital

Removing properties (physical capital) from harm's way is an important policy intervention for making communities resilient, but these measures are very difficult to implement. Studies suggest that communities will be less likely to suffer severe losses during a natural disaster if the local government makes the right choices with respect to land-use planning programs (Burby, Deyle, Godschalk, & Olshansky, 2000). The CRS program incentivizes *acquisition* (thereby creating open space preservation (CRS Series 300)) by providing points based on the number of buildings that have been removed from the floodplain with a maximum of 2250 credit points, the largest of any activity. The number of credit points is a function of the percentage of all the buildings in the SFHA that have been acquired or relocated.

Acquisition programs have been highly effective in reducing future flood losses. The State of Iowa acquired over 1500 properties in Special Flood Hazard Areas (SFHAs) over the past two decades to mitigate adverse effects of riverine flooding. A study of 12 communities that removed the structures revealed that $98.7 million in potential losses were avoided from floods that occurred during this time period (DHS, 2010). Disaster declarations in 1993 and 2002 led the city of Marion, adjacent to Cedar Rapids, to acquire 15 properties and convert them to open space. Cedar Rapids and Marion reduced their potential losses from the 2008 floods by $2.2 million by acquiring and demolishing over 1300 damaged properties in the Special Flood Hazard Area (SFHA) (DHS, 2010).

The City of Birmingham, Alabama is subject to flash flooding, notably along Village Creek where several residential areas were repeatedly flooded, displacing residents and creating community hazards due to sewage backups. Since the 1980s, with a cooperative effort by the community, the state, and the federal government, 735 structures were removed from the flood plain at a cost of $37.5 million, thus avoiding losses of over $60 million from floods in 2000. In fact, these floods replicated the 13.6 ft. flood level of 1996, when hundreds of properties in Village Creek were damaged. There was almost no residential property damage, no relocation of residents and no disaster assistance in Village Creek. Elsewhere in the city, the damage was serious enough to result in a Presidential disaster declaration (FEMA, 2011).

Acquisition projects in this way not only remove the physical capital of a community from harm's way but also add to its *natural capital*. For example, the acquisition project in Village Creek returned the floodplain to its natural state as a retention basin for floodwaters. Additionally, the financial savings realized by the community as the direct result of implementing the acquisition project can be put toward other civic improvements/projects adding to the *financial capital* of the community.[11] City of Birmingham, that joined the CRS in 1994, is currently classified as a Class 6 community based on its floodplain management activities, and city residents receive a 20% discount on their flood insurance premium.

4.5. Social capital

The benefits of social relationships that accrue to individuals and groups through membership in a social network is particularly important for a community's post disaster resilience (Breton, 2001; Kimhi & Shamai, 2004; Magis, 2010). Table 4 t shows that the largest number of links to the CRS series is via social capital.

Many examples could be provided of the importance of such social networks in promoting resilience. For example, in the aftermath of Hurricane Katrina, Elliott, Haney, and Sams-Abiodun (2010) interviewed 100 residents from two neighborhoods in New Orleans, Louisiana, USA the Lower Ninth Ward (primarily made up of African-Americans who lived below the poverty line), and Lakeview (a neighborhood made up primarily of affluent white residents), to understand how networks – especially bonding and linking social capital – played a role in recovery after the storm. In Lakeview, individuals were about 14% more likely to contact a neighbor compared to individuals in the Lower Ninth Ward. Overall, it took more than twice as long for residents of the Lower Ninth Ward than their counterparts in Lakeview to return to their homes.

Non-governmental organizations (NGOs) play a significant role in the development of social capital and community empowerment (Islam, 2014). These NGOs also support human recovery after disasters by connecting individuals who lost their homes with local agencies and services (Chandra & Acosta, 2009). In this regard, the St. Bernard Project in Louisiana (see: http://www.stbernardproject.org) currently supports families financially in rebuilding their homes and offers services to promote psychological healing, and local, neighborhood-driven service centers.

4.6. Political capital

4.6.1. Regulation to enhance resiliency
Political capital plays a major role in the establishment of new regulations related to enhancing resiliency. In 2013, Illinois State, USA Senator Sue Rezin of R – 38th District helped establish the Illinois Valley Flood Resilience Alliance (IVFRA) bringing communities, local governments, and emergency personnel together to prepare for floods through education, communication, and the purchasing of flood protection materials. Currently there are 24 Certified Floodplain Managers (CFMs) in the 38th District as a result of Senator Rezin's involvement in forming the IVFRA.[12]

4.6.2. Creating new funds for relief efforts
In New York City, USA, 21,000 donors were motivated to contribute more than $60 million to the Mayor's Fund for the City's emergency response needs and long-term restoration efforts following Hurricane Sandy. The grant program helped small businesses to restore their operations through the replacement of damaged inventory, supplies and equipment. Such activities through the Mayor's initiative and his decisions – and those of the donors – helped the communities affected by Sandy to bounce back.[13]

In the CRS program, the role of political capital is extensive and cuts across most of the activities – notably in relation to flood damage reduction activities (series 500). Food damage reduction activities in CRS entails activities such floodplain management planning, acquisition and relocation which require political will and support.

5. Community resilience case study: Cedar Rapids, Iowa, USA

In June 2008, the City of Cedar Rapids was severely affected by an extreme flood event that impacted 7198 parcels[14] including 5390 households, displacing more than 18,000 residents and damaging 310 city facilities (Cedar Rapids Flood Facts, 2008). Since then, the city has adopted a number of activities in an effort to make the city more resilient to future floods. In this section we highlight the Community Rating System (CRS) activities undertaken by Cedar Rapids and how the actions relate to the 6C and 4R framework.

Table 5. CRS activities in the City of Cedar Rapids, Iowa linked to 6C and 4R Framework.

CRS Activities	Points Earned (of maximum allowable) (%)	Capital	Properties
Elevation certificates	48	Physical	Robustness
Outreach projects	11	Social	Rapidity /Resourcefulness
Hazard disclosure	18	Social	Resourcefulness
Flood protection information	12	Social	Resourcefulness
Floodplain mapping	8	Human	Robustness
Open space preservation	11	Natural	Robustness
Higher regulatory standards	14	Physical	Robustness/Redundancy
Flood data maintenance	47	Human	Resourcefulness
Stormwater management	10	Physical	Rapidity
Acquisition and relocation	58	Physical	Robustness

Note: The points earned (as of 2014) is calculated as the percentage of maximum possible points.

5.1. CRS Activities in Cedar Rapids

The City of Cedar Rapids began participating in the CRS program in 2010 and was initially classified as a class 8 community earning a 10% flood insurance premium discount for policies within the Special Flood Hazard Area (SFHA). Due to the city's efforts to reduce the risk from future floods, it is currently a class 6 community earning a 20% discount for flood insurance policies in the SFHA. Table 5 lists the CRS activities that the Cedar Rapids is credited for (as of 2014) and links them to the 6C-4R framework. CRS activities in Cedar Rapids address each form of capital that contributes to resiliency.

We compared credit points earned in 2011 to those earned in 2014 across the CRS activities (Table 6). The city improved from CRS class 8 to class 6 due to its *acquisition and relocation* of property; CRS points increased from 143 to 1307 in this category, as shown in Table 6. All the other components in the CRS had very slight increases or decreases in points except for *map information services* where the number of points decreased from 140 to 0 and *flood protection information* where there was a 35 point decrease from 50 to 15. Below we characterize some of the changes in the evaluation of the CRS between 2011 and 2014.

Table 6. Comparison of Points for CRS activities between 2011 and 2014 for Cedar Rapids.

Activities	Activity #	Points 2011	Points 2014	% change
Elevation certificate	C310	56	56	0%
Map information services	C320	140	0	−100%
Outreach projects	C330	54	41	−24%
Hazard disclosure	C340	15	15	0%
Flood protection information	C350	50	15	−70%
Flood protection assistance	C360	4	0	−100%
Flood hazard mapping	C410	72	73	1%
Open space preservation	C420	262	264	1%
Higher regulatory standards	C430	337	347	3%
Flood data maintenance	C440	124	116	−6%
Storm water mgmt	C450	113	87	−23%
Floodplain mgmt planning	C510	0	0	
Acquisition and relocation	C520	143	1307	814%
Flood protection	C530	0	0	
Drainage system maintenance	C540	30	0	−100%
Flood warning and response	C610	0	0	
Levees	C620	0	0	
Dams	C630	0	0	

5.1.1. Acquisition and relocation

As a result of the 2008 floods, property acquisition has become a major component of flood risk mitigation and the long-term recovery process (reducing the community's *physical capital* but at the same time reducing its vulnerability and enhancing its resilience). The City bought 1356 properties in the inundated areas using funds from two types of mitigation grant programs – FEMA's Hazard Mitigation Grant Program (HMGP) and HUD's Community Development Block Grant Program (CDBG). Under both grant programs, property owners were offered 107% of the pre-flood assessed value, adjusted downward in cases with 'duplication of benefits,' such as funds already received through flood insurance payouts or FEMA Individual Assistance grants (Tate, Strong, Kraus, & Xiong, 2015).

The HMGP grants are deed restricted against structural improvement, the land reverting permanently to open space, recreation use, or natural floodplains (Conrad, 1998). Ninety-seven properties in Cedar Rapids were acquired and demolished using the HMPG grant with an average cost of $79,286 (Tate et al., 2015). The remaining 1259 properties were acquired using the CDBG grant and included businesses that were demolished or redeveloped depending on their location, use, and severity of flooding. Four percent of the properties bought using the CDBG grant were demolished, while the rest were either revitalized or reconstructed. The area where most of the demolition took place was converted into green space that not only prevents losses from future floods but also creates public recreation opportunities (Tate et al., 2015). Additionally, some of the area acquired under CDBG will be used by the City of Cedar Rapids to accommodate its Flood Control System (FCS), a series of levees, floodwalls, gates and pump stations being built to withstand a repeat of the 2008 flood event.

The remaining properties bought using CDBG funds became a part of a Neighborhood Revitalization Area. A majority of revitalization properties were outside the 500-year floodplain. The focus of the Neighborhood Revitalization Area was to develop affordable replacement housing incorporating walkability and a sense of place, while also providing opportunities for recreation and transportation (Tate et al., 2015).

5.1.2. Map information services and flood protection information

Both the map services and flood protection provide information about the potential flood hazard (adding to *human capital*). However, in 2014 we see from Table 6 that for these two activities the CRS points were down significantly from 2011 because a re-evaluation in 2014 of the initiatives undertaken in these two categories revealed that they were not adequately documented. The city plans to document all the initiatives when performing their recertification of the CRS program, required every 5 years.[15]

5.1.3. Higher regulatory standards

Standards which limit development in SFHAs, such as requiring buildings to be above base flood elevation (BFE)[16], protecting critical facilities, local drainage protection, and building codes are activities included in this category. The adoption by Cedar Rapids of the International Building Code in 2009 has led to improved protection standards for flood, wind, and other hazards over previous codes and aligns the city with the standard for Iowa. Additionally, the city has implemented a regulation requiring that new construction must have one foot of freeboard above the BFE. Well-enforced building codes strengthen the *physical capital* of the community.

5.1.4. Open space preservation

Of the 1356 properties that the City of Cedar Rapids acquired, 556 are located in a proposed greenway project that will add 110 acres to the city's open space at a cost of approximately $56 million through the FEMA and HUD mitigation grant programs, as noted above. Open space in a community not only creates an amenity value that is capitalized into property prices (Atreya, Kriesel, & Mullen, 2016) adding to the *financial capital* of a community through improved tax base, but also enhances hazard risk mitigation by avoiding future property losses.

Other environmental benefits of preserving open space include expanded ecological habitat, flood storage and conveyance, and recreational opportunities which adds to the *natural capital* of a community. In Cedar Rapids, the planned activities for the newly created open space via buyouts are: 8-block promenade, 4 miles of restored river edge, 8 acres of wetland, 15 acres of playing fields, and 12 miles of trail. The promenade, which includes restored river edge and trails, provide the recreational value while the preservation of wetlands can act as a natural defense structure to diminish the impacts of floods.

5.1.5. Stormwater management

Stormwater management reduces the quantity and improves the quality of the stormwater runoff. In the CRS there are four stormwater management elements: (a) stormwater management regulation that provides credit points for regulating development on a case-by-case basis to ensure that the peak flow of stormwater runoff from each site will not exceed the pre-development runoff; (b) watershed masterplan for regulating development; (c) erosion and sedimentation control regulations for land disturbed by construction or farming, and (d) water quality regulation to reduce stormwater runoff.

In Cedar Rapids, two cost-share programs for installation of Stormwater Best Management Practices (BMPs) are available to both residential and commercial/industrial property owners. The program provides technical and financial assistance for implementing stormwater BMPs. The financial assistance under the residential 'EZ program' includes reimbursement of up to 50% of the stormwater BMP project cost or $2000, whichever is less. The commercial/industrial program will also pay up to 50% of the BMP cost, but with no project cap. Fundable Stormwater BMPs must infiltrate stormwater, thus rainbarrels and cisterns are not currently funded through either of these programs. Common stormwater BMPs include the creation of rain gardens, rain barrels, redirecting downspouts, pervious pavement, and soil quality restoration (SQR).[17] All of these activities reduce the impact of urbanization that produce large fast moving stormwater runoff volumes. The stormwater BMP prevents stormwater pollution, improving the quality of *natural capital* and reducing the impact of floods to *physical capital*. The CRS does not provide credit for erosion and sediment control regulations or water quality unless those measures are enforced throughout the entire community.

5.1.6. Hazard disclosure

Hazard disclosure informs prospective property buyers of potential flood hazards before the lenders notifies them of the need for flood insurance. In Cedar Rapids, residents can see the anticipated inundation area based on river height up to 34.5 ft in a GIS map maintained by the city's Information Technology (IT) department. The map is searchable by address and has a number of basemap images including one taken shortly after the crest in 2008. A layer that indicates the floodplain boundaries based on FEMA's Flood Insurance Rate Maps is also available. The flood insurance premiums in Cedar Rapids are based on the Flood Insurance Rate Maps that became effective in 2010. The 2010 flood insurance rate maps changed the flood risk category of approximately 1900 properties in Cedar Rapids, half of which were designated in higher flood risk category and the other half in a lower flood risk category.

6. The effect on flood damages now and in the future: promising, but not yet proven

In 2014, several counties in eastern Iowa, including Linn County where the City of Cedar Rapids is located, experienced a severe flash flood that affected numerous areas across the city outside of the riverine floodplain. It was reported that the mitigation actions completed prior to 2014 resulted in an estimated $77 million in avoided damage to housing in 20 Iowa counties, $33.4 million of which was avoided losses in Linn County.[18], Whether the losses avoided were a result of the improvements in resilience discussed above has not yet been determined.

In October 2016, Cedar Rapids was hit by yet another riverine flood which led thousands of residents to evacuate their homes as floodwaters began to spill out of the rising Cedar River. The total economic losses to the business and industry area was estimated to be in excess of $25.7 million. Officials believe that Cedar Rapids sustained less damage because of long-term permanent flood control solutions.[19] Cedar Rapids is now a class 6 community and has implemented many flood reduction activities as discussed above. In each case they have earned CRS points, and each CRS activity can be linked back to the 6c/4R framework that more comprehensively characterizes community resilience to environmental hazards.

Although Cedar Rapids has yet not evaluated the benefits of the improvements as a result of joining the CRS program, there are studies that have assessed the effect the CRS program has on insured flood losses. Highfield and Brody (2017) determined that, on average, participating CRS communities experience a 41.6% reduction in flood claims compared to non-CRS participating communities. Kousky and Michel-Kerjan (2017) revealed that even communities that implement a minimal number of mitigation activities experience fewer individual flood claims. They found that Class 8 and 9 CRS communities experience approximately 13.5% fewer individual flood claims when compared to communities that do not participate in the CRS and that an increase of 100 CRS credit points reduces flood insurance claims in the community by approximately 2.5%. Although yet to be proven we see no reason to judge that similar effects would not be found in Cedar Rapids in proportion to what CRS measures have been implemented there.

7. Conclusion and suggestions for future research

The concept of *resilience* is gaining momentum in the wake of catastrophic events in the United States and around the world in recent years. This paper discusses the 6C-4R framework (6 capitals, 4 resilience properties) to assess community resilience holistically and how this framework relates to the NFIP's Community Rating System (CRS).

We demonstrate that several features of the 6Cs and 4Rs are captured by the CRS program that relate to the resilience status of a community: we interpret the correlation between the two systems as good. However, there is a need for a more formal quantitative analysis of the role the CRS has played in reducing flood losses by incentivizing the enhancing of resilience. In a working paper Michel-Kerjan, Atreya, and Czajkowski (2016) connect the CRS activities to 88 sources of resilience identified across 5 capitals that was proposed by the Zurich Flood Resilience Alliance (Keating et al., 2017) as a first step towards such a quantitative analysis.

One of the caveats of relying on CRS data to measure resilience comprehensively is that it does not cover all the aspects of a community's status and activities, such as the extent of social vulnerabilities and the level of engagement of diverse populations, such as those with special needs. These characteristics of communities can be key to preparing properly for disasters. Previous research has quantified the performance of CRS activities in terms of reported property damage (Highfield and Brody (2017); Kousky and Michel-Kerjan (2017)) but other measurements such as the proportion of population below poverty levels, or the proportion of population with special needs, may be necessary to capture resilience even more holistically.

Another important area for future study is to determine how the benefits of implementing these resilience-enhancing activities over time compared to the costs incurred. Communities are limited by resources to implement all the activities they would ideally like to undertake to improve their resiliency. In order to make informed decision and prioritize the implementation of resilience improving activities, a community – wide cost–benefit analysis of individual CRS activities and a combination of them would be useful in the future as inputs for developing further a strategy for reducing future flood losses.

Notes

1. Other definitions refer to resilience in ecosystems or social systems at the level of the individual, municiapality or region.
2. http://www.spur.org/policy-area/sustainability-resilience.
3. https://fema.ideascale.com/a/dtd/Baseline-Resilience-Indicators-for-Communities-BRIC/467674-14692#idea-tab-comments
4. https://huxley.wwu.edu/ri/resilus
5. https://toolkit.climate.gov/tool/coastal-resilience-index
6. Although the SL framework is built in the context of the third world, the concept of different capitals within the framework can be applied to the United States and other countries as well. Also, "community" in this paper refers to CRS community boundaries.
7. We thank Sandy Pumphrey, Flood Mitigation, Public Works Departmentof the City of Cedar Rapids for pointing this out.
8. We thank the National Academy of Sciences roundtable members and the participants of the National Association of Counties (NACo) conference in Colorado for suggesting that *political capital* should be added to the 5C framework, as it focuses on whether or not a community has the ability to move forward to build and implement resiliency.
9. The 2016 TMAC Annual Report can be found at https://www.fema.gov/media-library-data/1492803841077-57e4653a1b2de856e14672e56d6f0e64/TMAC_2016_Annual_Report_(508).pdf
10. See: http://www.cityofottawa.org/City_of_Ottawa_Flood_Information_2015.pdf
11. A community's financial capital can also be negatively impacted by acquisitions and demolitions by the loss of tax-base

12. See http://www.senatorrezin.com/ILValleyFloodResiliencyAlliance.aspx
13. See http://www1.nyc.gov/office-of-the-mayor/news/347-13/mayor-bloomberg-mayor-s-fund-advance-new-york-city-release-one-year-on-hurricane/#/0
14. Parcel is alternatively called a "lot" or a "plot" owned as a pat of a homeowner's property
15. We thank Sandy Pumphrey, Flood Mitigation, Public Works Department, City of Cedar Rapids for pointing this out.
16. The computed **elevation** to which floodwater is anticipated to rise during the **base flood.**
17. http://www.cedar-rapids.org/local_government/departments_g_-_v/public_works/stormwater_best_management_practices_cost-share_program.php
18. See http://homelandsecurity.iowa.gov/documents/misc/HSEMD_RecoveryTaskForce2014_FinalReport.pdf
19. http://www.cedar-rapids.org/Economic%20Development/Flood%20Recovery%20Report.pdf

Acknowledgements

This study was supported by National Science Foundation (NSF) grant EAR-1520683 and the Wharton Risk Management and Decision Processes Center. We thank Sandy Pumphrey for helpful comments on an earlier draft of the paper. We also thank the National Academy of Sciences' Resilient America Roundtable members for their comments. We appreciate all the assistance we received from Linda Langston and others in Cedar Rapids, Iowa.

Disclosure statement

No potential conflict of interest was reported by the authors.

References

Adger, N. W. (2000). Social and ecological resilience: Are they related? *Progress in Human Geography*, 24, 347–364.
Atreya, A., Kriesel, W., & Mullen, J. (2016). Valuing open space in a Marshland environment. *Journal of Agriculture and Applied Economics, Forthcoming.*
Berke, P. R., Kartez, J., & Wenger, D. (1993). Recovery after disaster: Achieving sustainable development, mitigation and equity. *Disasters, 17*(2), 93–109.
Breton, M. (2001). Neighbourhood resiliency. *Journal of Community Practice, 9*(1), 21–36.
Bruneau, M., Chang, S. E., Eguchi, R. T., Lee, G. C., O'Rourke, T. D., Reinhorn, A. M., & von Winterfeldt, D. (2003). A framework to quantitatively assess and enhance the seismic resilience of communities. *Earthquake Spectra, 19*(4), 733–752.
Burby, R. J., Deyle, R. E., Godschalk, D. R., & Olshansky, R. B. (2000). Creating hazard resilient communities through land-use planning. *Natural Hazards Review, 1*(2), 99–106.
Burton, C. G. (2015). A validation of metrics for community resilience to natural hazards and disasters using the recovery from Hurricane Katrina as a case study. *Annals of the Association of American Geographers, 105*(1), 67–86.
Cedar Rapids Flood Facts. (2008). Retrieved from http://www.cedar-rapids.org/government/departments/public- works/engineering/Flood%20Protection%20Information/Pages/2008FloodFacts.aspx
Chambers, R., & Conway, G. (1992). *Sustainable rural livelihoods: Practical concepts for the 21st century.* Institute of Development Studies (UK).
Chandra, A., & Acosta, J. D. (2009). *The role of nongovernmental organizations in long-term human recovery after disaster: Reflections from Louisiana four years after Hurricane Katrina.* Rand Corporation.
Community and Regional Resilience Institute (CARRI). (2013). Definitions of community resilience: An analysis. Retrieved from http://www.resilientus.org/wp-content/uploads/2013/08/definitions-of-community-resilience.pdf

Conrad, D. R. (1998). *Higher ground: A report on voluntary property buyouts in the nation's floodplains: A common ground solution serving people at risk, taxpayers and the environment.* National Wildlife Federation.

Cumming, G. S., Barnes, S., Perz, M., Schmink, K. E., Sieving, J., Southworth, M., ... Van Holt, T. (2005). An exploratory framework for the empirical measurement of resilience. *Ecosystems, 8,* 975–987.

Cutter, S. L. (2016). The landscape of disaster resilience indicators in the USA. *Natural Hazards, 80*(2), 741–758.

Cutter, S. L., Barnes, L., Berry, M., Burton, C., Evans, E., Tate, E., & Webb, J. (2008). A place-based model for understanding community resilience to natural disasters. *Global Environmental Change, 18*(4), 598–606.

Cutter, S., Boruff, B., & Shirley, W. L. (2003). Social vulnerability to environmental hazards. *Social Science Quarterly, 84,* 242–261.

Cutter, S. L., Burton, C. G., & Emrich, C. T. (2010). Disaster resilience indicators for bench-marking base-line conditions. *Journal of Homeland Security and Emergency Management, 7*(1), 1–22.

Czajkowski, J., & Simmons, K. (2014). Convective storm vulnerability: Quantifying the role of effective and well-enforced building codes in minimizing Missouri hail property damage. *Land Economics, 90*(3), 482–508.

Department of Homeland Security, Iowa. (2010). Retrieved from http://homelandsecurity.iowa.gov/documents/hazard_mitigation/HM_StatePlan_2-0_AnnexC.pdf

Elliott, J. R., Haney, T. J., & Sams-Abiodun, P. (2010). Limits to social capital: Comparing network assistance in Two New Orleans neighborhoods devastated by Hurricane Katrina. *The Sociological Quarterly, 51*(4), 624–648.

FEMA. (2011). Mitigation best practices. Retrieved from https://www.fema.gov/mitigation-best-practices-portfolio

Fronstin, P., & Holtmann, A. G. (1994). The determinants of residential property damage caused by hurricane andrew. *Southern Economic Journal,* 387–397.

Gunderson, L. H., & Holling, C. S. (2001). *Panarchy: Understanding transformations in human and natural systems.* Washington, DC: Island Press.

Highfield, W. E., & Brody, S. D. (2017). Determining the effects of the FEMA community rating system program on flood losses in the United States. *International Journal of Disaster Risk Reduction, 21,* 396–404.

Islam, M. R. (2014). Non-governmental organizations' role for social capital and community empowerment in community development: Experience from Bangladesh. *Asian Social Work and Policy Review, 8*(3), 261–274.

Jacobs, C. (2007). *Measuring success in communities: Understanding the community capitals framework.* White paper. South Dakota State University.

John Heinz III, H., & Center for Science Economics, and the Environment. (2002). *Human links to coastal disasters.* Washington, DC: H. John Heinz Center.

Keating, A., Campbell, K., Szoenyi, M., McQuistan, C., Nash, D., & Burer, M. (2017). Development and testing of a community flood resilience measurement tool. *Natural Hazards and Earth System Sciences, 17,* 77–101.

Keim, M. E. (2008). Building human resilience: The role of public health preparedness and response as an adaptation to climate change. *American Journal of Preventive Medicine, 35*(5), 508–516.

Kimhi, S., & Shamai, M. (2004). Community resilience and the impact of stress: Adult response to Israel's withdrawal from Lebanon. *Journal of Community Psychology, 32*(4), 439–451.

Kousky, C., & Michel-Kerjan, E. (2017). Examining flood insurance claims in the United States: Six key findings. *Journal of Risk and Insurance, 84*(3), 819–850.

Kousky, C., & Shabman, L. (2012). The realities of federal disaster aid: The case of floods. RFF issue brief. Washington. *DC: Resources for the Future.*

Kunreuther, H., & Michel-Kerjan, E. (2011). *At War with the weather: Managing large-scale risks in a New Era of catastrophes.* New York: MIT Press (Paperback edition).

Lindell, M. K., & Prater, C. S. (2003). Assessing community impacts of natural disasters. *Natural Hazards Review, 4*(4), 176–185.

Magis, K. (2010). Community resilience: An I1zndicator of social sustainability. *Society and Natural Resources: An International Journal, 23*(5), 401–416.

Mayunga, J. S. (2007). *Understanding and applying the concept of community disaster resilience: A capital-based approach.* Draft working paper prepared for the summer academy. Megacities as hotspots of risk: Social vulnerability and resilience building, Munich, Germany, 22–28 July 2007.

Michel-Kerjan, E., Atreya, A., & Czajkowski, J. (2016). Learning over time from FEMA's community rating system (CRS) and its link to flood resilience measurement. Wharton Risk Center Working Paper.

National Research Council (NRC). (2012). *Disaster resilience: A national imperative.* Washington, DC: National Academy Press.

News Tribune. (2012). Retrieved from http://newstrib.com/main.asp?SectionID=2&SubSectionID=28&ArticleID=22336

Norris, F. H., Stevens, S. P., Pfefferbaum, B., Wyche, K. F., & Pfefferbaum, R. L. (2008). Community resilience as a metaphor, theory, set of capacities, and strategy for disaster readiness. *American Journal of Community Psychology, 41*, 127–150.

Peacock, W. G., Brody, S. D., Seitz, W. A., Merrell, A. V., Zahran, S., Harriss, R. C., & Stickney, R. R. (2010). *Advancing the resilience of coastal localities: Implementing and sustaining the use of resilience indicators. Final report prepared for the coastal services center and the national oceanic and atmospheric administration.* College Station, TX: Hazard Reduction and Recovery Center.

Pfefferbaum, R. L., Pfefferbaum, B., & Van Horn, R. L. (2011). *Communities advancing resilience toolkit (CART): The CART integrated system.* Oklahoma City, OK: Terrorism and Disaster Center at the University of Oklahoma Health Sciences Center.

Pfefferbaum, R. L., Pfefferbaum, B., Van Horn, R. L., Klomp, R. W., Norris, F. H., & Reissman, D. B. (2013). The communities advancing resilience toolkit (CART): An intervention to build community resilience to disasters. *Journal of Public Health Management and Practice, 19*(3), 250–258.

Sadiq, A. A., & Noonan, D. S. (2015). Flood disaster management policy: An analysis of the United States community ratings system. *Journal of Natural Resources Policy Research, 7*(1), 5–22.

Tate, E., Strong, A., Kraus, T., & Xiong, H. (2015). Flood recovery and property acquisition in Cedar Rapids, Iowa. *Natural Hazards*, 1–25.

Tierney, K, and M Bruneau (2007). Conceptualizing and measuring resilience: A key to disaster loss reduction. *TR news*, p. 250.

Tobin, G. A. (1999). Sustainability and community resilience: The holy grail of hazards planning? *Environmental Hazards, 1*, 13–25.

Turnham, J., Burnett, K., Martin, C., McCall, T., Juras, R., & Spader, J. (2011). *Housing recovery on the gulf coast, phase II: Results of property owner survey in Louisiana, Mississippi, and Texas.* Washington, DC: Department of Housing and Urban Development, Office of Policy Development and Research.

UN/ISDR. (2002). *Living with risk: A global review of disaster reduction initiatives. Preliminary version prepared as an interagency effort co-ordinated by the ISDR secretariat.* Geneva.

Yip, W., Subramanian, S. V., Mitchell, A. D., Lee, D. T., Wang, J., & Kawachi, I. (2007). Does social capital enhance health and well-being? Evidence from rural China. *Social Science & Medicine, 64*(1), 35–49.

Research on disaster resilience of earthquake-stricken areas in Longmenshan fault zone based on GIS

Bin Liu, Xudong Chen, Zhongli Zhou, Min Tang and Shiming Li

ABSTRACT

Since the Ms 8.0 earthquake occurred in Yingxiu, a town in Wenchuan County in Sichuan Province, on May 12, 2008, frequent geological disasters ensued on the land affected by earthquakes in the Longmenshan fault zone have caused a large number of casualties and property losses. Therefore, measuring the disaster resilience of earthquake-stricken areas is the priority for disaster prevention and mitigation science. Based on the in-depth analysis of the hazard of disaster resilience, this paper constructs the compatibility coefficient of industrial and employment structure and per capita GDP growth rate from the socio-economic perspective to measure the disaster resilience. Considering seismic intensity, this paper, with the help of Geographic Information Systems (GIS) technology and the topographic complexity index described by the topographic information entropy method, analyses the temporal and spatial change rule of disaster resilience in the heavily damaged area. The results show that the impact of the "post-earthquake effect" on disaster resilience tends to decrease concussively over time, and the seismic intensity and topographic complexity are important internal factors that restrict the improvement in disaster resilience. This paper will provide a broader application scenario for the concept of resilience, especially in post-earthquake integrated risk management and reconstruction process.

1. Introduction

China is among the countries with the worst natural disasters in the world. Since the twentieth century, 28 provinces in China have been affected by earthquake disasters. Statistics show that there have been approximately 150 earthquakes causing casualties, with a total of more than 660,000 deaths, millions of injuries and hundreds of millions of people affected (Wen, Wenkai, & Zhonghong, 2019). Land-based destructive earthquakes in China account for one-third of the world's total, and more than 70% of them are shallow focus earthquakes. These occurrences are mainly due to the location of 23 major seismic zones in China and earthquake activity patterns, which feature a wide active area, high frequency, high intensity and shallow focus. In terms of basic intensity, 41% of the regional earthquakes reach 7 degrees or higher, and 79% of the regional earthquakes reach 6 degrees or higher. On 12 May 2008, Wenchuan County in Sichuan Province was struck by a magnitude Ms 8.0 earthquake. The casualties were as numerous as 70,000 and millions of people were missing, injured and homeless. This was yet another natural severe disaster of great intensity and scale in Chinese history after the Tangshan Earthquake. The disaster directly caused a total of 850 billion RMB in economic losses (Xu & Lu, 2012). On 20 April 2013, an earthquake of magnitude Ms 7.0 in Lushan County in Sichuan Province affected nearly 2 million people with direct economic losses reaching 50 billion yuan (Han, Tian, & Fan, 2018). The above-mentioned two earthquakes occurred in the middle and southern areas of the Longmenshan fault zone respectively. Although located in the same fault zone (with similar geological and geomorphological environments), the earthquake-causing losses in areas around the Longmenshan fault zone are quite different. In addition to the influence of different magnitudes, these differences are also related to the disaster-bearing and disaster-response ability in these areas (Timmerman, 1981). What's more, such geological disaster phenomena as collapse, landslide and debris flow ensue after strong earthquakes; these are known as post-earthquake effects. For many affected regions, destructive landslides and fault ruptures that tear through unlucky homes, flooding and liquefaction in recurring aftershocks, fires and disease can add up to sustained economic losses – let alone numerous casualties. The losses caused by earthquake-related geological hazards in earthquake-stricken areas have far exceeded those caused by earthquakes themselves (Stahl et al., 2017). Therefore, we have to consider how to measure the carrying capacity, disaster response capacity, and how the resilience of disaster-stricken areas under the influence of the 'post-earthquake effect'? Addressing these questions scientifically and reasonably is an important link for systematically evaluating the impact of seismic and geological hazards.

The emergence of the concept of 'resilience' provides a new perspective for us to understand the operation of complex systems and their sustainable development, especially systematic disaster prevention and mitigation (Meerow, Newell, & Stults, 2016). 'Resilience', originally a physical concept, refers to the ability of materials to absorb energy during plastic deformation and fracture. The evolution of the concept of 'resilience' has gone through three stages: engineering resilience (before 1973), ecological resilience (1973-1998) and social-ecological resilience (after 1998). These stages differ greatly in research scope, core attributes and other connotative characteristics. Engineering resilience defines the scope of research as a single and static system whose core attribute is the ability of the system to restore its original balance and maintain its stability after external shocks (Holling, 1996). In the 1970s, Holling, a Canadian ecologist, introduced the concept of resilience into the field of ecology and defined it as the amount of disturbance that can be sustained before system control or structure changes. The connotative characteristics of ecological resilience include researching the constantly changing complex systems and taking adaptability as its core attribute (Holling, 1973). Since the 1990s, resilience has been widely used in natural science, social science, humanities and in their crossover study (Holling, 1996). Covering the whole social-ecological system, social-ecological resilience emphasises that resilience is a dynamic system attribute that is closely related to continuous adjustment ability, which could be greatly influenced by human's subjective initiative. Thus variability becomes its core attribute (Walker, Anderies, Kinzig, & Ryan, 2006).

With the deepening of research on resilience, more and more geological hazard research bodies, experts and scholars have noticed the role of system resilience while researching earthquake disasters. In 1981, Timmerman defined disaster resilience as 'the ability of human societies to withstand external shocks or perturbations and to recover from such perturbations' (Timmerman, 1981). In 2005, Adger believed that resilience not only means restoration to the pre-disturbance state but also improved future response through learning current situations (Adger, 2005). In 2006, Bruneau, based on researching the post-earthquake situation of communities, defined seismic resilience as 'the ability of a community to recover quickly by absorbing damages'. Furthermore, he summarised the characteristics of resilience as firmness, redundancy, strategy and rapidity (Bruneau, Chang, & Eguchi, 2003). The United Nations Disaster Reduction Programme (UN/ISDR) redefined resilience in 2005 as ' the ability of a system, community or society to resist, absorb, adapt and recover from its impact in a timely and effective manner when exposed to danger, including the protection and restoration of its essential infrastructure and functions' (David, 2009). Manyena argued that disaster resilience can be regarded as an inherent ability of a system, community or society to rebuild itself, adapt and survive by changing its non-core attributes after being affected by shocks or pressures (Manyena, 2006). Additionally, in 2008, Cutter defined disaster resilience as 'the ability of the social system to recover from disasters' (Cutter et al., 2008).

For the study of disaster resilience, qualitative, quantitative or mixed research methods are usually adopted. Quantitative research mainly uses statistical and data mining methods. It includes two parts: one is the construction of resilience indicators and the other is the use of statistical methods for analysis, such as principal component analysis, cluster analysis and so on (Cai et al., 2018). Simpson put forward the model of the community resilience index in 2008. The index includes disaster risk, social capital, community assets, infrastructure quality, social services and population quality, etc., and it defines disaster resilience value as the ratio of preparedness to vulnerability (David, 2009). In 2008, Cutter and other scholars proposed the DROP model and constructed a community resilience evaluation index system (BRIC), including over 30 indicators that covered ecology, society, economy, infrastructure, system, community capacity, among others (Cutter et al., 2008). In 2010, Hongjian Zhou used a regression model to analyse 13 major socio-economic and demographic variables in 105 counties in the Wenchuan earthquake area. The principal component-based factor analysis method was used to integrate social and economic variables into four comprehensive factors, and K-means clustering method was used to divide sub-research areas to further analyse social resilience (Zhou, Wang, Wan, & Jia, 2010). In 2013, Orencio and Fujii used AHP analysis and Delphi methods to explore the post-disaster resilience of Philippine coastal cities (Orencio & Fujii, 2013). In 2014, Kusumastuti developed a resilience assessment framework for natural disasters; his work was guided by the definition of resilience by the United Nations International Strategy for Disaster Reduction (UNISDR) and Simpson's community resilience assessment method (Kusumastuti, Husodo, Suardi, & Danarsari, 2014). Later, Joerin established a Climate Disaster Resilience Index (CDRI) model, including economy, system, nature, material and society; he applied the model to an evaluation of the urban disaster resilience of Chennai, India (Joerin, Shaw, Takeuchi, & Krishnamurthy, 2014). In 2016, Xiaolu Liu established RIM model to quantitatively analyse community resilience in Wenchuan and its surrounding areas after the 2008 earthquake. Here, the cluster analysis method was used to divide the counties into four recovery levels according to the degree of exposure, damage and recovery. Discriminant analysis was used to quantify the impact of socio-economic characteristics on county resilience (Li et al., 2016). In 2017, Li Tongyue analysed the theoretical evolution and characteristics of resilient cities; he also constructed an evaluation index system of resilient cities in response to floods and a community resilience evaluation index system based on social, economic, institutional and infrastructure dimensions (Li, 2017). Generally, although resilience has been widely used in the field of disaster study and many results have been achieved, the current research on resilience is still at the initial stage of theoretical exploration, and there is no uniform research paradigm. Qualitative and semi-quantitative research remain the principal methodologies for studying disaster resilience, so research studies that utilise quantitative analysis are relatively limited.

Based on the geological disaster perspective, this paper constructs the index of disaster resilience research from the socio-economic dimension, considers the seismic intensity and topographical quantitative analysis, and combines GIS technology to analyse the spatial pattern and time trend of disaster resilience in earthquake-stricken areas, and reveal different spatial locations. Under the earthquake disaster area, considering the spatial distribution characteristics of disaster resilience, and the long-term impact of the 'post-earthquake effect', it reveals the key intrinsic factors for improving disaster resilience in the earthquake-stricken areas during post-disaster reconstruction. The predecessors constructed corresponding indicators from different dimensions such as ecology, society, economy and system when quantitatively analysing disaster resilience. Different from the above research, this study constructed the analysis index of disaster resilience from the macroscopic perspective on the basis of regional geographical characteristics. Then analyse the temporal and spatial evolution of disaster resilience.

The paper is organised as follows: the first part summarises the research progress of resilience; the second part gives the definition of disaster resilience and quantification method; the third part analyses the spatial and temporal evolution of disaster resilience from the perspective of regional seismic intensity and terrain complexity with 51 earthquake-stricken areas in Longmenshan fault zone as the research objects; finally, conclusions and discussions are presented in parts 4 and 5.

2. Disaster resilience from a geological disaster perspective

2.1. The definition of resilience

The concept of resilience is derived from ecology. Historically, the study of resilience has been fully developed and widely used in the research of security, engineering, social and economic fields (Zhou et al., 2010). The definition of the concept of resilience is very different even within the same research field. The same concept presents different connotation when used by different researchers. As a result, its connotation is accordingly enriched. However, we can conclude that resilience is related to the ability of a system to return to a stable state after being disturbed by the outside world (Klein,

Nicholls, & Thomalla, 2003). Research has deepened and studies have been conducted from different dimensions, which address the multidimensionality of social, economic, political and institutional components (Shim & Kim, 2015).In a word, resilience has been explored as a comprehensive economic, social and ecological complex (Walker et al., 2006).

2.2. The relationship between resilience and vulnerability

Resilience and vulnerability are two key concepts of natural hazard studies (Klein et al., 2003), but what is the relationship between the two? There is no unified standard answer. However, addressing this relationship is important in defining the meaning, implications and applications of resilience.

Vulnerability usually refers to the state of a system before disaster has struck, which can be seen as the foundation for resilience research. Vulnerability refers to the potential for loss (Zhou et al., 2010; Cutter, 1996). More specific definitions qualify the potential loss by factoring the likelihood of exposures and susceptibility into damage. However, in general, the concept of vulnerability focuses on the situation of a system before disaster, and it is helpful to preparedness for future hazard prevention.

Resilience refers to the capacity of the system to change, adapt and yet to be stable within certain thresholds (Shim & Kim, 2015; Berkes & Ross, 2013). Resilience, which is a process, mainly focuses on the stages of in- and post-disaster (when the loss occurs) and helps to enhance the ability of the system to resist disasters, recover from them and explore policy options for dealing with hazards.

2.3. Concept of disaster resilience

Disaster resilience can be defined as the ability of individuals, communities, organisations and states to adapt to and recover from hazards, shocks or stresses without compromising their long-term prospects for development (Ganderton, 2015). According to the Hyogo Framework for Action (UNISDR, 2002; Zhou et al., 2010), disaster resilience is determined by whether individuals, communities and public and private organisations can organise themselves to learn from past disasters and reduce the risks of future ones, at international, national, regional and local levels. Zhou defines disaster resilience as the 'capacity of hazard-affected bodies (HABs) to resist loss, to regenerate and to reorganise after disaster in a specific area in a given period' (Zhou et al., 2010; Mishra et al., 2017). In this paper, disaster resilience is defined as "the ability of earthquake-stricken areas to resist losses and recover from disasters during post-disaster reconstruction at specific times and in specific areas under the influence of 'post-earthquake effects'".

For any particular area, each regional system has its own resilience. When a system is subject to outside disturbance, the resilience may be related to two situations (Figure 1): (1) when the resilience of the system is lower than a certain threshold valve (λ), the system function is completely lost, which means that the system will no longer exist or (2) although the balance of the system is broken, under the influence of the endogenous power and external force of the system, resilience allows the system to gradually recover and perhaps reach or even exceed pre-disaster levels.

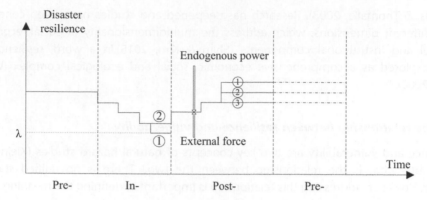

Figure 1. Schematic diagram of disaster resilience evolution.

2.4. Method for analysis of disaster resilience

In the process of studying disaster resilience, the diversification of local economy has been identified as an important policy objective for building resilience (Ullsten, Speth, & Chapin, 2004). Zhou et al.(2010) further provided evidence that income diversification can reduce vulnerability before loss and that regions with income diversification have more opportunities to recover from loss after disaster. At the same time, industrial structure is an important determinant of macroeconomic development and growth. Meanwhile, the employment structure reflects the configuration and use of regional labour resources, and the compatibility between the two is related to the sustainability of regional economic development. One of the ways that employment and industrial structures can become incompatible is if the industry and its factories and equipment are destroyed by a disaster. Then, there is no work for the workers: the employment structure (the employees, etc.) cannot be matched by the industrial structure (the factories, etc.) because the latter has been destroyed or partly destroyed. The rate of change in GDP growth will slow or even reverse. For example, after the earthquake, almost all industrial enterprises (phosphate mining enterprises) in the town of Qingping (one of the earthquake-stricken areas of the Wenchuan earthquake) stopped production, and the agricultural, animal husbandry and production industries were also severely damaged, resulting in widespread unemployment among a number of urban and rural populations. During post-disaster reconstruction in earthquake-stricken areas, the industrial and employment structures may be assisted by that reconstruction effort to the benefit of the regional economy, although they still may remain partially incompatible. Therefore, to quantitatively evaluate the disaster resilience of earthquake-stricken areas for earthquakes during the post-earthquake response based on the socio-economic perspective, this paper examines the socio-economic development status and development speed of earthquake-stricken areas from a macro perspective. To measure the recovery of earthquake-stricken areas after an earthquake, we developed two indicators to study the collected data: (1) compatibility coefficient of industrial and employment structure (H_{xy}) and (2) per capita GDP growth rate (R_{GDP}).

$$H_{xy} = \sum_{i=1}^{n} X_i Y_i / \sqrt{\sum_{i=1}^{n} X_i^2 \sum_{i=1}^{n} Y_i^2} \qquad (1)$$

The variable H_{xy} in Equation (1) is the index used to measure the relationship between industrial structure and employment structure, namely, the compatibility coefficient. The H_{xy} values range between 0 and 1. An H_{xy} value that is close to 0 indicates that there is poor compatibility between the regional industrial structure and employment structure and that there may be labour shortages and difficulties with employment. On the other hand, it is likely to mean that with higher compatibility of the employment structure and industrial structure, there may be full employment and no unemployment.

$$R_{\text{GDP}} = \frac{N_i}{N_{i-1}} - 1 \tag{2}$$

In the formula, N_i represents the current per capita GDP and N_{i-1} is the per capita GDP of the previous period.

3. Results

3.1. The study area

Located in the north-western edge of Sichuan Basin, the Longmenshan fault zone is part of the eastern margin of the Tibetan plateau (Figure 2). The location of the Longmenshan fault zone is very special. It is situated in the upper part of Sichuan Basin in China and is connected to the bottom of its north-western margin. As the uplift of the Tibetan plateau was hindered by Sichuan Basin, Longmenshan has become an active fault zone. The fault zone is generally NE45°, tending to NW, extending over 500 km in length and 25–50 km in width from east to west. There are three major faults: Longmenshan Qianshan fault (Dujiangyan–Hanwang–Anxian), Longmenshan central fault (Yingxiu–Beichuan–

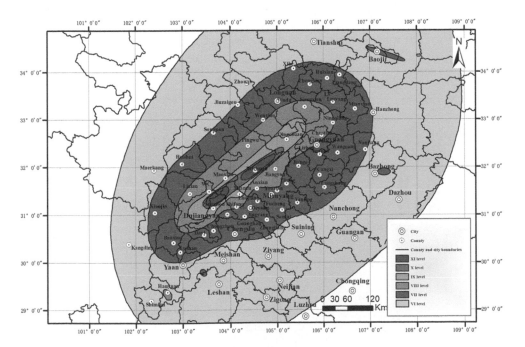

Figure 2. Distribution of seismic intensity in Longmenshan fault zone.

Guanzhuang) and Longmenshan Houshan fault (Wenchuan–Maoxian–Pingwu–Qing-chuan). Seismic activities on the fault zone are frequent and unevenly distributed.

After the Wenchuan Earthquake, approximately 90% of co-seismic landslides ensued along the slopes and in ravines, and those landslide deposits were prone to remobilisation during the sub-sequent rainy seasons (Zhang, Zhang, & Glade, 2014; Fan, Zhang, Wang, & Fan, 2018a). The general consensus is that the hillslope sediments produced by the co-seismic and post-seismic landslides were gradually transferred to channel deposits driven by abundant rainfall events (Zhang et al., 2014; Fan et al., 2018b; Lin, Chen, Chen, & Horng, 2008; Zhang, Zhang, Lacasse, & Nadim, 2016; Zhang & Zhang, 2017; Fan et al., 2018b), which led to continuous disasters in earthquake-stricken areas.

3.2. Spatial pattern of disaster resilience at the seismic intensity scale

Figure 2 shows that the 51 earthquake-stricken areas are distributed within the range of seismic intensity VI–XI in the Longmenshan fault zone. Considering the area of the seismic intensity in earthquake-stricken areas, we can get the results for study areas based on formula (1) and formula (2) as shown in Figure 3.

It can be seen from Figure 3, in the 3 years after the 12 May 2008 earthquake (until 2011), the H_{xy} compatibility coefficient performed better than before the earthquake and reached its peak (0.81) in 2009, mainly due to the impact of external intervention measures after the disaster, such as infrastructure reconstruction, which may promote local employment. After 2009, the external intervention measures gradually faded, and the post-earthquake effect gradually appeared, which led to the poor compatibility between the industrial structure and employment structure in the earthquake-stricken areas, such as the earthquake disaster zone. The town of Qingping was influenced by

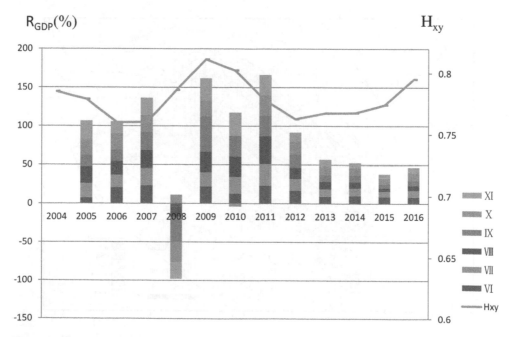

Figure 3. The trend of H_{xy} and R_{GDP} in earthquake-stricken area.

geological disasters, which created strong incompatibilities between the industrial structure and employment structure. Generally, H_{xy} of the 51 earthquake-stricken areas increased with time at first. Then, H_{xy} typically decreased, which was due to the large increase of geological disasters (landslides, etc.) such as the serious debris flow that occurred on 13 August 2010 and on 10 July 2013, in these areas. This proves that earthquake and geological disasters have a great impact on local economic omic development, and the 'post-earthquake effect' is obvious at this stage. After this stage, the gradual increase in H_{xy} was thanks to self-adjustment, and the impact of the 'post-earthquake effect' decreased accordingly. These overall trends match the previous research conclusion that post-earthquake geological disasters will periodically decrease in an oscillatory way. (In detail, geological disasters in post-disaster areas will last for 20–25 years. The period contains an influence peak of 4–5 years, then it decreases in an oscillatory way and finally restores to the pre-earthquake level).

The R_{GDP} histogram in Figure 3 shows that from 12 May 2008 (Wenchuan earthquake) to 2016, the R_{GDP} of regions with different seismic intensities generally trended downwards. R_{GDP} in the region with low earthquake intensity generally does not show negative growth, while the R_{GDP} in the region with high earthquake intensity may show negative growth due to the post-earthquake effect, which indicates the recovery speed in these regions is quite different. Furthermore, the percentage of the decline of IX and X seismic intensity was approximately 20%, and R_{GDP} became gradually stable at approximately 7% after 2012. At the same time, data show that the lower the seismic intensity level is, the more stable R_{GDP} becomes. This also reflects the influence of geological disasters on earthquake-stricken area in regions with low seismic intensity is relatively small while in regions with XI seismic intensity such as Beichuan, the influence is disasters is significant. The overall trend during 2008–2016 shows that the large increase after the earthquake in 2008–2012

The figure can be found in the original journal article

was mainly due to the impact of post-disaster assistance on the regional economy, but the growth rate of R_{GDP} after 2012 was relatively small, which was mainly due to the influence of the 'post-earthquake effect'. This indicates that the post-earthquake geological disaster has a great impact on the development speed of local economy; however, the degree of this influence gradually loses power and R_{GDP} stabilises accordingly.

3.3. Analysis of spatial distribution relationship between terrain complexity and disaster resilience

3.3.1. Terrain information entropy

Due to the weak regional geological environment caused by strong earthquakes, effects such as slope instability and rock mass destruction will occur, as well as all kinds of secondary disasters. At the same time, due to the long duration of seismic geological disasters, especially in mountainous areas with special topographic and geomorphic characteristics, post-disaster reconstruction is difficult. Topographic complexity analysis is an indispensable and important part of topographic and geomorphic research. Topographic complexity, which is an important parameter in the field of digital topographic analysis, describes the degree of surface relief and folding. Three indexes – elevation, slope and aspect – are always used to measure topographic complexity. Therefore, to quantitatively depict the topographical features of mountains based on previous mathematical model studies of topographic complexity (Zhou et al., 2010), this paper used information entropy to quantitatively describe local roughness and undulation of the surface and analysed the spatial distribution characteristics of disaster resilience in the areas that were most impacted by earthquakes.

Entropy, as a physical quantity, was first applied in thermodynamics. Subsequently, Shannon introduced entropy into information theory. Now, entropy, as an average information measure, has been widely applied in various fields (Hu, Guo, Yu, & Liu, 2010; Hosoya, Buchert, & Morita, 2004). In the face of complex topography and landforms with rough, irregular and generalised self-similarity, the topographic complexity index is calculated by selecting different scales, based on the resolution of topographic data in the research area. The calculation steps are as follows:

(1) Process the topographic data (DEM) in the study area in a regular grid
(2) Take r (number of grids) as the scale to define the sliding window to cover the research area
(3) Slide down the DEM grid unit from top to bottom, from left to right, traverse the whole terrain analysis area, and calculate the corresponding p(i) according to different scales r

$$p(i) = \frac{|h(i)|}{\left|\sum_{i=1}^{M} h(i)\right|} \tag{3}$$

wherein $h(i)$ is the elevation data of each point in the window
(4) Define topographic information entropy:

$$H_t = -\sum_{i=1}^{M} p(i)\log_2(p(i)) \approx -\sum_{i=1}^{M} p(i)(p(i) - 1) \tag{4}$$

wherein H_t is the entropy of terrain information and M is the total number of terrain data points in the sliding window. According to the definition of topographic information entropy, H_t reflects the average amount of information contained in the topography of the research area. Moreover, if the elevation value of the research area changes sharply, the more complex the topography is, the more abundant the information is, and the smaller the calculated value H_t is. On the other hand, H_t is bigger.

(5) Calculate topographic information entropy (namely, topographic complexity index) at different scales r.

3.3.2. The experiment of terrain complexity calculation

The altitude of the study area reaches up to 6500 m, and the largest topographic height difference reaches 5000 m. Within the region, the northwest part is higher and the southeast part is lower (Figure 5). In this paper, the terrain of the study area is used as experimental data (1:300,000 topographic data); the terrain complexity index is quantified with the help of topographic information entropy. First, the research area is regularly meshed, and the numbers of grids in horizontal and vertical coordinates are 251 and 360, respectively. Second, the different scales of 3×3, 5×5, etc., are selected as the 'sliding window'. The different scales represent differences in the calculation accuracy. Finally, the terrain complexity indexes at different scales come out by analysing the DEM data of the entire area, and then experimental results are shown in Figure 6.

From the analysis shown in Figure 6, it can be seen that the terrain complexity index changes from the microscopic to middle to macro scale. For the region, the discrimination degree increases with the growing scale. However, the smaller the scale is, the clearer the local topographical features become. However, from the perspective of regional integrity, choosing a moderate scale is more conducive to 'surface' analysis. The trend of the

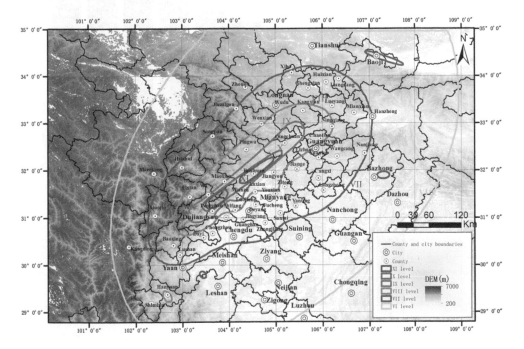

Figure 5. Isogram of terrain in study area.

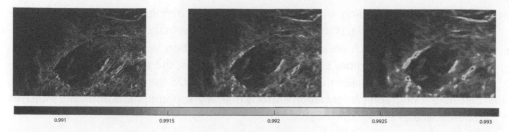

Figure 6. Experiment of diagram of terrain complexity index (5 × 5, 11 × 11, 19 × 19 from left to right scale).

average terrain complexity index of the whole region in Figure 7 also shows that the reasonable selection of the scale is of great significance to the expression of the topographic features of the study area. The experimental results conclude that if the scale is too small, the intrinsic continuity of the target would be broken; if the scale is too large, the inherent differences of the target would become weakened.

3.3.3. Spatial analysis of disaster resilience based on terrain complexity index

The scale 11 × 11 is selected from the DEM data of the Longmenshan fault zone as the sliding window, and the terrain complexity index of the whole area emerges. The calculation results are between 0.928 and 0.937. At the same time, with the combination of

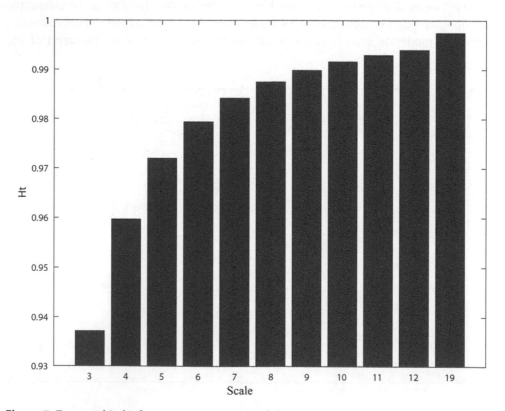

Figure 7. Topographical information entropy scale of the study area.

the seismic intensity and terrain complexity of the earthquake-stricken areas using GIS, we can obtain the visualisation results as shown in Figure 8. According to the definition of topographic information entropy, it could be inferred that the more complex the terrain is, the closer the value is to 0.928.

(1) The spatial distribution characteristics of terrain complexity in the study area

It can be seen from Figure 5 and Figure 8 that the terrain complexity of the higher elevations in the northwest is not higher. However, the terrain complexity of the lower elevations in the southeast is higher. The topographical features of the earthquake-stricken areas cannot be clearly seen in Figure 5. The terrain contour map is not intuitive enough to express the elevation, slope and aspect of the earthquake-stricken areas, and there is no direct relationship between the terrain complexity and the elevation height. However, it can be seen from the topographical complexity analysis in Figure 8 that the topographical features of the earthquake-stricken areas are fully expressed, and the description of the local spatial features is more intuitive. There are three seismic fault zones in the earthquake high-intensity area: the Longmenshan Qianshan fault (Dujiang-yan–Hanwang–Anxian), the Longmenshan central fault (Yingxiu–Beichuan–Guanzhuang) and the Longmenshan Houshan fault (Wenchuan–Maoxian–Pingwu–Qingchuan). These are consistent with the terrain complexity index in terms of spatial distribution, which indicates that the fracture zone has a higher degree of surface fragmentation. Combined with the disaster resilience analysis of the earthquake-stricken areas, the zones of higher topographic complexity are more readily affected by geological disasters during the process of

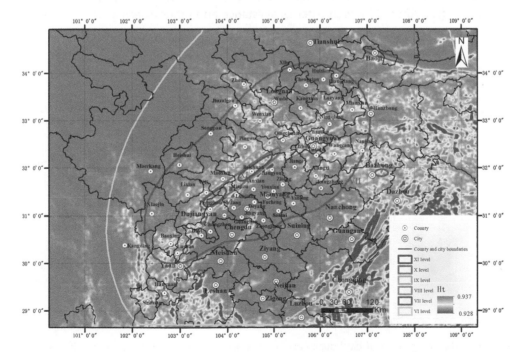

Figure 8. Spatial distribution map of the hardest hit area in terms of terrain complexity index.

reconstruction, meaning that the disaster resilience is often lower due to this frequent occurrence of additional geological disasters after the earthquake.

(1) Terrain complexity is an important internal factor that restricts the improvement of disaster resilience in earthquake-stricken areas

Among the 51 earthquake-stricken areas of Wenchuan earthquake in 2008, 10 counties and cities were greatly influenced by the earthquake, including Wenchuan, Beichuan, Mianzhu, Shifang, Qingchuan, Maoxian, Anxian, Dujiangyan, Pingwu and Pengzhou. Due to the huge difference in geographical location, the elevations of these 10 counties and cities are relatively low and the terrain complexity index is relatively high. Then, the M_{Hxy} (mean compatibility coefficient) and M_{GDP} (average GDP growth rate per capita) of the 10 earthquake extremely stricken areas and 41 earthquake-stricken areas are further analysed and the calculation results are shown in Figure 9.

It can be seen from Figure 9 that although the M_{Hxy} slowed down before the 2008 earthquake, the M_{GDP} was relatively stable. After the 2008 earthquake, the trend over time of the post-earthquake reconstruction investment caused a trend of 'sudden growth-return-rise' in the M_{Hxy} and M_{GDP} of the earthquake-stricken areas and the extremely earthquake-stricken areas. It can be seen from the trend of M_{GDP} that the extremely earthquake-stricken areas had dropped sharply in 2008 (−23.58%) and that the earthquake-stricken areas were less affected by the earthquake (9.36%). It can be seen from the trend in M_{Hxy} that the M_{Hxy} in the extremely earthquake-stricken areas and the earthquake-stricken areas had increased significantly within 3 years (before 2011), far exceeding the pre-earthquake level, but this result does not indicate the increased compatibility of the employment and industrial structures because the effect was mainly due to the 'external force' effect. After 2011, with external assistance and other external forces subsiding,

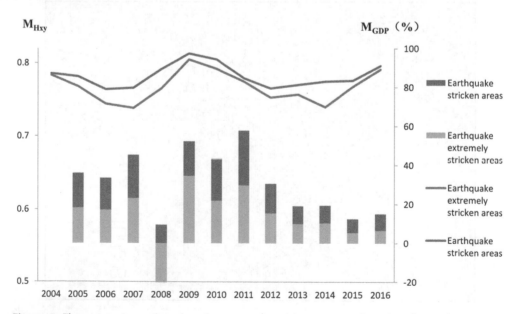

Figure 9. The comparison of earthquake extremely stricken areas and earthquake-stricken areas in terms of M_{Hxy} and M_{GDP}.

Table 1. The growth rate of M_{Hxy}.

Year	Earthquake extremely stricken areas	Earthquake-stricken areas
2005	−1.95	−0.52
2006	−3.12	−2.26
2007	−0.79	0.25
2008	3.66	3.37
2009	5.17	2.69
2010	−1.59	−1
2011	−2.21	−3.22
2012	−2.81	−1.76
2013	0.58	0.66
2014	−2.31	0.58
2015	3.77	0.16
2016	3.12	2.6

the regional 'internal force' played a greater role, indicating the difference between the geographical environment (the seismic intensity and terrain complexity of the areas) of the earthquake-stricken areas and the extremely earthquake-stricken areas. An 'external force' can improve M_{Hxy} and M_{GDP} in a short period of time, and the extremely earthquake-stricken areas will be more affected by the post-earthquake effect. At the same time, combined with the analysis of the M_{Hxy} change rate of the earthquake-stricken areas and the extremely earthquake-stricken areas (Table 1), the compatibility trends of the industrial and employment structures before the earthquake were similar. However, after the 2008 earthquake, the M_{Hxy} change rate of the extremely earthquake-stricken areas fluctuated greatly, indicating that the industrial structure and employment structure of the extremely earthquake-stricken areas showed large adjustments, while the earthquake-stricken areas were affected by the post-earthquake effect; however, the change in compatibility was relatively flat and the recovery was better. This result also further explains the relative importance of the geographical environment in the disaster area to the recovery process and its relationship with resilience.

4. Conclusion

In this paper, a quantitative analysis model of disaster resilience is established for the case study of the earthquake-stricken area in the Longmenshan fault zone. Furthermore, the disaster resilience is analysed from the perspective of earthquake intensity and topography. This is an actual evidence of learning from the Wenchuan earthquake in 2008. Adaptive strategies were undertaken to reduce vulnerability and enhance disaster resilience. Principal conclusions are as follows:

(1) Disaster resilience is important for understanding uncertainty and reducing losses from natural disasters. At the conceptual level, disaster resilience do not have an explicit definition due to the different characteristics of hazard-affected bodies. Disaster resilience analysis requires the assimilation of physical and socio-economic information from many sites each with a unique geographic location (Zhou et al., 2010; Shahid & Behrawan, 2008). Meanwhile, in order to achieve regional sustainable development, it is necessary to strengthen the application of the Internet of things, big data, GIS and other technologies to improve regional disaster resilience.

(2) Due to the continuous influence of post-earthquake effect, geological disasters in the post-earthquake areas occur frequently. At the same time, geological disasters are often extremely hidden and sudden, and it is difficult to identify and give early warning, this leads to an increase in the complexity of the natural disaster system. It can be seen from the analysis of the change rule of disaster resilience in earth-quake-stricken areas that, under the combined action of external and internal forces, the 'post-earthquake effect' in earthquake-stricken areas shows a trend of decreasing vibration. Therefore, in order to reduce the impact of the 'post-earthquake effect', it is necessary to further improve the construction of geological disaster moni-toring systems and strengthen the research and development of related technologies. In this way, technological development can be promoted in a targeted way, for the enhancement of regional disaster resilience.

(3) Disaster resilience is closely related to spatial location (the earthquake intensity area and topographic characteristics), while aid after disasters, and other external forces are not decisive for reconstruction. According to regional characteristics, we need to formulate a specific 'resilience' plan, to make the earthquake-stricken areas more resi-lient to future disasters in the process of recovery and reconstruction. At the same time, disaster resilience should also be taken as an important decision variable in the layout and construction of post-earthquake residential settlements, infrastructure and industrial bases.

5. Discussion

In post-disaster reconstruction, we should not only pay attention to the restoration or maintenance of the original social functions of the system but also consider whether it can survive and regenerate in future environmental changes. In this paper, the GIS method is used to analyse the disaster resilience of earthquake-stricken areas, and the problem of the integration of disaster resilience and geographic information under the unified space–time framework is solved. This study provides a new method for measuring the risk degree of seismic and geological disasters in the post-disaster reconstruction process, and a new perspective and method for the definition and quantification of disas-ter resilience. This is essential to enhance the resilience of disasters and reduce losses from natural hazards. However, many critical difficulties need to be resolved in the field of dis-aster resilience, which are described as follows.

First, due to the complexity of natural disaster systems, there are many difficulties in the quantitative analysis of disaster resilience. The composition of natural disaster systems is crucial to understand resilience. Natural disaster systems are composed of five elements: disaster-pregnant body, disaster-causing body, disaster-bearing bodies, emergency man-agement and disaster-resistant body. These five elements are distributed among physical, information and psychological spaces. The components, especially the disaster-causing body, the disaster-bearing bodies and the disaster-resistant body, may interact and trans-form each other. At the same time, each region of different types faces different disaster types and threats. It is necessary to build disaster simulation under the corresponding scenario and study the change rule of natural disaster systems to improve regional disaster prevention and reduction ability. As shown in Figure 10, earthquakes may cause a variety

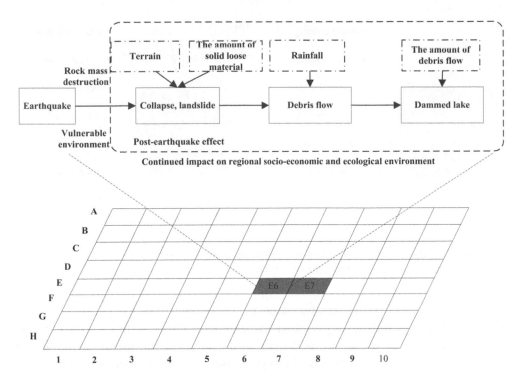

Figure 10. Space–time evolution mechanism of natural disaster system.

of possible secondary events. The E6 and E7 represent different geographical locations; among them, the E6 can produce a kind of disaster chain system over time and space, and may be extended to E7. Therefore, it is difficult to study the 'dynamic complex process' of disaster resilience within a unified research framework. It is necessary to understand the complexity of natural disaster system from a comprehensive geographical perspective, comprehensively analyse the elements and the relationships among them, and then understand the connotations of disaster resilience.

Second, due to the different types of disasters and disaster-bearing bodies, and the disaster resilience involves social, economic, ecological and other aspects, there is no uniform standard when establishing disaster resilience indicators. Therefore, different characteristics of the hazard-bearing body are being carried out. Disaster resilience analysis needs to be based on different perspectives. As shown in Figure 11, the input into geological disaster treatment after the Wenchuan earthquake accounted for approximately 17% of the total reconstruction investment, and the input into geological disaster treatment after the '8.13' massive mudslide accounted for approximately 71% of the total reconstruction input, which indicates that the mountainous earthquake-stricken areas were greatly affected by the 'post-earthquake effect'.

Furthermore, the disaster resilience analysis based on GIS technology is becoming more and more important, as it plays an important role for local, regional and national policy-makers. From the perspective of seismic intensity and terrain complexity, this paper uses GIS technology to conduct a quantitative analysis of disaster resilience in mountainous earthquake-stricken areas. However, building a regional 'soft system' from the results of disaster resilience analysis still represents a difficult problem. The construction

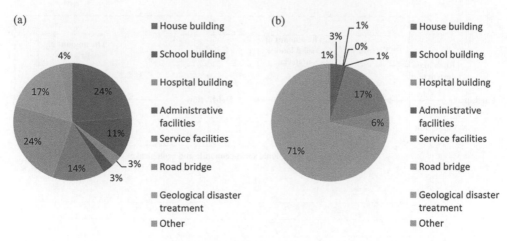

Figure 11. Proportion of investment in post-disaster recovery and reconstruction: (a) after the '5.12' Wenchuan earthquake in 2008, (b) after the '8.13' catastrophic mudslides disaster in 2010.

of a 'soft system' mainly includes two aspects: government organisation management and social security. Orderly organisation management and a perfect security system are critical to disaster prevention and reduction, which play an important role in the recovery of people's normal lives and the post-disaster development of economy and society. From the perspective of government organisation management, the improvement of laws and regulations can effectively guarantee the smooth progress of post-disaster reconstruction and recovery. In terms of social security, although remarkable achievements have been made in the post-Wenchuan earthquake reconstruction, social security is not perfect. For example, insurance coverage is very low (Xie et al., 2018). Therefore, future research is needed to explore the establishment of social security system in line with regional characteristics to improve the ability of different regions to cope with disasters.

Disclosure statement

No potential conflict of interest was reported by the authors.

Funding

This work was supported by National Key R&D Program of China: [Grant Number 2018YFC0604105]; Sichuan Science and Technology Program: [Grant Number 2019JDKY0017]; Opening Fund of Geomathematics Key Laboratory of Sichuan Province: [Grant Number scsxdz201601]; Scientific Research Fund of Sichuan Province Education Department: [Grant Number 18ZB0062].

References

Adger, W. N. (2005). Social-ecological resilience to coastal disasters. *Science, 309*(5737), 1036–1039.
Berkes, F., & Ross, H. (2013). Community resilience: Toward an integrated Approach. *Society and Natural Resources, 26*(1), 5–20.
Bruneau, M., Chang, S. E., & Eguchi, R. T. (2003). A framework to quantitatively assess and enhance the seismic resilience of communities. *Earthquake Spectra, 19*(4), 733–752.

Cai, H., Lam, N. S. N., Qiang, Y., Zou, L., Correll, R. M., & Mihunov, V. (2018). A synthesis of disaster resilience measurement methods and indices. *International Journal of Disaster Risk Reduction*, *31*, 844–855.

Cutter, S. L. (1996). Vulnerability to environmental hazards. *Progress in Human Geography*, *20*(4), 529–539.

Cutter, S. L., Barnes, L., Berry, M., Burton, C., Evans, E., Tate, E., & Webb, J. (2008). A place-based model for understanding community resilience to natural disasters. *Global Environmental Change-Human and Policy Dimensions*, *18*(4), 598–606.

David. (2009). UNISDR terminology on disaster risk reduction. http://www.universe.org/eng. terminology/terminology-2009-eng/html

Fan, X., Juang, C. H., Wasowski, J., Huang, R., Xu, Q., Scaringi, G., & Havenith, H. B. (2018b). What we have learned from the 2008 Wenchuan earthquake and its aftermath: A decade of research and challenges. *Engineering Geology*, *241*, 25–32.

Fan, R., Zhang, L. M., Wang, H., & Fan, X. M. (2018a). Evolution of debris flow activities in Gaojiagou Ravine during 2008–2016 after the Wenchuan earthquake. *Engineering Geology*, *235*, 1–10.

Ganderton, P. T. (2015). *Disaster resilience*. Social Science Electronic Publishing.

Han, P. F., Tian, S. J., & Fan, X. Y. (2018). Statistical analysis and forecasting of the secondary disasters induced by Lushan earthquake. *Journal of Natural Disasters*, *27*(1), 120–126.

Holling, C. S. (1973). Resilience and stability of ecological systems. *Annual Review of Ecology, Evolution, and Systematics*, *4*(1), 1–23.

Holling, C. S. (1996). Engineering resilience Versus ecological resilience. In P. C. schulze (Ed.), *Engineering within ecological Constraints* (pp. 31–44). Washington, DC: *National Academy Press*.

Hosoya, A., Buchert, T., & Morita, M. (2004). Information entropy in cosmology. *Physical Review Letters*, *92*(14), 141302–141800.

Hu, Q., Guo, M., Yu, D., & Liu, J. (2010). Information entropy for ordinal classification. *Science in China Series F: Information Sciences*, *53*(6), 1188–1200.

Huang, R. (2011). After effect of geohazards induced by the Wenchuan earthquake. *Engineering Geology*, *19*(2), 145–151. (in Chinese).

Joerin, J., Shaw, R., Takeuchi, Y., & Krishnamurthy, R. (2014). The adoption of a climate disaster resilience index in Chennai, India. *Disasters*, *38*(3), 540–561.

Klein, R. J. T., Nicholls, R. J., & Thomalla, F. (2003). Resilience to natural hazards: How useful is this concept? *Global Environmental Change Part B: Environmental Hazards*, *5*(1), 35–45.

Kusumastuti, R. D., Husodo, Z. A., Suardi, L., & Danarsari, D. N. (2014). Developing a resilience index towards natural disasters in Indonesia. *International Journal of Disaster Risk Reduction*, *10*, 327–340.

Li, T. Y. (2017). New progress in study on resilient cities. *Urban Planning International*, *32*(5), 15–25.

Li, X., Lam, N., Qiang, Y., Li, K., Yin, L., Liu, S., & Zheng, W. (2016). Measuring county resilience after the 2008 Wenchuan earthquake. *International Journal of Disaster Risk Science*, *7*(4), 393–412.

Lin, G. W., Chen, H., Chen, Y. H., & Horng, M. J. (2008). Influence of typhoons and earthquakes on rainfall-induced landslides and suspended sediments discharge. *Engineering Geology*, *97*(1), 32–41.

Manyena, S. B. (2006). The concept of resilience revisited. *Disasters*, *30*(4), 433–450.

Meerow, S., Newell, J. P., & Stults, M. (2016). Defining urban resilience: A review. *Landscape and Urban Planning*, *147*, 38–49.

Mishra, A., Ghate, R., Maharjan, A., Gurung, J., Pathak, G., & Upraity, A. N. (2017). Building ex ante resilience of disaster-exposed mountain communities: Drawing insights from the Nepal earthquake recovery. *International Journal of Disaster Risk Reduction*, *22*, 167–178.

Orencio, P. M., & Fujii, M. (2013). A localized disaster-resilience index to assess coastal communities based on an analytic hierarchy process (AHP). *International Journal of Disaster Risk Reduction*, *3*, 62–75.

Shahid, S., & Behrawan, H. (2008). Drought risk assessment in the western part of Bangladesh. *Natural Hazards*, *46*(3), 391–413.

Shim, J. H., & Kim, C. I. (2015). Measuring resilience to natural hazards: Towards sustainable hazard mitigation. *Sustainability*, *7*(10), 14153–14185.

Simpson, D. M. (2008). Disaster preparedness measures: A test case development and application. *Disaster Prevention and Management*, *17*(5), 645–661.

Stahl, T., Clark, M. K., Zekkos, D., Athanasopoulos-Zekkos, A., Willis, M., Medwedeff, W., … Jin, J. (2017). Earthquake science in resilient societies. *Tectonics*, *36*(4), 749–753.

Timmerman, P. (1981). Vulnerability, resilience and the collapse of society: A review of models and possible climatic applications. *Environmental Monograph*, *1*, 1–45. https://doi.org/10.1002/joc.3370010412.

Ullsten, O., Speth, J. G., & Chapin, F. S. (2004). Options for enhancing the resilience of northern countries to rapid social and environmental change. *Ambio: A Journal of the Human Environment*, *33*(6), 343–343.

UNISDR. (2002). *Living with risk: A global review of disaster reduction initiatives* Geneva United Nations: *BioMed Central Ltd*.

Walker, B. H., Anderies, J. M., Kinzig, A. P., & Ryan, P. (2006). Exploring resilience in social-ecological systems through comparative studies and theory development: Introduction to the special issue. *Ecology and Society*, *11*(1), 12.

Wen, L. I., Wenkai, C., & Zhonghong, Z. (2019). Analysis of temporal-spatial distribution of life losses caused by earthquake hazards in Chinese Mainland. *Journal of Catastrophology*, *34*(1), 222–228.

Xie, W., Rose, A., Li, S., He, J., Li, N., & Ali, T. (2018). Dynamic economic resilience and economic recovery from disasters: A quantitative assessment. *Risk Analysis*, *38*(6), 1306–1318.

Xu, J., & Lu, Y. (2012). Meta-synthesis pattern of post-disaster recovery and reconstruction: Based on actual investigation on 2008 Wenchuan earthquake. *Natural Hazards*, *60*(2), 199–222.

Zhang, S., & Zhang, L. (2017). Impact of the 2008 Wenchuan earthquake in China on subsequent long-term debris flow activities in the Epicentral area. *Geomorphology*, *276*, 86–103.

Zhang, S., Zhang, L. M., & Glade, T. W. (2014). Characteristics of earthquake-and rain-induced landslides near the Epicenter of Wenchuan earthquake. *Engineering Geology*, *175*(11), 58–73.

Zhang, S., Zhang, L. M., Lacasse, S., & Nadim, F. (2016). Evolution of mass movements near epicentre of Wenchuan earthquake, the first eight years. *Scientific Reports*, *6*(1), 36154.

Zhou, H., Wang, J., Wan, J., & Jia, H. (2010). Resilience to natural hazards: A geographic perspective. *Natural Hazards*, *53*(1), 21–41.

Zhou, Y., Tang, G. A., Zhang, T., Luo, M. L., & Jia, Y. N. (2010). Spatial pattern of terrain feature lines of Loess gully-hill region. *Arid Land Geography*, *33*(1), 106–111.

Coping and resilience in riverine Bangladesh

Parvin Sultana, Paul M. Thompson and Anna Wesselink

ABSTRACT
This paper investigates the impacts of two successive years of severe floods on households, their coping strategies and resilience to riverine hazards in northern Bangladesh. Based on focus groups and interviews with the same households after floods in 2016 and 2017, we found a cumulative decline in assets through sale of livestock and borrowing, and almost all households evacuated short term to higher places. Three notable recent ways that vulnerable households use socio-hydrological landscapes to enhance their resilience to hazards were revealed. Firstly, local flood protection embankments were the main destination for evacuation and were highly valued as safe places, although they breached and failed to protect the land. Secondly, community organisations, formed mainly for livelihood enhancement, took initiatives to provide warnings, to help households relocate during floods, and to access relief and rehabilitation services. Thirdly, seasonal migration by men, particularly to urban areas, is an important element of long-term coping and resilience based on diversified livelihoods for about 70% of these rural households. Although the unintended use of infrastructure, social capital and urban opportunities all form part of coping and resilience strategies in hazardous riverine landscapes, the high mobility that they are based on is not supported by enabling policies.

1. Introduction

This paper uses evidence from two recent years of exceptional floods (2016–2017) in a riverine area of Bangladesh to understand how vulnerable households use infrastructure, institutions and urbanisation to cope with hazards. Hydrological science has recently sought to take a more integrated view of societal interactions with floods (Sivapalan, Savenije, & Blöschl, 2012). This has been termed socio-hydrological, derived from the field of study and practice termed socio-hydrology, and is taken here to mean interdisciplinary study of the dynamic interactions and feedback between water and people. Arising from hydrologists the concept as recently adopted has a modelling focus that aims to capture the full range of human behavioural interactions with water resources. However, it can be argued (Wesselink, Kooy, & Warner, 2017) that socio-hydrology is a recent water system based incarnation of the much longer established field of socio-ecological systems (Berkes & Folke, 1998), which itself brought together aspects of systems approaches, human geography and institutional analysis related to natural resources. By contrast the concept of hydrosocial systems has its foundations in human geography with an emphasis more on hydrological cycles and analysing power relations (Linton, 2008; Wesselink et al., 2017). To empirically ground socio-hydrological propositions, Ferdous, Wesselink, Brandimarte, Slager, Zwarteveen & Di Baldassarre (2018) put forward the concept of socio-hydrological space as a geographical area in the landscape with distinct hydrological and social features that give rise to the emergence of distinct interactions and dynamics between society and water. In Bangladesh these spaces are distinguished by differing exposure to flood and riverbank erosion, delineating the active but inhabited floodplain from adjacent embanked lands (Ferdous et al., 2018). This paper shows that a riverine landscape that includes different socio-hydrological spaces is itself nested within, and increasingly dependent on, wider socio-economic institutions and systems, since people and water in rural and urban areas are becoming more and more connected. We find that people are more mobile than has been characterised in previous studies. Undoubtedly there is a strong attachment of rural people to home (village) locations, but there is evidence that those vulnerable to floods are not 'immobile' or 'trapped' where they currently live (Foresight, 2011). Rural people may take up work in urban areas while retaining a foothold in their floodplain origins. Urbanisation and inter-district communications in Bangladesh have developed at a fast pace. Also, the establishment of many community-based organisations (CBOs) in recent years has changed the situation of rural people bringing opportunities to help them withstand hazards. This paper enriches these insights on recent socio-economic developments affecting life in flood-prone rural areas in Bangladesh.

Bangladesh is one of the world's most flood-prone countries. Floods occur due to storm surges associated with cyclones in the Bay of Bengal affecting coastal areas, heavy rain over adjacent hills causing flash floods, and monsoon rains over the Ganges-Brahmaputra catchment that cause extensive riverine floods. Up to 80% of Bangladesh comprises of floodplain (Brammer, 1990), and about 25–33% of the country remains under water every year for 4–6 months during the monsoon (rainy season). 'Normal' floods have limited negative impacts and are even considered beneficial to agriculture, but the situation is very different in extreme floods (Paul, 1997). For example, in one of the last major riverine floods in 1998 about 60% of the country was flooded, about 1000 people died, about 2000 km of embankments were damaged and losses were valued at up to US$ 2.8 billion (Brammer, 2004). In addition, riverbank erosion causes loss of land and infrastructure almost every year.

Although evacuation in response to cyclones and coastal flooding has been well studied in Bangladesh (reviewed in Penning-Rowsell, Sultana, & Thompson, 2013), riverine floods and erosion lead to different responses due to the nature of the losses, the physical features of the area affected, and its settlement history. Superimposed on long-established hazards is a new popular discourse on climate change impacts which perceives large-scale migration as a potential source of instability. This has raised alarms for Bangladesh with fears of large-scale human displacement and mass migration from coastal areas which are predicted to become more regularly inundated by increasing sea levels (Mirza, 2002). With Bangladesh's high population density and high levels of poverty, the consequences of climate change for human life and society are often portrayed as cataclysmic and potentially leading to unmanageably large numbers of 'climate refugees' (Ahsan, Karuppanan, & Kellett, 2014; Salauddin & Ashikuzzaman, 2011). However, recent studies highlight the dynamic nature of the Bangladesh delta where major sediment deposition interacts with embankments to create waterlogged polders (Auerbach et al., 2015). Sedimentation may mitigate displacement of many people, while the impacts of a slowly-rising sea-level are less than those generated by increasing population pressure on natural resources (Brammer, 2014). Moreover, detailed studies reveal that drivers for migration are more complex and include several push factors such as environmental changes (Etzold, Ahmed, Hassan, & Neelormi, 2014; Joarder & Miller, 2013; Penning-Rowsell et al., 2013; Yasmin & Ahmed, 2013).

Since the 1960s flood mitigation in Bangladesh has been based on a structural approach of large-scale embankments constructed by the government. In the 1990s recognition was growing of adverse environmental and social effects of embankments that counteracted their benefits (Thompson & Sultana, 2000). Civil society debate grew over flood control and requirements for local participation and environmental mitigation, which challenged the technical-engineering dominance of policy (Sultana, Johnson, & Thompson, 2008). Public participation in water resource management in Bangladesh was formally adopted (Ministry of Water Resources, 2001), but active local participation in managing water is most often in smaller scale systems where communities manage irrigation and local leaders tend to dominate decisions (Sultana & Thompson, 2010). Studies on participatory water management in Bangladesh have reported local people collectively repairing damaged embankments (Penning-Rowsell et al., 2013), while Yu, Sangwan, Sung, Chen, and Merwade (2017) focused on idealised community initiatives to protect public infrastructure. However, these are exceptions rather than the norm, and practical public participation in larger-scale flood mitigation is limited (Dewan, Mukherji, & Buisson, 2015).

Despite engineering works, floods and riverbank erosion still cause substantial damage to land, crops, houses, and properties (Ayeb-Karlsson, van der Geest, Ahmed, Huq, & Warner, 2016; Islam, Hasan, Chowdhury, Rahaman, & Tusher, 2012; Thompson & Tod, 1998). Each year several thousand people become homeless and landless due to river bank erosion (Haque, 1988; Haque & Zaman, 1993; Hutton & Haque, 2004; Indra, 2000). Households living along the main rivers have to face the costs of frequently repairing or constructing new houses due to floods and riverbank erosion, which adds to their existing burdens (Yasmin & Ahmed, 2013). In the major floods of the 1970s and 1980s floods were a route to households falling into debt and impoverishment (Alamgir, 1980; Haque & Zaman, 1993).

Households who have lost both homestead and any cultivable land they owned due to riverbank erosion generally shelter on other people's land or on public land (including embankments) close to their original homes (Baqee, 1998; Elahi, Ahmed, & Mafizuddin, 1991; Gray & Mueller, 2012; Joarder & Miller, 2013; Rahman, 2010). The majority of flood-affected people try to stay near their houses in the hope that land may accrete along the river so that they can reclaim it (Brouwer, Akter, Brander, & Haque, 2007; Mamun, 1996). When these hopes are not fulfilled, they may then migrate to other districts where land or work might be available. Only a few migrate permanently to towns and cities (Ayeb-Karlsson et al., 2016; Gray & Mueller, 2012; Haque, 1988; Indra, 2000; Joarder & Miller, 2013).

This study updates knowledge and understanding of coping and resilience strategies in a changing socio-economic context, with a focus on how households use temporary evacuation and seasonal migration between rural flood-prone areas and urban areas. It also examines the role of local community-based organisations (CBOs) in strengthening resilience, and the role of structural flood mitigation (embankments) in connecting unprotected and protected socio-hydrological landscapes.

2. Methods

2.1. Study sites

This study focuses on three districts of northwest Bangladesh (Figure 1). The Jamuna–Brahmaputra River[1] in this area has been widening and shifting its course since 1830, becoming braided. During 1973–2000, the satellite images showed it widened by about 128 m per year (EGIS, 2000). River erosion affects lands that have been settled for generations and also the *chars* (islands of accreted sediment) within the braided river. In each of the three districts, three unions (the lowest administrative tier) were purposively selected to represent differences in exposure to riverine hazards and structural flood protection, and because CBOs were known to be active there.

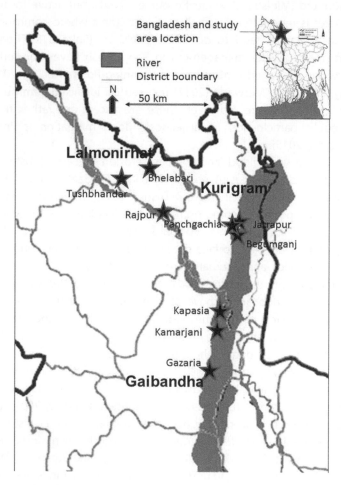

Figure 1. Locations of study sites.

Gaibandha District, the most southerly of the three districts covered, is situated to the west of the Jamuna–Brahmaputra River, with part of the district behind the Brahmaputra embankment and part in the riverine chars. The three unions studied represent lands stretching from the embankment out into the river: Gazaria Union includes a branch of the river and associated fishing communities; Kapasia Union, where most people live on the embankment or on land between the embankment and the river channels; and Kamarjani Union which mainly comprises chars.

Kurigram District, immediately to the north of Gaibandha, includes the most northerly reach of the Brahmaputra River within Bangladesh. Here the unions selected mainly represent chars separated from the mainland by branches of this braided river. Panchgachia Union borders the Brahmaputra and Dharla Rivers with some embankment-protected land and the rest is chars; Begumganj Union lies entirely in the riverine areas outside of embankments, and Jatrapur Union almost entirely comprises chars.

Lalmonirhat District, to the west of Kurigram, is located along the Teesta River just upstream of its confluence with the Brahmaputra River. The Teesta has a relatively narrow floodplain bordered by embankments on both sides. The study sites represented primarily mainland areas that have not accreted or eroded in recent times, protected by embankments from normal floods. The three unions are: Bhelabari Union located behind the Teesta embankment and equidistant from the Teesta and Dharla Rivers; Tushbhandar Union which is bisected by the Teesta embankment and has some unprotected land; and Rajpur Union where the Teesta River flows through the middle of the union, and active erosion of village land is common, but also with some embankment-protected areas.

2.2. Data collection

In these nine locations flood impacts and inhabitants' responses to floods and riverbank erosion were investigated initially through focus group discussions. Subsequently, surveys in one union in each district were undertaken. These unions were selected for their different exposure to flooding and riverbank erosion: an island char within the braided river with the highest exposure (Jatrapur Union in Kurigram district), a settlement on the embankment and adjacent riverside land with some exposure (Kapasia Union in Gaibandha district), and a village protected by embankments with least exposure (Rajpur Union in Lalmonirhat district). The survey focused on experiences during and after unusually severe floods in 2016 and 2017. We initially investigated the impacts of floods and erosion in 2016, and when 2017 also produced extreme flooding we repeated the surveys. In each of the three unions, within an area where CBOs were active, 40 households were randomly sampled. The head of household was interviewed in 2016, and the 107 households that remained in the three areas were re-interviewed in 2017, in both years during November soon after the flood season.

3. Results

3.1. Past flood experience and general economic trends

All respondents in the household survey had experienced floods and riverbank erosion during their lifetimes, and all respondents had moved home at least once. The recollection period by definition varies between respondents due to age, 58% of respondents were 31–50 years old so their memory extends back to at least 25 years. Their past experience of moving home due to flood or erosion is likely to influence coping strategies. Most households had moved home 2–8 times in their lifetime (Figure 2). The number of moves was associated with location ($\chi^2 = 23.2$, df $= 6$, $p < .05$), households had moved fewer times in Gaibandha, whereas even those currently living in locations at low erosion risk in Lalmonirhat were found to have moved there due to frequent river erosion elsewhere.

The households had recent experience of floods. Over the previous five years (2011–2015), 95% of surveyed households in all three study unions reported that they had experienced floods affecting their homes, with almost all also reporting loss of work and serious illness associated with floods. On average in all three unions, households had experienced two damaging floods in those five years. Most households had also been affected by erosion in one of these five years, although this varied between locations with the Gaibandha households worst affected (88% lost land and 83% lost housing to river erosion) compared with Kurigram (73% lost land but only 43% lost homes) and Lalmonirhat (65% lost land and 50% lost homes). Almost 80% of respondents believe that the incidence of floods and erosion is increasing.

During 2011–2015, the surveyed households reported that their net incomes had increased, based on the trends (increase, no change or decrease) that they reported for four main categories of possible income (agriculture, livestock, wage labour, seasonal migration). Differences between unions were consistent with differences in hazards, with the households in the most exposed union having the lowest incomes. In Lalmonirhat a majority of households reported that income from crop agriculture increased. Incomes from livestock rearing also increased; this is an important strategy in these areas where in the dry season there is plentiful grazing in the chars and cattle can be moved to the mainland before floods strike, and is discussed further in Section 3.3. Most households also reported increases in their income from local wage labour. In addition, 55–70% of households said household members undertook seasonal migration

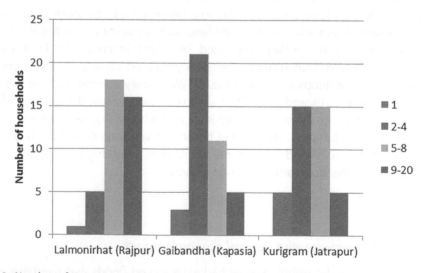

Figure 2. Number of times respondents moved home in their lifetime.

for work in the past five years involving more days worked and increasing wages. Over half of the households in Gaibandha and Kurigram reported that they had improved their house structure in this period. The resulting increase in total household asset value was diminished by declining landholdings in all three unions, with over 75% of respondents reporting that their landholding had declined, mainly due to erosion. Although incomes have increased, so have living costs, and hazard events in the same period caused loss of assets, which explains why about 80% of households reported running down their savings. Hence immediate coping ability may have been enhanced by rising incomes, but the implications for long-term resilience were unclear. The cumulative impacts of unusual floods in two successive years – 2016 and 2017 on resilience were a subject of investigation.

3.2. The 2016 and 2017 floods

In the whole study area, the 2016 and 2017 floods were more severe than recent years. The 2016 flood in Bangladesh in the Brahmaputra and its tributaries was estimated at the time to have a 1-in-100 year return period. In 2016 the water level at Bahadurabad gauging station (located in this area) exceeded the previous highest ever recorded water level of 1988 (Islam, 2016), but this level was exceeded again in 2017. The return period of the 2017 flood peak has been estimated as 1-in-30 to 1-in-100 years, but both estimates of return periods are affected by the short length of data series and changes in river mor-phology as well as uncertainties introduced by climate change (Sjoukje et al., 2018). In both years flood peaks occurred in both July and August. The flood peaks were relatively short in both years since they did not coincide with high flows in the Ganges, nor with spring tides in the Bay of Bengal.

From the focus groups data, it appears that in 2016 in Lalmonirhat only about 8% of the 20,000 households in the three unions had flood damage to their houses, fewer than in the other districts, and the affected households evacuated to embankments for about two weeks. In Gaibandha about 70% of 4300 households in the three unions were flooded, in some areas because embankment breaches occurred; the majority of those affected moved for up to a month to embankments or to protected areas. In Kurigram, about 56% of 8700 households in the three unions were estimated to have been flooded. In this district evacuation responses varied between locations: those affected by erosion evacuated to embankments, but many people sheltered on and near their homes. In 2017 all surveyed households in all three areas were flooded and temporarily evacuated, mostly to embankments. In 2017 erosion impacts were more severe than in 2016 in the unions in Lalmonirhat and Gaibandha districts, where almost half of the households lost some or all of their homestead land. Most households stayed in the area at the time of the surveys hoping that their land might reappear from the rivers, but a few households were reported to have already left the area permanently.

In 2016, up to 50% of houses were damaged in the floods, with widespread loss of poultry and some other livestock. In addition, most households lost income as daily labouring work was neither available nor possible locally, and they were busy coping with floods so could not look for work. The average reported losses per household in 2016 were about US$ 168 in Lalmonirhat, US$ 228 in Gaibandha and US$ 303 in Kurigram. Given the severity of the 2017 floods, losses were likely to have been higher, although some households had fewer assets to lose because they could not recover from 2016, and detailed data was not collected as the follow-up survey focused more on migration.

Ill-health was widely reported in both years' floods. This was a greater problem in the unprotected areas, particularly the chars (Table 1). In 2016, 70% of households lost on average 16 working days per household due to illness during the floods and spent almost US$ 37 per household on treatment. In 2017 households also lost on average about 16 working days per household due to illness during the floods. In 2017 they spent about US$ 50 per household on treatment, with higher costs in Kurigram and lower costs in Lalmonirhat.

3.3. Responses to floods

3.3.1. Range of coping actions

Households were asked what actions they had taken to cope with floods and riverbank erosion in 2016 and 2017. The incidence of coping actions reported was generally similar in the three unions and in the two years (Figure 3). In all three unions, all households changed their eating practices to cope in both years, reducing the number of meals and amount eaten, and ate lower quality or less preferred foods. Moreover 70% of households in 2016 and over 80% in 2017 borrowed food in addition to food aid that most of them received. A minority of households did casual labouring work for food in 2016; this increased in 2017. Very few households pledged labour for advance payments or sold expected harvests in advance. Both of these practices often lead to indebtedness.

Ownership of livestock can strengthen resilience, but also has disadvantages as an adaptation. Animals are movable and can be sold to raise cash, but are vulnerable to drowning and ill-health, also access to safe shelter during floods in the char areas was reported to limit their role in coping. Participants in a focus group in Gaibandha after the 2016 floods emphasised some of these vulnerabilities:

> There was no shelter for livestock in the chars during the flood, even though there was sufficient livestock feed. So we were forced to sell livestock. The char area had lots of buffalo before, but now the number decreased to almost none due to lack of space and feed. But the number of milk cows increased as a way of coping with flood risk.

Table 1. Reported health effects of 2016and 2017 floods.

	2016			2017		
	Lalmonirhat (Rajpur)	Gaibandha (Kapasia)	Kurigram (Jatrapur)	Lalmonirhat (Rajpur)	Gaibandha (Kapasia)	Kurigram (Jatrapur)
Number of households reporting illness	22	33	28	27	31	22
% households affected	55	83	70	68	78	55
Number of persons	27	45	34	44	43	37
Days income loss from illness per affected household	6.0	11.8	16.5	6.9	25.2	13.6
Cost of treatment (Tk per person)	461	1379	2381	922	2563	3888
Cost of treatment (Tk per household)	565	1880	2891	922	3555	6539

Note: Tk = Bangladesh Taka, approximately Tk 82 = US$1 in these years

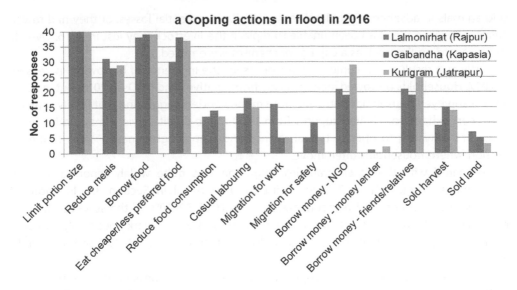

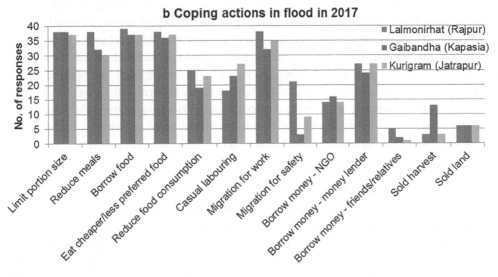

Figure 3. Coping actions.

In this area char households can sell milk to inhabitants of the protected 'mainland', revealing linked livelihood strategies between these areas. Livestock sale is to some extent forced upon households in the unprotected riverine landscape, but in 2016 sale of livestock, particularly goats and poultry, was also a common coping action. Challenges to feed cattle and move them to safety were reported to have been greater in 2016 than in previous years. In 2016 the Muslim festival of Eid-ul-Azha took place in September, in the late monsoon season, and many households in the study region decided to fatten cattle for sale as prices are usually high for the festival. In all three areas cattle were moved to higher ground including schools and embankments and remained there for up to a month, until the homesteads dried up. Nevertheless, 19 of the households lost cattle in the 2016 floods. In 2017 fewer households could raise cash from livestock sale. Eid-ul-Azha was about ten days earlier, and after the 2016 flood experience households either sold animals in advance of the flood season, to avoid potential losses, or they had fewer animals because they had been unable to replace the livestock they lost in 2016. Overall views regarding livestock as a coping mechanism were mixed.

Partly as a result of this cumulated loss of assets, the incidence of borrowing increased in 2017. Households borrowed money mainly from relatives and NGOs in 2016, but in 2017 25% of surveyed households borrowed from moneylenders suggesting that family sources had been exhausted. Limited support from NGO micro-credit services may be because NGOs are unwilling to operate typical savings and loan programmes in areas prone to erosion where households often move.

Compared with the other strategies, migration was quite widely reported as part of coping strategies (Table 2), and is discussed in detail in Section 3.4. In addition to being a component of livelihood strategies, in 2016 30–40% of the surveyed households said that household members undertook migration for work specifically to cope with flooding when no work was available locally, increasing to 50–70% of households in 2017.

The same broader elements of coping and resilience were considered by respondents to be important in all three unions in both years. They emphasised social capital in the form of help from relatives and neighbours, the availability of physical capital such as embankments for taking shelter, keeping livestock as a movable and saleable asset, and having savings. In addition, mobile phones were widely considered useful, for example to contact relatives and arrange evacuation. In 2017 credit from moneylenders and shops was more common than in 2016, which is consistent with fewer households having saleable assets left and having reduced savings because of the 2016 floods. Data from two successive years of unusual floods therefore show a pattern of widespread loss of assets, coping by reduced food intake, and short-term evacuation.

3.3.2. Role of embankments

Many of the nine unions are located in char areas or on the river side of existing embankments and are not protected by embankments. Yet, with the exception of one union, embankments are relatively close. In six out of eight unions these embankments breached in 2016, raising questions about their effectiveness in protecting crops and settlements from floods. Nevertheless, embankments were evaluated as very important in the surveys – as places of refuge. In the focus groups in Lalmonirhat and Kurigram it was reported that many households that lost their homes to erosion or had flooding inside their homes moved temporarily to find shelter on embankments. Participants in a focus group in Kurigram in 2016 reported that:

> Most people moved temporarily to embankment. . . . . These are very crowded places, some people even do not get space for sleeping. Due to mosquitoes and lack of clean drinking water we fell sick. For women latrine is a problem. Some NGOs helped people to raise land where they could stay. Some people lived on boats. Cooking food is a problem for lack of dry fuel and space. We could only cook once in a day. Getting work is a problem.

Thus although embankments provide a safe shelter for char people, there are also many difficulties faced by people taking shelter on embankments, which explain why they are mostly used short term until people can return to their homes.

The situation in Gaibandha is more complex, here several surveyed households already lived on the embankment, having previously lost their homes to erosion, but the embankment is itself threatened by erosion. Participants in a focus group here in 2016 said:

> Embankment is already inhabited by people and crowded. During flood when we want to move there a clash happens due to lack of drinking water and sanitation. People move with livestock which ruins the soft shoulders of the embankment. Ultimately breach happens when the water current increases during high monsoon.

Thus the embankment here had already become more than a short-term destination for char people displaced by erosion, and evacuees are well aware that their actions can compromise the functioning of embankments. Because the embankment was already crowded, in Gaibandha those households that were flooded in 2016 evacuated to other high land and public buildings. However, in 2017 about half of the surveyed households in this area were able to squeeze onto embankments for shelter. The opinion of people in Gaibandha about the importance of embankments reflects this change. In 2016 they asserted that embankments were not of use to them because they get no help there and have to make their own home and get no services; but a year later they recognised the value of embankments for shelter. Overall, embankments contribute substantially to resilience since they help households remain in the riverine landscape.

3.3.3. Role of collective action

Another key element of coping strategies is social capital in the form of trusted relatives and neighbours. This social capital was strengthened to some extent by community-based organisations (CBOs): local voluntary associations of households formally registered with government. Formed with the help of various past projects, they enable member households to cooperate to improve their livelihoods through water, agriculture and/or fishery management. None were formed specifically to provide hazard related services. However, in Gaibandha, for example, in all three unions the CBOs formed a basis for voluntary assistance by helping to warn and rescue people, and contributing some materials for the worst affected. The CBOs differ in their capacities and services in floods, but even those with limited capacity contributed to resilience. For example, the leader of Konai Brahmaputra CBO, which was formed in Gaibandha in the early 2000s for fishery management, said in 2016:

> Our members are very poor. Most of them fish for a living. But we helped people to move and when needed members helped each other to build shacks. Members collected bamboo from whoever had a bamboo grove for house rebuilding on the embankments.

CBOs differ from NGOs which are formed by one or a few individuals to provide services to target people, but CBOs formed a link with NGOs and other government agencies that supported flood warning dissemination through community volunteers.

In the study areas in Kurigram and Lalmonirhat, because the CBOs are federations with links to NGOs they have relatively wide coverage and put a priority on disaster risk reduction related activities. In the focus group discussions, CBOs were said to have played important local roles. For example, the leader of Satata Rajapur Federation in Lalmonirhat explained that

Table 2. Numbers of surveyed households involved in different types of movement in 2016 and 2017.

| District | 2016 | | | | 2017 | | | | | |
| | Temporary Evacuation | | Seasonal (for work) | | Temporary Evacuation | | Permanent move | | Seasonal (for work) | |
	no	%	no	%	no	%	no	%	no	%
Lalmonirhat (Rajpur)	39	97.5	22	55.0	34	100	6	15.0	14	41.2
Gaibandha (Kapasia)	28	70.0	27	82.5	36	100	4	10.0	23	63.9
Kurigram (Jatrapur)	39	97.5	30	75.0	37	100	3	7.5	27	73.0
Total	106	88.3	85	70.8	107	100	13	10.8	64	59.8

Percentages are out of sample size of 40 households in each district in 2016, but in 2017 are calculated out of those 107 households remaining

One of the main activities of our CBO is disaster management. The CBO members received training and formed volunteer group. We gave early warning through our own miking [loud-speaker] system. We relocate people by using members boats, and find out space for people and their livestock to move. We report to RDRS [a regional NGO] for relief and help. Also keep contact with government departments for rehabilitation, and cooperate with Union Parisad during relief distribution.

In Gaibandha externally based NGOs, rather than CBOs, provided similar services of general advice and flood warning systems with community volunteers in two areas. In Gaibandha the most erosion-prone union, NGOs (mainly the NGO Practical Action) helped people to move house, raise house plinths and establish new homes, and provided free replacement cattle.

In one union in Kurigram no NGO or government assistance was available. Despite almost all houses being flooded people did not move away and reported little damage. Here the CBO helped to warn people and move some vulnerable households. In the other two unions in this district, CBOs also helped people to move and rebuild or repair their houses, and also formed a link with NGOs and government to access relief for those worst affected by erosion or floods.

3.4. Migration in 2016 and 2017

There are several ways of categorising household movements and the relationship between these decisions and flood and erosion hazards. The distinction between tempor-ary evacuation and a permanent move can be fuzzy, since households may evacuate when they are flooded, only to find their homestead land has eroded and then they either remain for a longer period living on embankments or make a house on other available land, often in the same area. Also, longer-term migration often only applies to one person in the household, who migrates for work 'seasonally' but often for a substantial part of the year.

3.4.1. Evacuation and longer-term shifts in residence

Many flooded households evacuated temporarily to safer locations. All households traced in 2017 evacuated, and almost all evacuated in 2016 (Table 3), except for Gaibandha where already some of the households lived on the embankment. Embankments and villages protected by embankments were the main evacuation destinations, although in 2016 fewer people moved to embankments in Gaibandha because they already had many

Table 3. Percentage of sample households evacuating in floods by origin and destination.

Location	Lalmonirhat (Rajpur)		Gaibandha (Kapasia)		Kurigram (Jatrapur)	
	2016	2017	2016	2017	2016	2017
Number of households	39	35	28	36	39	37
Origin (place evacuated from)						
Island char	15.4	17.1	0.0	44.4	94.9	40.5
Riverside (not protected)	10.3	42.9	92.9	52.8	0.0	16.2
On embankment	0.0	2.9	3.6	2.8	0.0	0.0
Village area behind embankment	74.4	37.1	3.6	0.0	5.1	43.2
Destination (place evacuated to)						
Island char	2.6	2.9	3.6	0.0	0.0	0.0
Riverside (not protected)	0.0	2.9	46.4	0.0	0.0	0.0
On embankment	69.2	41.2	3.6	50.0	87.2	97.3
Village area behind embankment	25.6	44.1	46.4	50.0	10.3	0.0
Other more distant rural area	2.6	8.8	0.0	0.0	2.6	2.7

households living there from previous displacement. Moreover, 25% of households that evacuated their homes in 2016 had to do this twice, and in 2017 58% of households that evacuated did this twice. In Lalmonirhat in 2016 evacuation was mainly for 1–5 days along the Teesta River, and for 6–10 days for the first event and an additional 16–30 days for those households that evacuated twice along the Brahmaputra-Jamuna River. There was a similar pattern in 2017 in the three areas of brief evacuation in the first flood peak and evacuation for 11–30 days in the second flood peak.

Permanent movement in the 2016 and 2017 floods was a response to erosion of homes and was short distance usually to embankments or other nearby places in order to stay close to relatives and be on hand for land possibly re-emerging from the river. Hence most households keep a physical and social base in the riverine landscape (on embankments, behind them, or in unprotected areas) according to opportunities and social links.

In all three districts, people who evacuated in the 2016 and 2017 considered social capital in the form of relatives and other known and trusted people was the main advantage of living in their home locations compared with relocation places (Table 4). Resilience was also strengthened by similarly affected households cooperating and moving together (for example with CBO help). Safety, communications, productive land, and availability of land were highlighted as advantages of home areas. The main reported disadvantage of their home location, especially in 2017 in all locations, was the risk there followed by a lack of work opportunities.

Whether movements should be seen as temporary evacuation or longer term is debatable. In many cases the households moved back to their homes shortly after the floods, but in other cases due to changes in the river and char morphology they remained for longer in 'temporary' places, expecting that their land might become inhabitable. In 2017 12% of the households surveyed in 2016 left the area and were untraceable, they were reported by former neighbours to have left permanently (with the highest emigration from Lalmonirhat, and lowest from Kurigram).

3.4.2. Seasonal migration for work
Seasonal migration for work is an important part of livelihood strategies for almost 70% of households surveyed (highest incidence in Kurigram, lowest incidence – half of households in Lalmonirhat). Migration is reported to be increasing as a way of earning money

Table 4. Main advantages and disadvantages reported for home location (% of households).

Attribute	Lalmonirhat (Rajpur)		Gaibandha (Kapasia)		Kurigram (Jatrapur)		All	
	2016	2017	2016	2017	2016	2017	2016	2017
Advantages								
Relatives and known people	38.5	54.3	46.4	50.0	38.5	64.9	41.1	56.5
Safe location	5.1	28.6	28.6	19.4	25.6	37.8	19.8	28.7
Good communications	12.8	20.0	21.4	36.1	12.8	13.5	15.7	23.1
Productive land	23.1	28.6	7.1	5.6	23.1	13.5	17.8	15.7
Land to live on	2.6	17.1	7.1	38.9	0.0	21.6	3.2	25.9
Good public services (e.g. school, health)	12.8	14.3	3.6	13.9	10.3	10.8	8.9	13.0
Good environment	7.7	5.7	3.6	0.0	12.8	5.4	8.0	3.7
Plenty of work	7.7	11.4	0.0	13.9	0.0	2.7	2.6	9.3
Disadvantages								
Risky location	7.7	42.9	21.4	58.3	30.0	54.1	19.7	51.9
Little work	10.3	34.3	7.1	30.6	7.5	35.1	9.1	33.3
Poor communications	28.2	22.9	7.1	13.9	32.5	16.2	22.6	17.6
Theft/crime/tolls	7.7	25.7	21.4	27.8	2.5	5.4	10.5	19.4
Poor public services (e.g. school, health)	2.6	17.1	0.0	11.1	10.0	27.0	4.2	18.5
No/poor sanitation	2.6	14.3	14.3	19.4	15.0	5.4	10.6	13.0
Poor environment	2.6	2.9	0.0	16.7	2.5	10.8	1.7	10.2
Unproductive land	0.0	5.7	0.0	8.3	2.6	16.2	0.9	10.2
No land to live on	0.0	2.9	0.0	11.1	2.5	8.1	0.8	7.4
People not helpful	0.0	5.7	0.0	8.3	2.5	0.0	0.8	4.6

Note: Multiple responses possible so each % is from those surveyed.

to cope after floods. Migration for work mostly involved one man per household aged 16–40 with little or no education, hardly any women migrate for work. They moved either to cities and their edges (mostly Dhaka but also from Lalmonirhat to towns in neighbouring districts) for work in construction, labouring, pulling rickshaws, and rarely in factories; or to distant districts as teams/groups to plant and harvest crops during peak demand periods (notably from Kurigram to Munshiganj). Migration strategies are influenced by travel distance, contacts and opportunities. Migrants often move in groups and have built up personal contacts in destinations, which have been enhanced in the last decade by use of mobile phones. This 'seasonal migration' is for six or more months of the year, usually in several trips. Migrants commonly visit home at approximately monthly intervals. The lowest incidence of such migration was in July-August in both years when men came back to their villages to help their families cope in the flood season, and was high in the dry season and also in September–October as part of recovery after floods (Figure 4).

Conditions faced by these migrants are difficult – they have no proper shelter, face poor sanitation and drinking water, and are vulnerable to exploitation in the cities. In focus groups with people involved in seasonal migration they highlighted the hardships involved, for example a participant in one focus group in Gaibandha said:

> We go in a group and stay there until a specific job is finished. But getting work every day is uncertain. We have to work hard to earn money. Moreover, we are always worried about the family left at home. Sometimes people fall in a trap and get involved in drugs and other crime. Construction labourers don't get wages on time and never get full payment. We all face water, sanitation, food and shelter problems. Besides we have to pay local mastans [musclemen] and police.

Although such migration appears to improve coping and resilience, migrating men have to borrow to provide initial funds for their wives and families that remain behind,

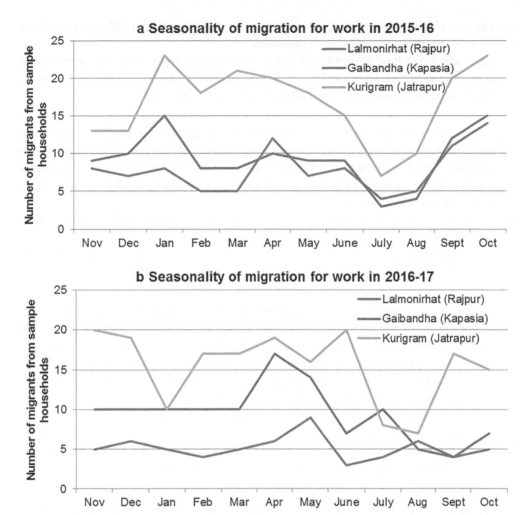

Figure 4. Seasonality of migration for work.

to cover travel costs, and to cover initial living costs when they move; all of which add new vulnerability and reduce the actual benefit received.

4. Discussion and conclusions

The three districts and study locations selected for this study are representative of the socio-hydrological landscapes along the Brahmaputra-Jamuna River in Bangladesh, but the similarity of these areas with other locations along the main rivers, and the social linkages found with distant urban areas mean that the findings are of more general relevance. In the floods in 2016 and 2017, complex decisions on evacuation and longer-term movements were made by rural floodplain inhabitants. These decisions are not just a function of immediate flood and erosion impacts, or opportunities for shelter and for livelihoods nearby. Decisions were also found to be influenced, for example, by improved road communications that enable repeated trips to rapidly growing urban centres to take advantage of economic opportunities. Decisions are balanced by the risks and disadvantages of life in towns and cities. Hence migration of different types is no longer a last resort for flood-prone households.

Our study shows how households living in rural riverine areas make use of the socio-hydrological landscapes they inhabit, but also of enhanced physical and socio-economic connections to other landscapes including more distant urban areas:

- *Physically* they inhabit recently accreted and older floodplain lands that have high flood risks and where erosion forces frequent shifts of home, they also use the embankments that form the interface between local 'unprotected' and 'protected' socio-hydrological landscapes as temporary flood shelter, and for those affected by erosion as longer term shelter close to their lost land.
- *Economically* they use work opportunities in expanding urban areas and other districts to compensate for limited risky rural livelihoods in riverine areas.
- *Socially* they cooperate with others from their community informally and through community organisations to cope locally and to find work.

Living conditions in places of shelter (embankments, schools and other high places) were reported to be very difficult due to overcrowding, poor water and sanitation, and the difficulty of cooking. People from one char union preferred not to evacuate and instead stayed on boats or raised platforms to protect their homes. This local perspective highlights a difference between highly flood and erosion-prone people regarding their use of the socio-hydrological landscapes compared with that of 'experts'. Water resource engineering in Bangladesh and elsewhere has conceived of structural flood mitigation in the shape of embankments to enable hazard-free livelihoods in the protected areas without considering use by people affected by floods and erosion. The normal engineering perspective is exemplified by a comment from a district level Bangladesh Water Development Board official in Kurigram.after the 2017 floods:

> Embankments are mostly built for flood defence. In Bangladesh they are also meant to be used for transportation. However, when flood affected people move to embankment and make a permanent settlement there it becomes vulnerable to breaching, and that is what is happening here.

This study shows how embankments are an important structural resource for resilience of inhabitants of the active floodplain and its riverine islands, even when they are breached or not continuous. Meanwhile embankments do not ensure a risk-free environment for inhabitants of protected areas because of frequent breaching, but they also provide places of shelter for inhabitants of the 'protected' or 'mainland' socio-hydrological landscape. This raises the question whether the government's structural responses to flood and erosion hazards considering the broader riverine landscape should focus on expensive rehabilitation, reinforcement and extension of embankments along the main rivers, or on other measures such as raised shelter areas above the level of exceptional floods. These raised shelter areas could be connected by moderately raised but flood resilient embankment-roads that would protect crops in 'normal' floods. Such alternatives deserve consideration by water resource engineers. Government programmes tend to treat socio-hydrological spaces separately, but if connections and migration were seen more positively as a component of resilience then costs saved from structural flood protection could, for example, be used to improve public services for urban migrants and for char communities.

Past studies on flood impacts and coping, such as Sultana and Rayhan (2012), have highlighted the importance of individual coping including borrowing and asset sales during and after floods. This study confirms how successive years of severe floods can erode personal assets accumulated over time and reveals how individual coping is complemented by social capital, and access to distant employment opportunities, which in turn is enhanced by urbanisation. We find that temporary evacuation during floods as well as seasonal migration for work involves groups of households and individuals, who organise to exploit these opportunities together. In addition to such informal arrangements, CBOs established for other purposes spontaneously responded to hazards and help local people to organise to improve their coping strategies. Examples of CBO initiatives included disseminating warnings, helping people to evacuate, and helping displaced households to access relief services.

There is scope for CBOs, assisted by NGOs and government, to further help vulnerable households to work together to enhance their resilience. Whether seasonal migration is finally accepted by policy makers as part of a resilient rural livelihood strategy or not, CBOs might advocate better living conditions for migrants and help migrants to obtain better-paid work. CBOs could also provide group loans to fund seasonal migration, including the living costs of families left in the riverine areas, on less exploitative terms than traditional money lenders. At the same time balanced collective action could help the women left behind when men migrate, women could organise in parallel groups to men in order to develop income sources and strengthen cooperation to protect their homes and families. These opportunities are enhanced by wider societal changes. Mobile phone technology now offers the scope for CBOs to improve information collection and distribution, and the flows of funds between households and their migrant workers.

If government and NGOs took a more integrated socio-hydrological perspective of riverine hazards and rural and urban linkages, this could pave the way to innovations that start from the perspective of rural inhabitants. This would help build longer term resilience to hazards that are likely to become more severe with climate change. Such an approach would involve technical and institutional changes such as designing embankments along main rivers that are more effective as flood shelters and improving water supply, sanitation and housing conditions for seasonal migrants in urban areas. The potential shown by local community organisations could be enabled to improve flood coping and resilience for mobile riverine households. Lastly the study showed how migration for work is a male phenomenon, revealing an opportunity to strengthen and improve cooperation among women for both hazard coping and resilient livelihoods that has so far been unaddressed.

Note

1. The river known as the Brahmaputra in India continues to be known by that name in Bangladesh as far as the southern tip of Kurigram District, but further downstream is called the Jamuna River. The former course of the river (now a minor off-shoot to the east) continues to be named the (Old) Brahmaputra River.

Acknowledgements

This paper is an output of the project 'Hydro-Social Deltas': Understanding flows of water and people to improve policies and strategies for disaster risk reduction and sustainable development of delta areas in the Netherlands and Bangladesh. We are grateful to our colleagues, the leaders of the participating Community-Based Organisations, and the flood-affected households interviewed for their assistance and cooperation. We thank Md. Ruknul Ferdous for contributions to the literature review and Michelle Kooy for guidance and comments on the study. We are grateful to Professor Edmund Penning-Rowsell, Dr Brian Cook, and an anonymous reviewer for their comments and suggestions which have improved this paper. The views expressed are those of the authors alone.

Disclosure statement

No potential conflict of interest was reported by the authors.

Funding

This work was supported by the Netherlands Organisation for Scientific Research (NWO) Urbanising Deltas of the World programme under grant number W 07.69.110.

References

Ahsan, R., Karuppanan, S., & Kellett, J. (2014). Climate induced migration: Lessons from Bangladesh. *The International Journal of Climate Change: Impact and Responses, 5*(2), 1–15.

Alamgir, M. (1980). *Famine in South Asia: Political economy of mass starvation*. Cambridge, MA: Oelgeschlager, Gunn & Hain.

Auerbach, L. W., Goodbred, S. L., Mondal, D. R., Wilson, C. A., Ahmed, K. R., Roy, K., ... Ackerly, B. A. (2015). Flood risk of natural and embanked landscapes on the Ganges-Brahmaputra tidal delta plain. *Nature Climate Change, 5*, 153–157.

Ayeb-Karlsson, S., van der Geest, K., Ahmed, I., Huq, S., & Warner, K. (2016). A people-centred perspective on climate change, environmental stress, and livelihood resilience in Bangladesh. *Sustainability Science, 11*(4), 679–694.

Baqee, A. (1998). *Peopling in the land of Allah Jaane. Power, peopling and environment: The case of char-lands of Bangladesh*. Dhaka: The University Press.

Berkes, F., & Folke, C. (Eds.). (1998). *Linking social and ecological systems: Management practices and social mechanisms for building resilience*. Cambridge: Cambridge University Press.

Brammer, H. (1990). Floods in Bangladesh: Geographical background to the 1987 and 1988 floods. *The Geographic Journal, 156*(1), 12–22.

Brammer, H. (2004). *Can Bangladesh be protected from floods?* Dhaka: University Press.

Brammer, H. (2014). Bangladesh's dynamic coastal regions and sea-level rise. *Climate Risk Management, 1*, 51–62.

Brouwer, R., Akter, S., Brander, L., & Haque, E. (2007). Socioeconomic vulnerability and adaptation to environmental risk: A case study of climate change and flooding in Bangladesh. *Risk Analysis, 27*(2), 313–326.

Dewan, C., Mukherji, A., & Buisson, M. C. (2015). Evolution of water management in coastal Bangladesh: From temporary earthen embankments to depoliticized community-managed polders. *Water International, 40*, 401–416.

EGIS. (2000). *Riverine chars in Bangladesh environmental dynamics and management issues*. Dhaka: Environment and GIS Support Project for Water Sector Planning (EGIS) and The University Press Limited.

Elahi, K., Ahmed, K. S., & Mafizuddin, M. (1991). *Riverbank erosion, flood and population displacement in Bangladesh*. Jahangirnagar University Riverbank Erosion Impact Study, Dhaka.

Etzold, B., Ahmed, A. U., Hassan, S. R., & Neelormi, S. (2014). Clouds gather in the sky, but no rain falls: Vulnerability to rainfall variability and food insecurity in Northern Bangladesh and its effects on migration. *Climate and Development*, *6*(1), 18–27.

Ferdous, M. R., Wesselink, A., Brandimarte, L., Slager, K., Zwarteveen, M., & Di Baldassarre, G. (2018). Socio-hydrological spaces in the Jamuna River floodplain in Bangladesh. *Hydrology and Earth System Science*, *22*, 5159–5173.

Foresight. (2011). *Migration and global environmental change*. London: The Government Office for Science.

Gray, C. L., & Mueller, V. (2012). Natural disasters and population mobility in Bangladesh. *Proceedings of the National Academy of Sciences*, *109*(16), 6000–6005.

Haque, C. E. (1988). Human adjustments to river bank erosion hazard in the Jamuna floodplain, Bangladesh. *Human Ecology*, *16*(4), 421–437.

Haque, C. E., & Zaman, M. Q. (1993). Human responses to riverine hazards in Bangladesh: A proposal for sustainable floodplain development. *World Development*, *21*(1), 93–10.

Hutton, D., & Haque, C. E. (2004). Human vulnerability, dislocation and resettlement: Adaptation processes of river-bank erosion-induced displacees in Bangladesh. *Disasters*, *28*(1), 41–62.

Indra, D. (2000). Not just dis-placed and poor: How environmentally forced migrants in rural Bangladesh recreate space and place under trying conditions. *Rethinking Refuge and Displacement: Selected Papers of Refugees and Immigrants*, *8*, 163–191.

Islam, A. K. M. S. (2016). Aspects of 2016 flooding in Bangladesh. *New Age*, August 5.

Islam, M. S., Hasan, T., Chowdhury, M. S. I. R., Rahaman, M. H., & Tusher, T. R. (2012). Coping techniques of local people to flood and river erosion in char areas of Bangladesh. *Journal of Environmental Science and Natural Resources*, *5*(2), 251–261.

Joarder, Md. A. M., & Miller, P. W. (2013). Factors affecting whether environmental migration is temporary or permanent: Evidence from Bangladesh. *Global Environmental Change*, *23*, 1511–1524.

Linton, J. (2008). Is the hydrologic cycle sustainable? A historical-geographical critique of a modern concept. *Annals of the Association of American Geographers*, *98*(3), 630–649.

Mamun, M. Z. (1996). Awareness, preparedness and adjustment measures of river-bank erosion-prone people: A case study. *Disasters*, *20*(1), 68–74.

Ministry of Water Resources. (2001). *Guidelines for Participatory Water Management*. Dhaka: Ministry of Water Resources, Government of the People's Republic of Bangladesh.

Mirza, M. M. Q. (2002). Global warming and changes in the probability of occurrence of floods in Bangladesh and implications. *Global Environmental Change*, *12*(2), 127–138.

Paul, B. K. (1997). Flood research in Bangladesh in retrospect and prospect: A review. *Geoforum; Journal of Physical, Human, and Regional Geosciences*, *28*(2), 121–131.

Penning-Rowsell, E., Sultana, P., & Thompson, P. (2013). The 'last resort'? Population movement in response to climate-related hazards in Bangladesh. *Environmental Science & Policy*, *27*(Suppl 1), S44–S59.

Rahman, M. R. (2010). Impact of riverbank erosion hazard in the Jamuna floodplain areas in Bangladesh. *Journal of Science Foundation*, *8*(1–2), 55–65.

Salauddin, M., & Ashikuzzaman, M. (2011). Nature and extent of population displacement due to climate change-triggered disasters in the south-western coastal region of Bangladesh. *Management of Environmental Quality*, *22*(5), 620–631.

Sivapalan, M., Savenije, H. H. G., & Blöschl, G. (2012). Sociohydrology: A new science of people and water. *Hydrol. Process*, *26*, 1270–1276.

Sjoukje, P., Sparrow, S., Kew, S. F., van der Wiel, K., Wanders, N., Singh, R., … van Oldenborgh, G. J. (2018). Attributing the 2017 Bangladesh floods from meteorological and hydrological perspectives. *Hydrology and Earth System Sciences Discussion*, doi:10.5194/hess-2018-379

Sultana, P., Johnson, C., & Thompson, P. (2008). The impact of major floods on flood policy evolution: Insights from Bangladesh. *International Journal of River Basin Management*, *6*(special issue), 1–10.

Sultana, N., & Rayhan, M. I. (2012). Coping strategies with floods in Bangladesh: An empirical study. *Natural Hazards*, *64*(2), 1209–1218.

Sultana, P., & Thompson, P. M. (2010). Local institutions for floodplain management in Bangladesh and the influence of the Flood Action Plan. *Environmental Hazards*, *9*(1), 26–42.

Thompson, P. M., & Sultana, P. (2000). Shortcomings of Flood Embankment strategies in Bangladesh. In D. J. Parker (Ed.), *Floods (vol. 1)* (pp. 316–333). London: Routledge.

Thompson, P., & Tod, I. (1998). Mitigating flood losses in the active floodplains of Bangladesh. *Disaster Prevention and Management: An International Journal, 7*(2), 113–123.

Wesselink, A., Kooy, M., & Warner, J. (2017). Socio-hydrology and hydrosocial analysis?: Toward dialogues across disciplines. *WIRES Water, 4*, 1–14. doi:10.1002/wat2.1196

Yasmin, T., & Ahmed, K. M. (2013). The comparative analysis of coping in two different vulnerable areas in Bangladesh. *International Journal of Scientific and Technology Research, 2*(8), 26–38.

Yu, D. J., Sangwan, N., Sung, K., Chen, X., & Merwade, V. (2017). Incorporating institutions and collective action into a sociohydrological model of flood resilience. *Water Resources Research, 53*(2), 1336–1353.

Urbanisation and disaster risk: the resilience of the Nigerian community in Auckland to natural hazards

Osamuede Odiase, Suzanne Wilkinson and Andreas Neef ⓘ

ABSTRACT

Community resources and the ability to organise them in times of challenges are essential in resilience. This study investigated the resilience of the Nigerian community in Auckland to the risk of natural hazards. Auckland has a multi-hazard landscape and an increasing rate of urbanisation which encompasses diverse communities and nationalities. The study examined five resources, the social, economic, communication, disaster competency and physical resources to determine the resilience of the Nigerian community to natural hazards in Auckland. The study collected data from both secondary and primary sources. The primary source included information from surveys and interviews; and secondary data from existing literature and documents in the public domain. The research used both parametric and non-parametric methods and themes identification for analysis. A community resilience index was created to calculate the current resilience status of the Nigerian community. Research findings suggested policy interventions to enhance indicators of low resilience. Enhancing resilience in the community depends on how the government addresses these indicators in future community resilience planning.

1. Introduction

Disasters arisen from natural hazard events have resulted in the loss of life and decreased economic livelihoods. In New Zealand, Parker and Steenkamp (2012, p. 14) estimated the impact of the Canterbury Earthquake to have affected 460,000 people in the cities of Christchurch, Selwyn and Waimakariri. In addition to the human impact of natural hazards, the cost of rebuilding the affected communities was estimated to be around NZ$ 40 billion (Reserve Bank of New Zealand, 2016, p. 3). Although there has been a reduction in both human and material loss from natural hazards since 2012 (Guha-Sapir, Hoyois, & Below, 2016), concerns regarding urbanisation and human vulnerability have increased (Blaikie, Cannon, Davis, & Wisner, 2014). Apart from the development of hazard-prone areas, urbanisation has contributed to resource inequality (Behrens & Robert-Nicoud, 2014). Emerging urban communities have become sensitive as a result of an ethno-cultural perception of risk (Perry & Green, 1982) and the migration from a place of known risk and familial support to the unfamiliar environment (Mitchell, 1999).

Efforts to reduce risk have centred on the traditional risk assessment approach (Saunders & Kilvington, 2016) and vulnerability reduction (Bogardi & Birkmann, 2004). Although vulnerability assessment is essential in understanding disaster causation and impact, its primary focus is on resource deficiency rather than capacity. Notwithstanding the level of vulnerability in a community, resilience exists in the context of adversity. Many studies on community resilience have shown that some communities can thrive regardless of inherent challenges. Such capacities have been investigated in New Zealand by Thorney, Ball, Signal, Lawrence-Te Aho & Rawson (2014); Paton, Millar, and Johnston (2001); Wilson (2013) and Smith, Davies-Colley, Mackay, and Bankoff (2011). However, the primary focus of their studies was on the social dimension of resilience, and they addressed resilience from a post-disaster perspective. Notwithstanding, the susceptibility of Auckland to natural hazards, available literature suggest a dearth of study on the resilience of the African communities to the risk of natural hazards in Auckland.

This study examines the resilience of the Nigerian community in Auckland to the risk of natural hazards. The research adopted holistic and pre-disaster approaches to assess the resilience of the Nigerian communities. These approaches help to identify the strengths and weaknesses of resilience indicators before a disaster event. To achieve the primary objective of this research, this research answers the following questions:

- What indicators of social resilience exist in the Nigerian community?
- How economically resilient is the Nigerian community?
- How competent is the community in disaster risk reduction?
- What communication resources exist in the Nigerian community?
- What physical resources exist in the community to improve resilience?

This paper proceeds with the theoretical and conceptual framework for the study, followed by the method of data collection and analysis and then the study's findings and discussion and lastly, the conclusion of the study.

2. The theoretical and conceptual framework for the study

2.1. Urbanisation and disaster risk

Cross (2001) argues that the wealth of an urban community enables it to be better prepared and mitigate potential hazard impacts. Central to Cross' argument was that urbanisation in itself does not constitute a disaster risk if proper mitigation strategies are designed and enforced. However, urbanisation becomes a concern if the process contributes a significant variable in the exposure, sensitivity, and vulnerability of urban communities (Jones & Kandel, 2004). Urban risk arises from the location, development and subsequent expansion of communities into hazard-prone areas (Mitchell, 1999). The population is more at risk of natural hazards if there is intentional and institutional failure to integrate informal settlement into mainstream development of the community (Satterthwaite, 2007).

The social and economic strata that emerge as a result of urbanisation have implications for community resilience to natural hazards. The nature of economic inequality within an urban community (Behrens & Robert-Nicoud, 2014) and cultural diversity influence risk perception (Lindell & Perry, 1992) and the capacity of communities to cope with disasters. (Behrens & Robert-Nicoud, 2014). Addressing issues of urban vulnerability to disaster risk requires an investigation into the capacities of communities to reorganise during challenges.

2.2. Conceptualising community resilience to natural hazards

Although resilience has been used over many decades ago, cross-disciplinary application form mechanical to ecology has been credited to C S Holling seminal work in 1973 (Alexander, 2013. Holling (1973, p. 14) defines resilience as a 'measure of the persistence of systems and their ability to absorb change and disturbance and still maintain the same relationships between populations or state variables'. Although Holling's definition laid the groundwork for the conceptual development of resilience, it has been criticised for its allusion to previous equilibrium. The reason being that resilience as argued by Leitch and Bohensky (2014) is not about returning that the initial vulnerable but continuous interaction between the physical and social elements and the ability of the later to cope with the stress imposed by the physical forces. The subsequent development of resilience has conceptualised resilience as socio attribute of a social system (Adger, 2000), as a socio-ecological attribute (Folke, 2006), and as an attribute of a place (Cutter et al. 2008a). Recently, discourse on resilience has shifted to the practical application of the concept. The work of Weichselgartner and Kelman (2015) is in that direction. They suggested a synthesis of resilience ideas to produce a generic blueprint for building resilience. The essence of harmonisation of ideas is to produce an operational definition and resilience baseline condition that will apply to different spatial and temporal scales. The possibility of this achievement could be herculean epistemological and context differences in resilience. To move resilience forward, a link between risk, sustainability and resilience should be established (Tobin, 1999) and vulnerability reduction incorporated in resilience planning process rather than a separate entity (Weichselgartner & Kelman, 2015). Resilience has been distinguished by how a community addresses its ecological vulnerability. While a community could strengthen its emergency and hard engineering capacity to address a potential disaster (Dovers & Handmer, 1992) it may also opt for a proactive approach based on the understanding of the ecological system and social adaptation. These typologies underscore resilience as an outcome of intervention or a process of adaptation.

The relationship between resilience, vulnerability and adaptive capacity have ensued as confounding issues in resilience. While it is accepted that both concepts have oppositional characteristics, the existence of vulnerability does not preclude resilience as both concepts exist in society. What determines resilience baseline condition is the preponderance of vulnerability over resilience and vice-versa. The difficulty in defining the relationship between the concepts arises from the multi-disciplinary usage and definitional overlapping between vulnerability and resilience (Manyena, 2006). Nonetheless, vulnerability and resilience are contextually related (Weichselgartner & Bertens, 2000) and conceptually linked (Cutter et al. 2008) by the adaptive capacity of a community (Engle, 2011). Adaptive capacity encompasses the use of social learning and existing resources to manage current challenges (Longstaff, Armstrong, Perrin, Parker, & Hidek, 2010). Although adaptive capacity is essential in resilience, it is an element in resilience (Tierney & Bruneau, 2007). Addressing the sensitivity and exposure of elements at risk are also essential in resilience (Murray & Ebi, 2012). While resilience may be related to a high adaptive capacity, access and distribution of resources among the population are central to community resilience (Klein, Nicholls, & Thomalla, 2003).

Notwithstanding the conceptual development of resilience, it remains a fluidic concept that is understood and applied differently by stakeholders. An agreement exists that resilience is characterise by diversity, redundancy and robustness of community resources (Tierney & Bruneau, 2007), sustainable behaviour (Tobin, 1999) and social network (Frankenberger, Mueller, Spangler, & Alexander, 2013); and community engagement in disaster risk reduction activities. From the preceding, this research conceptualises resilience as the ability of a community to reorganise itself and use its available resources to anticipate, cope, adapt and function despite the challenges imposed by natural hazard event.

3. Construction of community resilience index (CRI)

3.1. Resilience domain and indicator selection

The index approach is widely used for assessing baseline indicators of community resilience (Ainuddin & Routray, 2012) and also, to identify variables that contribute or diminish the capacity of a community to cope with a disaster event (Cutter et al., 2008b). Although researchers used nomenclatures or the stressors under consideration to distinguish their community resilience index, the construction of the index follows the same procedure.

In developing a (CRI), this research used a content analytical process to identify relevant and recurring domains in similar resilience assessment research. The analytical process identifies the physical, natural, social, economic, institutional, communication and community competence. The research modifies the community resilience index to develop a (CRI) for the Nigerian community based on the domains of social, economic, physical, competence and communication and information aspects of the community. This research excluded the institutional and natural domains from the Nigerian (CRI) because the population was the unit of analysis and the capacity to predict the resilience of a population is located at personal resources rather than geographic space.

Following the previous works by Cutter, Burton, and Emrich (2010); Ainuddin and Routray (2012); Qasim et al. (2016) and Yoon et al., (2015), the study selected indicators of resilience from prior resilience research. The selection of indicators was based on the following criteria namely: (1) relevance to the objectives of the research; (2) indicators have been tested and are consistent in similar study; (3) the ability of selected indicators to influence disaster management cycle activities Peacock et al. (2010). In the social domain, this study identified indicators that measure individuals' involvement in a social group, communal affairs and network within the community (Peacock et al., 2010; Putnam, 2000). Economic attributes that support individual during challenges were selected to measure the economic domain. The study measures the physical domain by the possession of transport and house that enhanced individual coping with a disaster event. The variables that assess expertise in disaster management were used to measure competency domain. Private possession of the means of receiving and communicating risk information was selected as measurement variables for information and communication domain. An essential part of the selection process was the pre-testing of the indicators and expert opinion on the relevance and reliability of the indicators. The essence was to have a plural input into the indicator selection process.

3.2. Calculation of (CRI) for the Nigerian community

In calculating the (CRI) for this study, the researchers arranged the indicators under their respective domains (Parsons et al., 2016) and adjusted to be on a standard scale to ease the mathematical combination of indicators (Cutter et al., 2010; Mayunga, 2007). The study adjusted the indicators by population size of the communities and converted to percentages for uniformity. The adjustment and conversion enabled the research to know the percentage distribution of the population on a given indicator. Although studies, i.e. Mayunga (2007) and Cutter et al. (2010) normalised indicators to avoid extreme values dominating the index development (Freudenberg, 2003), this study did not follow that procedure because the indicators have been expressed in percentages. This choice of research was in line with studies by Qasim et al. (2016) and Ainuddin and Routray (2012). Whereas the capital-based approach to resilience attached weight to indicators to acknowledge differential contributions within a domain (Mayunga, 2007), this study opted for an equal weight for all indicators both at sub and composite index levels for two reasons. Firstly, there is no theoretical preference underpinning the importance of differential weight allocation (Cutter et al., 2010), and secondly, differential weight allocation undermines interdependence among resilience domains (Gutierrez-Montes, Emery, & Fernandez-Baca, 2009). In calculating (CRI); therefore, the researchers used a simple average method which is based on equal weight. The method has also been used in studies conducted by Qasim et al. (2016); Ainuddin and Routray (2012) and Cutter et al. (2010). This method provides for simple comprehension and transparency in calculating the resilience index of the Nigerian community. A Weighted Mean Index (WMI) method was used to compute the Aggregate Weighted Mean Index (AWMI) for each of the resilience domains. The value of the (AWMI) constituted the resilience contribution of that domain to the overall resilience of the Nigerian community.

The equation below summarises the mathematical formula for combining variable indicators of a domain to generate individual indices for the domains under consideration:

$$d_i = \sum (X1\,w1 + X2\,w2 + X3\,w3 + \dotsfill X\,n\,w\,n) \qquad (1)$$

Where d_i = Domain index; X = Indicator; W = Weight; n = Number of indicators or weight considered; i = indicator number

In calculating each of the domain resilience indexes, this study computed the resilience index of each domain as the mean values of resilience index of the variable indicators under the domain. The formula below depicts the construction of domain resilience index for this study.

$$\text{Domain resilience index (DRI)} = \sum_{i=1}^{n} \frac{d_i}{n}. \qquad (2)$$

Where; di = domain index; n = the number of indicators of that domain

$$\text{Aggregate resilience index for all domains (ADRI)} = \sum_{i=1}^{5} \frac{DRIi}{N} \qquad (3)$$

Where; DRI = domain resilience index; N = total numbers of resilience domain under consideration.

Based on the dimensions selected to assess the resilience of the Nigerian community in Auckland, this study calculated the overall disaster resilience index; thus:

Community disaster resilience index can be calculated = M (E) +M (S) +M (P) +M (Ci) +M $(C)/5$

Where M (E), M (S), M (P) M, (Ci) and M (C) represented the mean values for economic, social, physical and communication & info.; and competence resilience; and 5, the numbers of community domains.

4. The study's method of data collection

4.1. Area of study

The location of Auckland City is on the North Island of New Zealand, and it is the fastest-growing multi-cultural city in the country. More than 90% of Auckland's population lives in urban areas. Unlike the South Island city of Christchurch and its environs, Auckland is least susceptible to an earthquake because of its location on the Australian tectonic plate, 300–500 km north-west of the active plate boundary of the Australian and Pacific plates (Auckland Council 2015, p. 6). However, it is most prone to volcanic eruptions and coastal erosion because of its location on the Auckland Volcanic Field (AVF) covering 100 km^2 of the urban areas and approximately 3000 km length of the coastal shoreline (Auckland Council, 2015, p. 6). Apart from earthquakes and volcanic eruptions, Auckland is also vulnerable to a wide range of multiple hazard impacts from severe weather events, floods, tsunamis, and landslides.

4.2. The community of study

This research conceptualises a community as a local entity near the source of risk. Apart from sharing the same characteristics as a group, members of the community share the same hazard exposure and diverse risk perception, but they are united by community risk assessment and the desire to reduce disaster risk (Pearce, 2003). The Nigerian community in Auckland is dispersed but united by common ancestry, ethnicity and culture. A major platform for community interaction is independence and cultural day celebrations. The resilience of the Nigerian community is under consideration because the community is small, relatively new and exposed to multiple risks of natural hazards in Auckland. As a growing community, risk perception may be culturally underpinned, and access to various resources may be limited.

According to the data available from Statistics New Zealand (2015, p. 1), the Nigerian population in Auckland was estimated to be 294 and out of which 53.1% were male and 46.9% were female. The population of Nigeria constituted about 0.021% of Auckland population (1,415,550) and 0.007% of New Zealand population (4,242,048) as at 2013 census. About 81.6% of Nigerian lived in the North Island and 18.4% in the South Island (Statistics New Zealand, 2013, p. 1). Almost 89.7% of the Nigerian lived in the main urban areas, out which 52.6% of the population lived in the Auckland Region, followed by the Canterbury Region (14.4%), and the Wellington Region (11.3%) (Statistics New Zealand, 2013, p. 1). In Auckland Region, the majority lives in the Otara-Papatoetoe Local Board Area (17.3%), Mangere-Otahuhu Local Board Area (9.6%), and Howick Local

Board Area (7.7%) (Statistics New Zealand, 2013, p. 1). Since Auckland is a fast-growing and diverse city, the study could act as a proxy for other small and recently integrated communities in Auckland.

The socio-demographic results of the survey data that was administered to adult participants in the Nigerian community are presented in Table 1 below:

4.3. Method of data collection and analysis

The primary objective of this study was to investigate the resilience of the Nigerian community to the risk of natural hazards in Auckland. In achieving this objective, this study used a mixed-method approach for data collection. The primary sources involved survey and interviews. Existing literature and official documents constituted secondary

Table 1. Socio-demographic characteristics of the Nigerian communities in Auckland.

Socio-demographic variables	f (n = 90)	%
Gender		
Male	56	62.2
Female	34	37.8
Age category		
18–25	5	5.6
26–45	56	62.9
46–65	25	28.1
66 and above	3	3.4
Education level		
Primary	3	
Secondary	7	8.1
Trade, cert. moreover, diploma	9	10.3
Undergraduate	15	17.3
Postgraduate	53	60.9
Immigration status		
Citizen	41	45.6
Permanent resident	14	15.5
Study permit	27	30.0
Work permit	8	8.9
Years in Auckland		
0–5	43	48.3
6–10	20	22.5
11–15	13	14.6
16–20	7	7.9
21–25	6	6.7
26 above	–	–
House ownership		
Others	10	11.2
Rented accommodation	57	64.1
Owned a house	22	24.7
Employment		
Employed	57	64.9
Unemployed	4	4.5
Retired	1	1.1
Social benefit	–	–
Student	26	29.5
Annual income (NZD)		
0–20,000	32	41.6
21,000–40,000	20	26.0
41,000–60,000	16	20.7
61,000–80,000	8	10.4
80,000 above	1	1.3

data. The study added the Auckland / Red Cross-Application as one of the indicators during an informal meeting with members of Civil Defence Emergency Management in Auckland. The survey protocol consisted of participants' socio-demographic background and questions regarding the objectives of the research. The researchers distributed 200 surveys to adult Nigerians during the Independence Day celebration as the occasion attracted a majority of Nigerians in Auckland. The remaining survey was distributed with the aid of research assistants. Participants returned a total of 90 questionnaires. The response rate was above the minimum number recommended by Onwuegbuzie and Collins (2007) for a quantitative correlational design and two-tailed hypotheses testing. Further to the survey administration, five face to face and two telephone interviews were conducted by the researcher with respondents that indicated their interest in further discussions. The essence of the interviews was to seek more details on the responses. Parametric and non-parametric tests involving independent sample t-test and Spearman's rank correlation were used to analyse quantitative data with the aid of a Statistical Package for the Social Sciences 23 (SPSS/IBM). Thomas's (2006) General Inductive Approach to qualitative analysis and a 3-step coding cycle, as explained by Saldana (2013) were used to analyse interviews. A content analytical process of theme formation was used to analyse secondary materials. The research used qualitative findings to bolster the survey's findings, and overall discussions were in line with existing literature.

5. Findings and discussions

The resilience scores of each domain are presented in Table 2 below. The result shows the resilience contributions of each indicator to the aggregate resilience index of the Nigerian community.

The results of the Nigerian (CRI) are displayed in a radar chart on a scale of 1–5, which shows the resilience of each resource in Figure 1 below. One represents a very low resilience and 5, a very high resilience.

The purpose of this study was to investigate the resilience of the Nigerian community to potential natural challenges in Auckland. Variable indicators related to community social, economic, physical, competency and communication resources were assessed to determine the resilience of the Nigerian community in Auckland. The practical resilience of the Nigerian community was 2.88 on the scale of 1–5. The current resilience baseline condition of the community is presented in Figure 2 below.

Regarding the social aspect of the Nigerian community resilience, the participants were questions regarding community affection, network and social participation. The domain resilience index of 3.26 indicated that social capital would play an essential role during natural challenges because of the high mean value ($M = 2.96$; SD 1.21) for charitable and community activities and also, volunteerism and community meeting attendance. However, only 11.4% of the participants indicated that they would be fully committed to those endeavours during a crisis. Social capital will contribute more to the resilience of the community if members of the community are encouraged to be proactive in participating in community activities. The percentage of community members that could be relied upon in time of crises was too small to contribute meaningfully to social resilience. The current percentage fell short of the minimum 50% contribution threshold required from social indicators to effectively enhance resilience (Ainuddin & Routray, 2012). The

Table 2. Resilience scores of the Nigerian community.

Domain / category	Variable indicators	Resilience mean score	Source
Social resource			
1 High school education & above	% of people above high school education	4.24	Peacock et al. (2010)
2 Economic population	% of people with employment	2.30	Cutter et al. (2008)
3 Social relationship*	% of the population participating in social, charity, meetings and volunteer activities	2.96	Cutter et al., (2008); Paton and Johnston (2001); Becker, 2010
4 Community affection*	% of the population willing to help, visit others, share risk information and encourage preparedness activities	2.95	Cutter et al. (2008); Norris, Stevens, Pfefferbaum, Wyche, and Pfefferbaum (2008); Kulig, Edge, and Joyce (2008); Paton et al. (2001); Paton and Johnston (2001); Becker, 2010
5 Participation in disaster management programme*	% of the population participating in community projects, attending preparedness fair, disaster training and exercise	2.60	Cutter et al., (2008); Norris et al. (2008); Paton and Johnston (2001); Becker, 2010; Thornley, Ball, Signal, Lawson-Te Aho, and Rawson (2015)
6 Membership of association	% of the population in association	4.50	Longstaff et al. (2010)
Average resilience mean index		*3.26*	
Economic resource			
7 Insurance coverage	% of the population with disaster insurance	1.18	McBean and Falkiner (2005).
8 Income above the poverty line	% of the population above minimum wage	2.04	Sherrieb, Norris, and Galea (2010).
9 Individual ability save	% of the population with the ability to save	2.90	Ainuddin and Routray (2012).
10 Creditability	% of the population with creditability	2.88	Mayunga (2007).
Average resilience mean index		*2.25*	
Physical resource			
12 House ownership	% of the population with home	2.13	Mayunga (2007).
13 Car ownership	% of the population with a vehicle	3.33	Colten et al., (2008); Masozera et al. (2007)
Average resilience mean index		*2.73*	
Communication resource			
15 Risk communication app.	% of the population with risk App.	1.52	Auckland Civil Defence (2018)
16 Social media platform	% of people on Facebook	4.70	Tobin and Whiteford (2002).
17 Sources of information*	% of the population with a mobile phone, radio and television	4.46	Peacock et al. (2010).
Average resilience mean index		*3.56*	
Individual competence			
18 Disaster Mgt. competence	% with knowledge of early warning evacuation procedures	2.73	Norris et al. (2008).
19 Understanding local risk	% of people with risk awareness	2.51	Cutter et al. (2008)
Average resilience mean index		*2.62*	
Aggregate resilience index		*2.88*	

*Social variables for assessing the social resilience of the Nigerian community were categorised to create indices of community social relationship; affection and participation in disaster risk reduction programme (DRR). Index of community affection encompasses helping others in the community, visitation, sharing risk information and encouraging preparedness behaviour. The social relationship sub-theme constitutes contributions to social activities, community charity, meeting attendance and volunteering. The community participation index in (DRR) was a summation of participation in a community project, preparedness fair and training and exercise. Index of the source of information includes a mobile phone, television and radio.

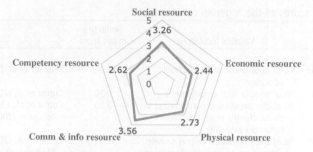

Figure 1. Domain rating of resilience resources of the Nigerian community.

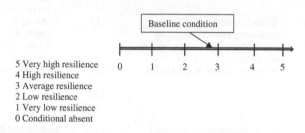

Figure 2. Resilience baseline of the Nigerian community in Auckland.

mean value of *(M = 2.96)* indicated the existence of a social network that the community could utilise during challenges. The importance of pre-existing social network was pivotal in the resilience of the Manawatu community to flood hazard (Smith et al., 2011). The community was eager to communicate risk and advise one another to be prepared. This finding was boosted by the high percentage of people that were willing to visit other community members. The high percentage of visitation was not a surprise because interviewees from the community regarded visitation as a community norm, especially in the time challenges.

A mean values of *(M = 2.96 and M = 2.95)* social relation and affection in the community were indications that the Nigerian community could link one another to resources and livelihood opportunities during challenges (Lo, Xu, Chan, & Su, 2015). About 70% of the community members belong to the Nigerian association. Members of the community can utilise the trust and network that are associated with membership of social associations. However, the opportunity for diverse assistance was minimal as the percentage of people in the community that belong to more than one association was low. About 37% of the population belong to religious, ethnic and continental associations. The importance of resource diversity in enhancing and assessing resilience has been expressed by Frankenberger et al. (2013); Longstaff et al. (2010) and Tierney and Bruneau (2007). A community with diverse resources and persons with multiple sources of assistance are more likely to cope and recover faster than counterparts with a mono resource or support.

Further analysis of the 37% of the population that belong to more than one association, 16% was religious. An indication that religious institutions will play a pivotal role during an inherent challenge. The percentage of community members that belong to disaster and humanitarian agencies was 3% and out of which 1% were members of paramilitary agencies. The negligible enlistment of community members in response agencies indicated a lack of interest in pre-disaster response activities. Low participation in response activities will impair the Nigerian community as the first responder during in future disaster event because of the percentages of the population that have participated in the emergency drill, training and exercise were below the 60% required for a resilient community (Ainuddin & Routray, 2012). While the importance of community participation was crucial in responding to a forest fire outbreak in Haifa, Israel (Simonovich & Sharabi, 2013), the same participation may not be expected from the Nigerian community because of lack of interest in response activities. Nonetheless that the community has no formal procedure for mobilising the community in the time of assistance, the community was capable of organising members through existing networks, and cultural affiliations within the Nigerian community toward any endeavours.

Regarding the economic resilience of the Nigerian community, the research assessed variables relating to employment, income, insurance, savings and credit availability. The high employment rate in the community may contribute to the community coping and recovery capacities. However, the median income (NZ$22,000) of the community was a significant concern as it fell short of the national median income by NZ$ 8,500 (Statistics New Zealand, 2013). Equally worrisome was that all the research participants were mono income and less than 50% of the community earned above the median income. The research participants attributed the prevalence of mono income to double taxation and family responsibilities. The role of income in addressing post-disaster poverty and subsequent vulnerability was doubtful. There is a need to encourage the community to consider multiple sources of income to enhance their resilience. A community is resilient if the population has a diverse resource (Adger, Kelly, Winkels, Huy, & Locke, 2002). Despite the low level of income, the propensity to save was high in the community. 64.3% were inclined to savings, a mean value ($M = 2.89$; $SD = 1.01$) shows that a little above half of the population can save for contingencies. Equally high in the community was the ability to obtain a credit facility from financial institutions. The mean value for savings ($M = 2.90$; $SD = 1.02$) was high. An indication that 60% of the population can address the problem of the scarce resource through a credit facility besides personal savings. While all participants had health cover, they showed little interest in disaster-related insurance. The low level of interest in disaster-related insurance was attributed to negligible house ownership and disaster experience in the community. The current level of 2% investment in disaster insurance fell behind the required 50% recommended by Ainuddin and Routray (2012) as the threshold for resilience. More people were likely to seek alternative assistance for mitigation and post-disaster reconstruction. This research, using a Spearman coefficient analysis correlated personal income with personal savings, creditability and purchase of disaster insurance because of the ability of income to influence other variables. The results of the correlation are presented in Table 3 below.

Table 3. A summary of the correlation between individual income and disaster insurance, savings and credit facility

Dependent variable	Individual Income
Savings	.350**
Credit facility	.286**
Disaster insurance	−.156

**Correlation is significant at the 0.01 level (1-tailed).

The positive correlations between individual income and savings and credit facility indicated that income would positively influence the roles of savings and credit facility in the resilience of participants. Income was not positively related to disaster insurance because of the negligible percentage of house ownership and lack of disaster experience in the community.

The mean index value of the community physical dimension was above average. A substantial contribution to the domain was the high percentage of car ownership by participants. The community did not consider car ownership as a luxury but a necessity. The traffic situation in the city and the need to make a quick escape from danger were the main reason for car ownership in the community. The community view on car ownership collaborated the assessment of Masozera, Bailey, and Kerchner (2007) on the importance of personal means of transportation in evacuation and responding to Hurricane Katrina in New Orleans. The percentage of car ownership in the community was an indication that the majority of the population will respond to early warning and be able to rescue personal resources for coping with impending disaster. The contribution of house ownership to physical resilience was low because only 24% of the study participants owned a house in Auckland. About three-quarters of the participants will struggle with post-disaster accommodation in Auckland. The lack of affordable accommodation, inadequate emergency shelter and the small number of vacant accommodation in Auckland will complicate the post-disaster coping ability of the community.

The use of social media and mobile telephones were prevalent among participants. Most people in the community will receive and disseminate risk information among themselves because of the possession of means of communication and the existence of social capital in the community. The community had less interest in real-time risk and emergency information. About 11% of the study's participants subscribed to Auckland's emergency alert. The members of the community attributed the low patronage for alert application to lack of information and publicity regarding its existence. The importance of risk application in resilience is that participants are always abreast with real-time information on preparedness and response regarding imminent risk (Auckland Civil Defence, 2018). The high mean value for trusted information *(M = 3.74; SD = 1.24)* indicated that the community trusted in hazard information from the local emergency agency. The high level of trust will positively influence disaster response because the community lacked knowledge of hazard and disaster management activities.

The research accessed community competence in disaster management because a community is regarded as the first responders to a disaster (Twigg & Mosel, 2017). Indicators of risk knowledge and Understanding of preparedness, response and recovery activities were used to assess community competency. The community risk awareness was related to dread and information regarding hazards. The community was more aware of the risk of earthquakes and flood than other local hazards. The community overall competence in hazard management was above average. However, competence was related to the personal dread for a hazard. Competency was higher in earthquake and tsunami than other hazards. The community misconstrued competency as adherence to risk information. The moving to higher ground and relocating household properties from the floor to a more top position in case of flood and tsunami was synonymous with competence.

Table 4. Results of Spearman correlation coefficient analysis: the Nigerian community.

Dependent variables	Nigerian community (Independent social variables)			
	Age	Education level	Yrs. in Auckland	Income
Communal help	.206*	.237*	.076	.121
Project participation	.151	.059	.232*	.038
Social participation	.158	.091	.214*	−.037
Charity	.405**	.033	.312**	.017
Meeting attendance	.293**	.156	.098	−.026
Volunteerism	.253*	−.009	.321**	.092
Visitation	−.235*	.054	.065	.004
Organisations' membership	.219*	−.014	.115	−.137
Attending preparedness fair	.084	.057	−.012	−.118
Share risk information	−.052	.075	−.068	−.175
Encouraging others to prepare	.033	.180	−.056	−.114
Participate in training and exercise	.031	.116	−.109	−.091
Individual emergency competence	.212*	−.049	.305**	.248*

Ns = not significant ($p > .05$), significant = $p < .05$.
**Correlation is significant at the 0.01 level (1-tailed).
*Correlation is significant at the 0.05 level (1-tailed).

To further analyse the resilience of the Nigerian community to natural hazards, an independent Sample T-Test and Spearman correlation coefficient analyses were conducted to examine the influence and direction between independent and dependent variables. The mean difference between genders in the community was not significant ($p > .05$) to constitute a gender difference in disaster resilience. However, some correlations existed between the socio-demographic variables of the community and the proxies of social resilience. The results of the relationships are presented in Table 4 below.

Age was positively related to community participation in charitable and volunteering activities, association enlistment and attend a meeting. The length of years community members stay in Auckland was influential in their willingness to participate in a community project, the act of charity and volunteerism. Individual competence in emergency improves with the length of stay in Auckland. The relationship between income and competence was unexpected in the study because nowhere in the literature to suggested such a relationship.

5. Conclusions

This paper investigated the resilience of the Nigerian community to a potential natural hazard event in Auckland. This research was necessitated by the hazard landscape and the rate of urbanisation in Auckland. Aside from the contribution of urbanisation to the exposure and vulnerability of the elements at risk, urbanisation has implication for risk perception and access to resilience resources. While cultural diversity has a positive dimension to resilience because of pluralities of ideas in resilience decision and social cohesion within communities, it could also be a source of vulnerability as risk is understood and interpreted differently.

The current resilience baseline condition of the Nigerian community was arrived at using the index method. This research acknowledged the shortcomings of the index method because of the tendency for assessors to ignore pluralism of ideas from qualitative data that may not lend themselves to measurement and concentrate on top-down indicators that might not be contextual. The role of the index method in data summation and visualisation of the resilience baseline condition cannot be underestimated. The findings of this research projected indicators of strength and weakness regarding the resilience of the community. The current findings suggested the need to take resilience to an operational level. At the policy level, we argue for a discussion that will address the economic indicators of the community in future resilience planning. The Nigerian community, in conjunction with official agencies, could implement policies aimed at increasing the economic resilience through diversification of livelihood, equity and encourage community members to take out insurance against natural eventualities. The research supports raising community awareness of pre-response activities. Although these measures are preparatory, they have an added value for enhancing coping capacity and long term resilience.

It should be noted that neither the indicators of resilience used this research nor the current baseline condition is static because they are underpinned by the inherent vulnerability and resilience of the community and prior interventions to this assessment. This research, therefore, recommends the need to update resilience through a longitudinal study because of the changing relationship between indicators and ecological system and implement interventions to meet changing ecological situations.

A limitation of this research was the exclusion of the natural and institutional domains from the assessment framework. The purpose was to predict resilience at an individual scale rather than the geography of a community.

Disclosure statement

No potential conflict of interest was reported by the authors.

Funding

This work was supported by Ministry of Business, Innovation and Employment: [Grant Number 9072/3711735].

ORCID

Andreas Neef ⓘ http://orcid.org/0000-0002-5079-3323

References

Adger, W. N. (2000). Social and ecological resilience: Are they related? *Progress in Human Geography*, *24*(3), 347–364.
Adger, W. N., Kelly, P. M., Winkels, A., Huy, L. Q., & Locke, C. (2002). Migration, remittances, livelihood trajectories, and social resilience. *AMBIO: A Journal of the Human Environment*, *31*(4), 358–366.
Ainuddin, S., & Routray, J. (2012). Earthquake hazards and community resilience in Baluchistan. *Natural Hazards*, *63*(2), 909–937.
Auckland Civil Defence. (2018). Red Cross hazard application. Retrieved from http://www.aucklandcivildefence.org.nz/alerting
Becker, J. (2010). *Understanding disaster preparedness and resilience in Canterbury: Results of interviews focus groups and a questionnaire survey*. Wellington, New Zealand: GNS Science.
Behrens, K., & Robert-Nicoud, F. (2014). Survival of the fittest in cities: Urbanisation and inequality. *The Economic Journal*, *124*(581), 1371–1400.

Blaikie, P., Cannon, T., Davis, I., & Wisner, B. (2014). *At risk: Natural hazards, people's vulnerability and disasters* (2nd ed). London: Routledge.

Bogardi, J., & Birkmann, J. (2004). Vulnerability assessment: The first step towards sustainable risk reduction. *Disaster and Society—From Hazard Assessment to Risk Reduction, Logos Verlag Berlin, Berlin*, 75–82.

Cross, J. A. (2001). Megacities and small towns: Different perspectives on hazard vulnerability. *Environmental Hazards, 3*(2), 63–80.

Cutter, L. S., Barnes, L., Berry, M., Burton, C., Evans, E., Tate, E., & Webb, J. (2008b). *Community and Regional Resilience to Natural Disasters: Perspective from Hazards, Disasters and Emergency Management.*

Cutter, S. L., Barnes, L., Berry, M., Burton, C., Evans, E., Tate, E., & Webb, J. (2008a). A place-based model for understanding community resilience to natural disasters. *Global Environmental Change, 18*(4), 598–606.

Cutter, S. L., Burton, C. G., & Emrich, C. T. (2010). Disaster resilience indicators for benchmarking baseline conditions. *Journal of Homeland Security and Emergency Management, 7*(1), 1–22.

Dovers, S. R., & Handmer, J. W. (1992). Uncertainty, sustainability and change. *Global Environmental Change, 2*(4), 262–276.

Engle, N. L. (2011). Adaptive capacity and its assessment. *Global Environmental Change, 21*(2), 647–656.

Folke, C. (2006). Resilience: The emergence of a perspective for social-ecological systems analyses. *Global Environmental Change, 16*(3), 253–267.

Frankenberger, T., Mueller, M., Spangler, T., & Alexander, S. (2013). *Community resilience: Conceptual framework and measurement feed the future learning agenda.* Unpublished manuscript.

Freudenberg, M. (2003). *Composite indicators of country performance: A critical assessment. (NO 2003/16).* Paris: OECD Publishing. Retrieved from http://econpapers.repec.org/paper/oecstiaaa/2003_2f16-en.htm

Guha-Sapir, D., Hoyois, P., & Below, R. (2016). *Annual disaster statistical review: Numbers and trends 2015.*. Brussels, Belgium: Centre for Research on the Epidemiology of Disasters. Retrieved from http://www.cred.be/sites/default/files/ADSR_2015.pdf

Gutierrez-Montes, I., Emery, M., & Fernandez-Baca, E. (2009). The sustainable livelihoods approach and the community capitals framework: The importance of system-level approaches to community change efforts. *Community Development, 40*(2), 106–113.

Holling, C. S. (1973). Resilience and stability of ecological systems. *Annual Review of Ecology and Systematics, 4*(1), 1–23.

Jones, B. G., & Kandel, W. A. (2004). Population growth, urbanisation, and disaster risk and vulnerability in metropolitan areas: A conceptual framework. *The Economics of Natural Hazards, 2*, 427–452.

Klein, R. J. T., Nicholls, R. J., & Thomalla, F. (2003). Resilience to natural hazards: How useful is this concept? *Global Environmental Change B: Environmental Hazards, 5*(1), 35–45.

Kulig, J. C., Edge, D., & Joyce, B. (2008). Understanding community resilience in rural communities through multimethod research. *Journal of Rural and Community Development, 3*(3), 77–94. Retrieved from http://hdl.handle.net/10133/1265

Leitch, A. M., & Bohensky, E. L. (2014). Return to 'a new normal': Discourses of resilience to natural disasters in Australian newspapers 2006-2010. *Global Environmental Change, 26*, 14–26.

Lindell, M. K., & Perry, R. W. (1992). *Behavioural foundations of community emergency planning.* Washington, U.A: Hemisphere Publ. Co.

Lo, A. Y., Xu, B., Chan, F. K. S., & Su, R. (2015). Social capital and community preparation for urban flooding in China. *Applied Geography, 64*, 1–11.

Longstaff, P. H., Armstrong, N. J., Perrin, K., Parker, W. M., & Hidek, M. A. (2010). Building resilient communities: A preliminary framework for assessment. *Homeland Security Affairs, 6*(3), 1–35.

Manyena, S. B. (2006). The concept of resilience revisited. *Disasters, 30*(4), 434–450.

Masozera, M., Bailey, M., & Kerchner, C. (2007). Distribution of impacts of natural disasters across income groups: A case study of New Orleans. *Ecological Economics, 63*(2), 299–306.

Mayunga, J. S. (2007). Understanding and applying the concept of community disaster resilience: a capital-based approach. Paper presented at the *Summer Academy of Social Vulnerability and Resilience Building*, 1–16.

McBean, G, Murphy, B, & Falkiner, L. (2005). *Enhancing local level emergency management*. Toronto, Canada: Institute for Catastrophic Loss Reduction.

Mitchell, J. K. (1999). Natural disasters in the context of mega-cities. *Crucibles of Hazard: Mega-Cities and Disasters in Transition, New York, United Nations University*.

Murray, V., & Ebi, K. L. (2012). No title. *IPCC special report on managing the risks of extreme events and disasters to advance climate change adaptation (SREX)*.

Norris, F., Stevens, S., Pfefferbaum, B., Wyche, K., & Pfefferbaum, R. (2008). Community resilience as a metaphor, theory, set of capacities, and strategy for disaster readiness. *American Journal of Community Psychology, 41*(1), 127–150.

Onwuegbuzie, A. J., & Collins, K. M. T. (2007). A typology of mixed methods sampling designs in social science research. *The Qualitative Report, 12*(2), 281.

Parker, M., & Steenkamp, D. (2012). The economic impact of the Canterbury earthquakes. *Reserve Bank of New Zealand Bulletin, 75*(3), 13–25. Retrieved from http://www.econis.eu/PPNSET?PPN= 727122371

Parsons, M., Glavac, S., Hastings, P., Marshall, G., McGregor, J., McNeill, J., & Stayner, R. (2016). Top-down assessment of disaster resilience: A conceptual framework using coping and adaptive capacities. *International Journal of Disaster Risk Reduction, 19*, 1–11.

Paton, D., & Johnston, D. (2001). Disasters and communities: Vulnerability, resilience and prepared-ness. *Disaster Prevention and Management: An International Journal, 10*(4), 270–277.

Paton, D., Millar, M., & Johnston, D. (2001). Community resilience to volcanic hazard consequences. *Natural Hazards, 24*(2), 157–169.

Peacock, W. G., Brody, S. D., Seitz, W. A., Merrell, W. J., Vedlitz, A., Zahran, S., & Stickney, R. (2010). *Advancing resilience of coastal localities: Developing, implementing, and sustaining the use of coastal resilience indicators: A final report. Hazard reduction and recovery centre*. College Station, Texas: Hazard reduction and recovery centre.

Pearce, L. (2003). Disaster management and community planning, and public participation: How to achieve sustainable hazard mitigation. *Natural Hazards, 28*(2–3), 211–228.

Perry, R. W., & Green, M. R. (1982). The role of ethnicity in the emergency decision-making process. *Sociological Inquiry, 52*(4), 306–334.

Putnam, R. D. (2000). *Bowling alone: The collapse and revival of American community*. New York: Simon and Schuster. Retrieved from http://parlinfo.aph.gov.au/parlInfo/search/summary/summary.w3p; query=Id:%22library/lcatalog/10152780%22

Qasim, S., Qasim, M., Shrestha, R. P., Khan, A. N., Tun, K., & Ashraf, M. (2016). Community resilience to flood hazards in Khyber Pakhtunkhwa province of Pakistan. *International Journal of Disaster Risk Reduction, 18*, 100–106.

Reserve Bank of New Zealand. (2016, February). The Canterbury rebuild five years on from the Christchurch earthquake. *Bulletin*. Retrieved from https://www.rbnz.govt.nz/-/media/ ReserveBank/Files/Publications/Bulletins/2016/2016feb79-3.pdf

Satterthwaite, D. (2007). *The transition to a predominately urban world and its underpinning*. London: lied.

Saunders, W. S., & Kilvington, M. (2016). Innovative land-use planning for natural hazard risk reduction: A consequence-driven approach from New Zealand. *International Journal of Disaster Risk Reduction, 18*, 244–255.

Sherrieb, K., Norris, F. H., & Galea, S. (2010). Measuring capacities for community resilience. *Social Indicators Research, 99*(2), 227–247.

Simonovich, J., & Sharabi, M. (2013). Dealing with environmental disaster: The intervention of com-munity emergency teams (CET) in the 2010 Israeli forest fire disaster. *Journal of Sustainable Development, 6*(2), 86.

Smith, W., Davies-Colley, C., Mackay, A., & Bankoff, G. (2011). Social impact of the 2004 Manawatu floods and the 'hollowing out' of rural New Zealand. *Disasters, 35*(3), 540–553.

Statistics New Zealand. (2013). Ethnic group profile: Nigerian. Retrieved from http://archive.stats. govt.nz/Census/2013-census/profile-and-summary-reports/ethnic- profiles.aspx?url=/Census/ 2013-census/profile-and-summary-reports/ethnic-
profiles.aspx&request_value=24787&tabname=Income&sc_device=pdf

Statistics New Zealand. (2015). *Subnational population projection: 2013 (base)-2043*. Wellington: Statistics New Zealand. Retrieved from http://www.stats.govt.nz/browse_for_stats/population/ estimates_and_projections/SubnationalPopulationProjections_HOTP2013base.aspx

Thornley, L., Ball, J., Signal, L., Lawson-Te Aho, K., & Rawson, E. (2015). Building community resilience: Learning from the Canterbury earthquakes. *Kotuitui: New Zealand Journal of Social Sciences Online*, *10*(1), 23–35.

Tierney, K., & Bruneau, M. (2007). Conceptualising and measuring resilience: A key to disaster loss reduction. *TR News*, (250).

Tobin, G. A. (1999). Sustainability and community resilience: The holy grail of hazards planning? *Global Environmental Change Part B: Environmental Hazards*, *1*(1), 13–25.

Tobin, G. A., & Whiteford, L. M. (2002). Community resilience and volcano hazard: The eruption of Tungurahua and evacuation of the Faldas in Ecuador. *Disasters*, *26*(1), 28–48.

Twigg, J., & Mosel, I. (2017). Emergent groups and spontaneous volunteers in urban disaster response. *Environment and Urbanization*, *29*(2), 443–458.

Weichselgartner, J., & Bertens, J. (2000). Natural disasters: Acts of god, nature or society? On the social relation to natural hazards. *WIT Transactions on Ecology and the Environment*, *45*, 3–12.

Weichselgartner, J., & Kelman, I. (2015). Geographies of resilience: Challenges and opportunities of a descriptive concept. *Progress in Human Geography*, *39*(3), 249–267.

Wilson, G. A. (2013). Community resilience, social memory and the post-2010 Christchurch (New Zealand) earthquakes. *Area*, *45*(2), 207–215.

The French *Cat' Nat'* system: post-flood recovery and resilience issues

Bernard Barraqué and Annabelle Moatty

ABSTRACT

Successive French governments have progressively decentralised flood control policy to increase the role of local authorities in planning and crisis management. In 1982, a law mandated local risk maps for 5 types of exceptional natural hazards and set up Cat' Nat', a national system of damage compensation based on an insurance super-fund. While this system clearly improved the situation of victims of extreme events through subsidies in housing and infrastructure reconstruction, it did not necessarily foster a parallel reduction of vulnerability: insurance is more tuned to the past than to the future, and the tacit rule supports identical reconstruction so as not to increase the pre-disaster vulnerability (but not reducing it either). Yet indirectly, the recognition of a state of natural disaster triggers vulnerability reduction later, through various measures at various scales, from housing level (*build back better*) to the PAPI (action programmes for flood prevention); we describe them and present a case study before presenting hypotheses to explain the resistance of private landowners as well as potential improvements to better bridge recovery and resilience.

Introduction

In France, flood control (the most important natural hazard both in terms of the number of events and the scale of economic impact) has been framed by specific central-local relationships, like most territorial policies. The Constitution of the Vth Republic adopted in 1958 increased the centralised nature of the regime, but met with resistance from local government, giving rise to what Michel Crozier and his disciples, Pierre Grémion (1976), Jean-Pierre Worms (1966) called 'cross regulation': central government developed a policy to modernise the economy, but at local level, society would resist the proposed change, unless the government would bring financial support. Without this support it would have to postpone the reforms or give dispensations. Bargaining took place between the 'prefect' (*préfet*, head of central administration at *département* – county – level), and the county's political leaders and aldermen. International comparisons fostered by the rise of European water policy then triggered decentralisation as a way to make territorial authorities more responsible.

The decentralisation policy initiated in 1982 included an important reform of the control of 5 natural hazards: floods, landslides, avalanches, earthquakes and volcano eruptions (the latter in overseas counties). It created a national funding mechanism generated from additional insurance premiums paid by all households and vehicle owners, to be mobilised to cover damages resulting from natural disasters (exceptional events). As a counterpart, local authorities in risk areas were required to draft 'risk exposure maps' to ban construction or subject it to vulnerability reduction measures in the disaster prone parts of their territory. Repeated severe droughts more recently led to add a sixth eligible disaster: housing damages due to clay soil subsidence.

According to Lamond and Proverbs (2009), the notion of resilience encompasses pre-disaster planning and warning systems, emergency handling procedures and post-disaster reconstruction. In this article we focus on the link between recovery and resilience; we first present the context of flood control and land-use policies in the country; then we discuss the way post-disaster recovery procedures operate, and how they improve the situation for victims of natural disasters; and follow this by questioning whether the recovery system also triggers resilience measures, in particular we ask 'at household level, is recovery leading to "building back better"?' Finally, we illustrate the presentation with a field case study, before concluding with potential improvements to the policy.

In other words, we address four questions: how far was decentralisation of flood control driven and why was it incomplete? Is post-disaster recovery now improved and how much does it cost? Does the insurance-based recovery funding improve resilience? And how can landowners be better involved in vulnerability reduction?

Decentralisation of water policy and flood control in France

As early as the 1960s, a regionalisation policy initiated a new form of governance, made more transparent and open to civil society through public participation mechanisms (Duran & Thoenig, 1996). In the water policy sector, this regionalisation included the creation of 6 *Agences de l'eau* and 6 *Comités de bassin* (institutions at the river basin district level) covering the country, and operating under a mutualised version of the polluter-pays principle: i.e. that water users degrading the quality or the quantity of water in the environment pay a collectively agreed upon levy to the institution, which uses the money to subsidise environmentally friendly projects proposed by pro-active water users. Over the following decades, these river-basin institutions helped the learning process about valuing water and turning water policy into one focussed on usership rather than on ownership, *i.e.* separating the right to use water from landownership rights, an important change in the country which invented the Civil Code. Interestingly, French river-basin institutions were not entrusted with flood control measures: they remained a prerogative of central government (Barraqué, 2014) despite the overlap between the recovery of the aquatic environment and flood control, as some land-based measures support both.

Decentralisation was promoted further with laws voted in 1982–83, under the left wing government. New planning laws gave competence for granting building permits to local authorities, on the condition that they first set up a land-use plan, for review by the government's services at county level. The idea was to decentralise urban planning at local level, but to limit the possible subjections of elected councils to vested interests and land-use based speculation, as well as to check the incorporation of national regulations. But rarely would a prefect take a local authority to court for illegal planning decisions.

In the same spirit in 1982, a law was passed to launch a new recovery and vulnerability reduction policy for the above-mentioned 5 types of hazards.[1] To fund the recovery, rather than use the budget of the *Agences de l'eau*, the law mobilised insurance companies to provide compensation to victims from a special insurance fund called Cat' Nat' (abbreviation for *Catastrophe Naturelle*) (Figure 1), set up thanks to an addition on all insurance premiums at national level; for vulnerability control, the law mandated all appropriate local authorities to draft risk maps, the *'Plans de prevention du risque'* (risk prevention plans, PPR, and for floods, PPR-*inondation* or PPRi); however, the plans were not done well or quickly; in addition long term precaution and environmental issues could not compete with short term added value of urban development. In 1987, through additional legislation, the prefects recovered the responsibility to draw the PPR, but difficulties remained in their completion and they were not rapidly incorporated into local town plans.

Central government services eventually found themselves with flood events responsibilities, as illustrated by the Xynthia disaster (47 casualties in total): in February 2010 on the Atlantic coast in La Faute-sur-Mer, close to La Rochelle, a severe depression provoked a sea surge which broke the dykes and submerged housing estates on floodable land (Vinet, Boissier, & Defossez, 2011). Families of the victims brought a lawsuit against the municipality of La Faute-sur-Mer and central government. The municipality was condemned, criminally for the mayor; but central government was also condemned for its inaction against a town plan which favoured development in the hazard prone areas, and for insufficient maintenance of the dykes.

Today, the implementation of the Floods Directive[2] (FD – 2007/60/EC) is the responsibility of the prefects and local authorities, as in the past, i.e. at the level of administrative territories. However, in the years before the FD's adoption in 2007, catchment-based institutions called *Etablissements Publics Territoriaux de Bassin* (Public Territorial Catchment Institutions – EPTB) were created at a smaller scale than the 6 river basin districts, and some of them developed a tool called *Programmes d'Action pour la Prevention des Inondations* (Flood Prevention Action Programs – PAPI), to financially support projects aimed at reducing vulnerability and 'returning space' to the rivers; which has become an essential part of flood risk management policy, as illustrated below.

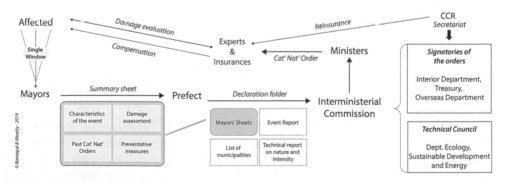

Figure 1. The Cat' Nat' damage compensation system (source: authors' elaboration).

More recently in 2014, and possibly following the Xynthia lawsuit, the government decided to transfer the competency for aquatic environment management and flood protection at local level, by organising a grouping of tiny local authorities into larger and more sustainable units called *Etablissements Publics de Cooperation Intercommunale* (Intercommunal Cooperation Public Institutions – EPCI). Again local authorities complained that these new tasks would be an unfunded mandate. The government allowed them to tax local residents and landowners up to 40 €/cap/yr to cover this new expense. But on 1 January 2019, only one third of EPCI had adopted this new tax (AdCF, 2019), with limited rates, since they are already pressed by the Cour des Comptes (national accounting office) to reduce budget spending in general …

The progressive decentralisation of the control of risks as presented here should be challenged with key questions related to recovery and resilience: does this evolution improve post-disaster recovery, and in addition does it foster vulnerability reduction? Before discussing this, a presentation of the recovery system and its effectiveness is needed.

Recovery with Cat' Nat': how does it work?

As mentioned above, Cat' Nat' works thanks to an additional fee on all insurance premiums in the country, whether policy holders are at risk or not. The resulting fund covers only the damages of extreme events, which were not covered before the 1982 law: insurance companies would then argue a case of *force majeure* (*i.e.* beyond what is statistically foreseeable, sometimes called 'an act of God' in the English speaking world).

In their typology of flood coverage systems at international level, Lamond and Penning-Rowsell have picked Cat' Nat' as the typical example of 'bundled insurance backed by the State' (Lamond & Penning Rowsell, 2014). The Cat' Nat' fund is created from 12% extra on the premiums paid to cover homes, businesses and some agricultural assets, and 6% on automobile premiums. 53.1% of the total comes from domestic residences, 37.3% from business risks, 3.4% from agriculture risks, and 6.2% from automobiles *i.e.* €100 million. If an insurance company has difficulties in facing its liabilities, it can get the support of the Caisse Centrale de Réassurance (Central Reinsurance Agency – CCR).

When an extreme event occurs, mayors of municipalities hit by a natural disaster can make a Cat' Nat' claim to the prefect (Figure 1); the prefect in turn reports to an inter-ministerial commission, responsible for deciding on the 'naturalness' and 'exceptional intensity' of the phenomenon, and whether 'the usual measures to avoid such damage could not prevent them to happen or could not be taken'.[3] Upon the commission's advice, the relevant ministers will eventually sign a Cat' Nat' order, acknowledging a state of natural disaster. Only then, are insurance companies liable to mobilise the additional premiums they have set aside through the super-fund, and cover the damages.

Post-disaster coverage includes the replacement of damaged infrastructure and buildings, as well as the complete restoration of services and economic revitalisation. This support through the insurance system can be supplemented with direct subsidies from the State: The government supports local authorities in dealing with the distress of victims on top of the insurance system, including compensation for uninsured local public assets. For instance, the *Fonds de Solidarité en faveur des Collectivités Territoriales* (Solidarity Fund in Support of Local Authorities – FSCT)[4] can be mobilised if the amount of damage to the local authority is between €150,000 and 6 million; beyond this level the *Fonds de Calamités Publiques* (Public Calamity Fund) can be called in. These funding measures, which are separate from Cat' Nat', play a decisive role, since they better encourage the adoption of preventive adaptation measures within reconstruction (Moatty, Gaillard, & Vinet, 2017).

The Cat' Nat' system prevents many insured families and businesses from being ruined, and in theory allows them to obtain quickly (within 3 months) the funds they need to repair their belongings, at least for those who are insured. In addition, after the event, insurance companies are in principle not allowed to raise the premiums of concerned insurees, given that it is exceptional. This can be compared with the situation in England where insurers may change the contract conditions for those flooded (Lamond, Proverbs, & Hammonds, 2009). Although based on a small set of examples, the following Tables 1 and 2 illustrates the improved damage coverage allowed by Cat' Nat' compared with 2016 data collected by Munich Re:

In 2017, the last year with available consolidated statistics, the total amount of additional premiums raised by insurance companies reached 1.641 bn €, *vs* 1.601 bn € in 2016. This budget is being used chiefly to cover two exceptional events: the first, cyclone Irma which hit the Caribbean islands of St Martin and St Barthélemy, and cost around 2 billion € of insured losses according to CCR; the second, the 2018 drought following the one of 2016 will certainly cost even more to the system.

Altogether in its 35 years' existence, the Cat' Nat' system has managed to finance the physical reconstruction of damaged assets at national level whilst increasing additional premiums paid to insurance companies by only 10%. Confidence in the system relies on public reinsurance system which supports insurance companies facing hardship, and collects the data allowing a good follow-up (CCR, 2018). On top of it, the whole system is guaranteed by central government budget which would typically be mobilised in case of a more devastating event like an earthquake hitting the Nice area.

Now, a potential accumulation of extreme events, related to climate change, questions the sustainability of the recovery system. The fee on insurance premiums has already been

Table 1. Cost and coverage rate of some 2016 disasters (Munich Re; French CCR).

Area	Event	Cost estimate	Coverage rate
World	Total natural disasters	€ 167 bn	30%
Japan	Earthquake April 14–16	€ 29.6 bn	19%
Europe	May-June storms	€ 5.7 bn	47%
France	May floods*	€ 1.4 bn	64%

*this was the second costliest event since the inception of Cat' Nat', with 150,000 victims and 5 casualties.

Table 2. Cat' Nat' key 2017 data (from CCR, 2018).

	2017	From 1982 (2000 for vehic.)	Yearly average
Number of insurance contracts (– including vehicles)	90 million		
Cat' Nat' income from additional premiums	1.64 Md€		
Cost of Cat' Nat' events on non-vehicle insured damages	IRMA 1.97 Md€ Drought 800 M €	33 Md€ (56% floods, 33% droughts)	936 M€
Cost of insured -vehicle damages		707 M€	40 M€

growing steadily since its inception: initially, the additional rate on non-vehicle premiums was set at 5.5%, but it was raised to 9% in 1985, to restore an interannual balance between income and expenses; and it rose again to 12% in 2000, because more severe and repeated droughts resulted in soil subsidence provoking damage to overlying properties. Today with the possibility that climate change increases the number and severity of extreme events, the system could potentially become bankrupt. In 2016, the total disbursement of insurance companies to victims of floods and droughts already exceeded the total budget gathered from insurees by 4.5%. For 2017, the deficit is even worse. It is still difficult to assess because the rate of insurance coverage is much lower in the French West Indies than on the mainland: the rate of housing insurance penetration is only 52% in the *départements d'outre-mer* (overseas counties – DOM) vs 99% in mainland France; but the cost of Irma alone represents more than 6% of total disasters' damage cost since the creation of Cat' Nat' in 1982. In 2018 an exceptional drought on most of France (and Northern Europe), together with a dramatic flash flood in Aude county in October will also probably leave the fund in deficit. It remains to be seen if implementation of the FD will lead to reviewing the PPRi and increasing land-use limitations, or if Cat' Nat' will need a new increase of the fee on insurance premiums.

From recovery to resilience: how and when

Economic resilience depends on the capacity of a government to fund recovery and reconstruction through a large span of public and private mechanisms, like budget reallocations, tax increase, reserves mobilisation, national and international bonds, international (European) aids, insurance and re-insurance indemnities and financial obligations like 'catastrophe obligations'. (Mechler, Linnerooth-Bayer, Hochrainer, & Pflug, 2006; UNISDR, 2013)

The reconstruction period offers a 'window of opportunity' to reduce vulnerability (Christoplos, 2006; Moatty, 2017) This can take two different shapes, which can *a priori* appear contradictory, and yet in fact follow each other and set resilience as a dynamic process. Indeed, during the crisis, resilience means resisting, at household or territorial level; while in the longer post-disaster run, it means a capacity to bounce back to normal or to non-degraded functioning, eventually with adaptation (Moatty, Vinet, Defossez, Cherel, & Grelot, 2018).

But in France, the Cat' Nat' relative success has a negative counterpart: insurance companies unintendedly reduce resilience since they reimburse victims on the basis of actual damages incurred, which does not foster different and more resilient reconstruction: additional costs of vulnerability reduction are not well covered, as an official inquiry on post-disaster victims' compensation puts it: 'The risk prevention policy and the compensation of natural disasters are juxtaposed but they largely ignore each other' (Moatty & Vinet, 2016). On top of this, many damages are not considered by insurance companies: those on the public domain, outdoor housing damages, non-monetisable values (casualties, health impacts, emotional loss), etc. (Moatty et al., 2018). Vulnerability reduction has to be funded differently.

In the French system, resilience then relies more on prevention than on crisis management and post disaster measures, and prevention is chiefly introduced through the risk maps: the PPR. The PPR act as a counterpart of national reconstruction financing by making a local obligation to zone risk areas. Typically PPRi are maps showing areas exposed to a reference flood hazard (it is often the 100-year flood that is used, or the highest known water levels when they are higher), and include bans on building (in areas where the risk is considered strong), or restrictions on building (*e.g.* no valuables on ground floor) in other areas. This remains a touchy issue due to the political weight of private property, and PPR meet land-owners resistance.

And, since Cat' Nat' funding is national but PPR are local, obviously local authorities are tempted to underestimate the risks and to support economic and urban development at the expense of prevention, all the more so when they expect that losses will be covered by Cat' Nat'. Zoning regulations are feared by local representatives as impacting negatively on land values and the attractiveness of the commune, particularly where other natural or industrial risks are identified. According to business daily *Les Echos*,

> this very protective regime also has a perverse effect: it delays setting up efficient prevention policies, and even in some cases, it relieves actors from their responsibilities. Too many mayors allow house building in flood prone areas; too few coastal communes adopt plans on sea surge risks; and too few in general impose a geological survey before granting a building permit. (Maujean, 2018 – our translation)

Various experts question a perverse effect: 'automatic reimbursement' of disasters tends to reduce the victims' responsibility (Bourrelier, 1997; Ledoux, 2000; Lefrou et al., 2000).

This is also why the drafting of the plans was transferred from municipalities to prefects' services at county level in 1987, but implementation difficulties remain frequent, due to the 'crossed regulation' politics mentioned at the beginning of this paper.

Additional measures to reduce vulnerability

Floods however do trigger vulnerability reduction efforts, even though not necessarily in the recovery phase. Despite local resistance, the above-mentioned PPRi progressively cover all concerned communes. At the end of 2017, the number of approved PPR and other risk plans resulting from previous legislation exceeded respectively 20,000 (all risks) and 14,000 (floods and mudslides), covering 10,400 communes for flood risks. This means that in most communes at risk of flooding, there are some limitations on building in areas at risk. In addition, a 2004 law mandated municipalities to set up, within 2 years after the PPRi is approved, what is called a *Plan Communal de Sauvegarde* (Communal Safeguard Plan – PCS). This PCS establishes alert and crisis management procedures[5] such as evacuation and emergency resettlement. The law also mandates local authorities to issue a *document d'information communal sur les risques majeurs* (Communal Information Document on Major Risks – DICRIM), to help the local population be prepared if a disaster is announced. At the end of 2017, respectively 8000 and 6000 communes had set up their PCS and their DICRIM, which means that local knowledge about the level of risk has probably improved, even though in a variable manner: realising a PCS implies that the local authority either has staff qualified in environment and urban planning, or outsources to consultants, who make the vulnerability diagnosis and draft the immediate action sheets needed. In the latter case there is a chance that PCS remain unknown or non-appropriated by elected representatives, reflecting lack of interest or excessive standardisation due to economies of time and money.

Altogether, however, the post-disaster period is a time of increased consciousness of vulnerabilities, and thus of better acceptance of prevention's additional costs (Quarantelli, 1999). Improved knowledge of hazards and vulnerabilities lead central government and the insurance system together to update and reinforce regulations. To give an example, a review on *Build Back Better* incentives (MRN, 2018; MRN & et al., 2017) was made after the costly flood of May-June 2016 south of Paris. It concluded that the pre-existing PPRi should be reinforced: they had been designed after the worst event ever recorded, the flood of 1910, and yet in 2016 some cities recorded higher levels still (+40 cm in Nemours, +30 cm in Montargis). Detailed observation at building level showed that 95% of the damage occurred inside houses, and could have been reduced by installing temporary flood barriers, protections on basement light wells, anti-flood backflow valves on sewer connections, and by moving various in-house appliances like electricity, above the highest known water level. Many of these measures, which undoubtedly reduce the vulnerability of residential housing, were in fact mandated before the 2016 disaster: it is typically the case with sewer backflow valves which are mandated in the sewerage regulations, or with moving up electric appliances, mandated in most PPRi. However, as usual in France, regulations do not apply retrospectively so these rules apply primarily to new buildings and eventually to post-disaster reconstruction. Insurance companies are not obliged to fund additional costs, but they tend to do it more, for instance by paying for replacing wood floors with tiling at ground level.

Another type of measure indirectly supporting vulnerability control is the reduction of insurance coverage in the case of repeated disasters. As we wrote above, insurance companies were not allowed to raise premiums after a disaster. But this rule was indirectly relaxed when insurance and reinsurance companies realised they were compensating the same landowners in the same communes, for the same works several years in a row! With the reform adopted in 2000, where communes do not have a PPR, the (residential and business) deduction on insurance reimbursement[6] (*franchise*) is raised when events are repeated: since 2001, if in the last 5 years there were three events, the deductible ('excess' in the UK) is doubled, with 4 events it is trebled, and with 5 events it is quadrupled. In the period between 2000 and 2017, only 5% of contractual deductibles were thus modulated, and less than 1.5% were more than doubled. What then appears as a credible threat should incentivise the most vulnerable local authorities to adopt a PPRi faster, and in turn residents to invest in vulnerability reduction.

The most innovative measure was mentioned at the beginning of this article: the PAPI. PAPIs were initiated by communes and catchment institutions which realised that drafting the PPR did not in itself reduce the vulnerability of existing construction, and something more had to be done. For all types of flooding (river overflowing, groundwater rising, stormwater flooding and sea surge), local authorities are eligible to tap a special fund derived from Cat' Nat' called Barnier fund or *fonds de prévention des risques naturels majeurs* (Prevention Fund on Major Natural Risks – FPRNM) (Figure 2), to reduce vulnerability before a disaster takes place. This began when environment minister Michel Barnier sought a solution to move the population of a village before a neighbouring cliff would collapse on it. Since houses could not be sold, a solution had to be found to buy their property amicably and allow them to relocate. Several other cases were then identified, and finally in 2003 it was decided to call for tenders of local projects to reduce vulnerabilities; and a yearly percentage of the Cat' Nat' fund was diverted to

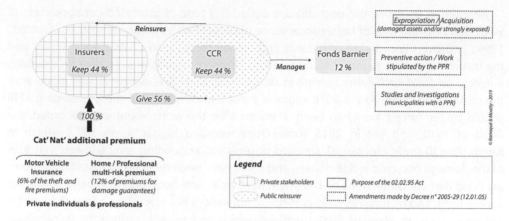

Figure 2. Barnier fund origin and eligible operations (authors' elaboration).

subsidise these projects. The call for tender was a success since more than 100 PAPI were developed, covering respectively 40% and 60% of the population exposed to risks from flooding and from sea surges.

In 2017, the transfer to the Barnier fund had reached 12% of the Cat' Nat' budget, *i.e.* providing 200 mln € at national level (representing 38% on average of the PAPI investments). Although insufficient compared to all the vulnerability reduction projects, as is now being revealed with the FD implementation, it is difficult to increase the funding from Cat' Nat' for fear that insurance companies would not be able to cover post-disaster damages, whether in case of a major event, or in case of cumulated events of lesser magnitude repeated over several years.

The most recent measure concerning resilience was a decision of the government in 2014 to decentralise the management of flood control infrastructure at local level. On the one hand it is a good decision to give local authorities the responsibility for both the structural measures and for land-use based solutions. But on the other hand, observers suspect that after the Xynthia disaster in 2010, the government wanted to put an end to its own responsibility for dyke maintenance, so as to reduce its deficit. Four new responsibilities were devolved to local authorities, together labelled *Gestion de l'Eau, des Milieux Aquatiques et Prévention des Inondations* (management of water and aquatic environment and flood prevention – GEMAPI). To allow them to face resulting financing needs, local authorities are allowed to tax local residents and landowners up to 40 €/cap/yr to cover this new expense. After 4 years on 1 January 2019, only one third of EPCI had adopted this new tax (ADCF, 2019), and with rates much below 40 €, since they are already under pressure from the national accounting office for overspending. On top of it, GEMAPI tax is set in part on local housing taxes which the government intends to phase out progressively, as part of general taxation reduction on modest revenues. So it remains to be seen how this further decentralisation move will improve resilience.

A local example

To illustrate this situation of limited progress, we now cover a case in an area south of Paris, where a small river called Yvette is a tributary to Orge river, which is in turn a tributary to the Seine upstream of Paris. Yvette suffered a serious flooding episode in May-June 2016, (CCR, 2016), which was eligible to Cat' Nat' funding.[7] But in fact, flooding had previously taken place repeatedly, in particular downstream where the Yvette and the Orge meet before merging into the Seine: 17% of the 121 communes in the catchment were supported by Cat' Nat' 9 times or more between 1983 and 2016! Saint-Rémy-les-Chevreuse, upstream on the Yvette, was also badly flooded several times, in particular in 2016 when a nearly 100 year flood occurred and flooded 300 homes, 10 public buildings and 6 shops. Little was done to stop this repeated flooding directly, while some wetland restoration recently improved the situation.

One has to remember here that the *agences de l'eau* can support the implementation of the programme of measures of the Water Framework Directive (WFD – 2000/60/EC), including measures on wetlands that have indirect, positive impacts on floods; but cannot support direct flood control measures in the flood risk management plans of the FD. In the case study, the area is partly covered by a regional natural park,[8] the institution of which funded important efforts towards sustainable landscape governance. The catchment institutions of the Orge and Yvette rivers' and the park together initiated a *Schéma d'Aménagement et de Gestion des Eaux* (catchment plan – SAGE) as soon as 1997. A *Commission Locale de l'Eau* (local water committee – CLE) was set up in 1999, under participative democracy principles: the committee includes 30 elected representatives from the (116) municipalities and the (two) counties, 24 user representatives (farming union, fishing, consumer and environment associations, a university, the regional park, etc.) and 9 representatives of the Government (regional and county prefects, the *agence de l'eau*, regional services of various ministries), etc. Once the SAGE was drafted and adopted by the CLE in 2006, and approved by the prefect, resulting investments were made by the local river institution which is competent on floods and also on waste water: a joint board acronymed SIAHVY.[9] The plan had to be updated following the water law of 2006, and the new SAGE was approved in 2014. After the 2016 flood, the decision was made to develop a PAPI to try reducing vulnerability.

In the SAGE approved in 2006, there were already three major targets dealing with floods: protection of people and goods in lower areas (Flood Risk Management Plan, integration in urban plans, flood expansion areas); protection of people and goods from flooding due to runoff (specific stormwater control measures); and developing a risk culture in the population. These projects were picked up again in the revised plan in 2014, and included in the integrated catchment planning carried out by the SIAHVY. Some projects were being prepared before the serious flooding of May 2016, and could be implemented rapidly. In particular, the re-naturalisation of a small tributary of Yvette, the Mérantaise, was carried out to accommodate and store flood waters in a rural site with no damage to properties. Purchase of 800 ha and restoration of the wetland, a 4.5 million € project, was eligible for funding from the Agence de l'Eau Seine Normandy, as part of the programme of measures of the WFD. This project could improve resilience directly, but also indirectly through improving the legitimacy of the PAPI project.

The Mérantaise operation was a success and is now a show case for the *agence de l'eau*, since in the 2018 flood, there was no damage in that area; and elected representatives are now better convinced of the merits of land-based flood control. The PAPI Orge-Yvette was then drafted: it contains 8 types of actions, covering different phases of risk management (prevention and crisis management) and targeting different scales, from the watershed to the citizen:

- Improving risk knowledge
- Improving citizens' consciousness of the risk
- Monitoring and forecast of flooding episodes
- Alert and crisis management
- Integration of risk maps in town plans
- Reducing vulnerability and improving resilience
- Channelling the high flows
- Maintenance of flood protection structures

In addition, a transversal axis for governance includes a web site to share information between the institutions. In terms of vulnerability reduction and resilience improvement, the chief action is to help the most impacted municipalities in being prepared: drafting vulnerability assessments on public buildings, workplaces and shops, and setting up a strategy to cope with dramatic episodes in real time and hasten recovery. The cumulated cost of the 35 forecast actions of the plan is estimated at 1.57 M€ plus VAT (Mérantaise operation excluded), two thirds being devoted to flood knowledge and consciousness, and 200,000 € to improving resilience. The expected PAPI funding would mobilise the Barnier fund for 45%, the *Agence de l'eau* for only 4%, the County council for 26%, with the rest being self-funded by the builders/managers of infrastructure.

However, local managers of the Yvette catchment are worried about the area being eligible for the Barnier fund, due to the small size of the issue compared to other cases in France. They fear delays in implementation due to bureaucratic control of their projects, leading in the end to a 'temporal dilution' of the initial good will of local authorities. Staff in charge of the PAPI are concerned that it does not include more precise anticipation measures to improve resilience in post-disaster recovery.

In addition, vulnerability reduction funding focuses almost entirely on public and business infrastructure, and not on housing: the staff in charge of the PAPI also fear that individual landowners at risk will not invest in vulnerability reduction due to lack of incentives. They propose residents install anti-backflow valves to prevent flooding from sewers, but it is not mandatory, and people are reluctant due to the high cost. In other words, residents are rich enough to be ineligible for benefits, but cannot afford to invest on additional resilience measures, which are not included in the Cat' Nat' indemnity. The worst case is when insurance companies dramatically increase the reimbursement deduction if vulnerability reduction investments are not made. An example was given of an insurance company threatening to raise the *franchise* substantially if buildings were not equipped with removable light walls to block ground floor entrances, and flood barriers. This may ultimately lead people to quit their insurance. Conversely, these requests are met by the managers of public buildings like schools despite the cost, as being part of the PAPI.

Concluding remarks

Post-disaster recovery can be either addressed by collective or individual coping strategies, planned or spontaneous, with varying objectives depending on the stakeholders.

The French Cat' Nat' system combines individual and collective coping; it clearly represents an improvement in terms of recovery. Its success is due to its funding by all policy holders, be they at risk or not. This globally allows supporting recovery of extreme events without subsequently raising the insurance premiums of the victims, which obviously limits the risk of policy holders exiting. It can be compared with the American National Flood Insurance Program (NFIP) which for Lamond and Penning Rowsell (2014) is a typical example of 'add-on or separate policy, state-backed': insurees are eligible for subsidised extra flood insurance in areas which are largely self-designed as hazard areas. This leaves large areas outside of the scheme, and in addition, the price of extra insurance may be too high, leading to some policy holders exiting and potential fund bankruptcy.

If we compare with Italy, which has one of the worst disaster vulnerabilities in Europe, there is a clear difference: there the risk map is made and updated at national level, but priorities to develop preventative measures are left to local authorities, which are often too small to really control land-use developments at risk. The focus is on emergency management, and not on mid- and long-term territorial resilience. At least in France, Cat' Nat' binds local authorities and the Government through the insurance system and the PPRi together.

However, it seems that Cat' Nat' does not really encourage vulnerability reduction, in particular as concerns private landowners. Considerable amounts of money are invested in recovery but most of them finance an 'identical' reconstruction. This is probably nothing new for flood insurance experts at international level. But we want to highlight the difficulty in bridging flood prevention and global river and aquatic environment policy (FD and WFD directives' implementation). This is partly due to the ongoing economic crisis and the excessive national deficit compared to the Maastricht 3% target. Thus, in line with the spirit of the decentralisation laws, the concern of the present Government is to transfer its previous responsibilities on to local authorities. But they also want to regulate the use of public funds to make sure that they are not used, after disasters, to make up for the maladaptation or even the lack of maintenance of public assets before disasters (*e.g.* dykes, the responsibility of which falls to local authorities since 2014); and that this support does not increase the pre-existing vulnerability. In addition, Cat' Nat' reimbursements are reduced if floods occur repeatedly in an area in which stakeholders have not taken any preventive or mitigation measures. Facing this relatively negative / punitive approach to flood control, integrated river policies at catchment scale frequently include more positive measures to improve social and territorial resilience. But they also frequently stop at the gate of private properties.

This can be illustrated with the difficulties in drafting the PPRi. They were first decentralised to municipalities, but soon recentralised in the prefectoral services. So decentralisation was only partial, with local elected representatives mainly continuing to support the urban development potential of flood prone areas at the expense of vulnerability reduction. Resilience anticipation is thus limited by this typical characteristic of French political system: the State is above the citizens but private property is well defended. While landowners often manage to undermine the building bans or limitations in PPRi, the State is tempted to underestimate the impact of floods on private property and the need for adaptations, or leave it to the insurance companies.

Of course, it is much easier to draft a more severe PPRi and a more resilient land-use plan in the aftermath of a disaster: in La Faute-sur-Mer today, in the place of the housing estates where 29 people drowned in the February 2010 Xynthia event, there is now a golf course. Yet anybody can ask why wasn't it zoned that way beforehand?

One answer is to reconsider land-use planning from the perspective of the aquatic environment, and derive planning rules from a catchment plan, which is elaborated more in terms of water usership than of landownership. Participative democracy, which was particularly developed in the water sector in the last 50 years, may then help mitigate the well-known State *vs* private property French antagonism. In places where local stake-holders convene to draft a SAGE (catchment plan) there is a better potential involvement of the population; yet flood control measures directly concerning landowners as such are not frequent, as our case study illustrates. This is also because in France, catchment insti-tutions and the *agences de l'eau* are not encouraged to include flood control measures in their tasks: they would need to get a corresponding funding mechanism beyond what is made available through the Barnier fund, which is reckoned as largely insufficient to trigger improved resilience in most vulnerable areas, while remaining complex to obtain.

Insurance companies are aware of the difficulties encountered after floods by impacted landowners. Yet they have to admit that, paradoxically, the Barnier fund remains partly unspent year after year, and unfortunately any positive fiscal cash flow attracts the envy of the Treasury! After taking 10% of the 200 million € since 2016, this year instead the total transfer from the Cat' Nat' fund is capped at 137 million €. Facing this unsatisfactory bottleneck, the Government and the CCR (2016) are considering an evolution of the regime, at least with two measures: subsidising temporary relocation of victims, since recovery/reconstruction duration is frequently much longer than expected; and reimbur-sing some additional costs corresponding to vulnerability reduction in the damages payment calculations by insurance companies.

But again this raises the general question of the trade-off between after and before dis-asters particularly in Cat' Nat': the more one spends on more resilient recovery, the more is needed for the PAPI and Barnier funding, and the less is left to cover disasters after they occurred. And all this in a context where climate change increases the uncertainty on financial sustainability, not only for flooding but also for drought events which seem to multiply in France.

Some observers consider it would be a good idea to transfer the Barnier fund to the 6 *agences de l'eau*, which could then merge this budget with their own budget aimed at recovering the good status of the aquatic environment, and help develop more integrated WFD and FD policies. This would be a reversal of the previous policy leaving the *agences* out of the issue of flood control. But the way chosen in 2014 is somewhat different: con-solidated local authorities are allowed to raise a new tax on landowners, the above men-tioned GEMAPI. It remains to be seen how the new legal responsibility on flood planning and infrastructure devolved to local authorities will help boost this important but yet fragile task: a more resilient recovery thanks to more active policies on land-use planning and on buildings restoration.

In the Netherlands, after the 1953 disaster (1830 casualties), insurance companies refused to cover the flood risks, and then the government developed a strong collective protection system based on dykes at two levels (national and regional). Today most of the flood control policy is implemented at the level of the former polder waterboards, which were dramatically consolidated since this event into what they call Regional Water Authorities (RWAs). But many people think that this focus on flood protection through structural measures has been overly disconnected from land-use management by local authorities, resulting in an aggravation of vulnerability (Doorn-Hoekveld, 2018). So they tend to promote a new 'multi-layered' policy, where improvement of the aquatic environment would be combined with alternative flood control policies: returning space to the rivers, mobilising private land to temporarily store and discharge excess water (with due compensation of the losses); and above all, developing better links between RWAs, and provinces and local authorities in charge of land-use management so that they are incited to reduce spatial vulnerability. Local zoning plans and some building permits are now subjected to a 'water test' to clearly inform local authorities of their consequences in terms of flooding. This new 'living with water' approach even includes an experiment on insuring some assets against floods: the 'Neerlandse' insurance was initiated in 2011 but has not attracted many subscribers yet.

If the French would draw on the Dutch multi-layered policy, they could support a more systematic co-ordination between the consolidated local authorities (EPCI) and the catchment institutions (EPTB), so as to plan vulnerability reduction through land-use control, including with compensated storage of excess water on private property; and funding could combine the Barnier fund, the GEMAPI tax levied by the EPCI, plus some specific mutual levy system to be developed at the EPTB level, in the aim at better bridging the implementation of the FD and the WFD.

Notes

1. Act n°82-600 of 13 July 1982 for the compensation of victims of natural disasters. Floods represent more than 90% of events.
2. Directive 2007/60/EC of the European Parliament and of the Council of 23 October 2007 on the assessment and management of flood risks.
3. Quoted from the 1982 Cat' Nat' law. Our translation.
4. The FSCT is a financial package announced in the Decree in the Official Journal of 27 August 2008. So far, we have limited information on its performance.
5. Act n° 2004–811 of 13 August 2004 dite « loi de modernisation de la sécurité civile » : civil security modernisation law, art. 13.
6. In some insurance contracts there is a fixed deductible on reimbursement, in exchange with a premium reduction.
7. Severe storms followed by an extreme rainfall episode provoked floods in several parts of France between May 25 and June 6, resulting in damages on insured properties up to 1.4 bn €
8. Parc Naturel Régional de la Haute Vallée de Chevreuse. https://www.parc-naturel-chevreuse.fr/
9. SIAHVY : Syndicat Intercommunal pour l'Aménagement Hydraulique de la Vallée de l'Yvette

Disclosure statement

No potential conflict of interest was reported by the authors.

References

Association des Communautés de France (AdCF). (2019, February 8). Gemapi: 35% des communautés ont désormais institué la taxe affectée. In AdCF-direct, Newsletter n°457. Retrieved from https://

www.adcf.org/contenu-article-adcf-direct?num_article=4533&num_thematique=1&id_ newsletter=382&source_newsletter=457&u=MTc3Ng

Barraqué, B. (2014). The common property issue in flood control through land use in France. *Journal of Flood Risk Management*. doi:10.1111/jfr3.12092. (Virtual Special Issue: *The European Flood Risk Management Plan*, Guest Editors: Asst. Prof. Dr. Thomas Hartmann and Prof. Dr. Robert Juepner).

Bourrelier, P. H. (Dir.) (1997). *La prévention des risques naturels : rapport de l'instance d'évaluation*. Paris: la Documentation Française, 702 p.

Caisse Centrale de Réassurance (CCR). (2018). *Les Catastrophes naturelles, en France, bilan 1982–2018*. Self-Published Report. Retrieved from http://www.side.developpement-durable.gouv.fr/ EXPLOITATION/ACCIDR/doc/IFD/IFD_REFDOC_0550808

CCR – Service R&D modélisation – Direction des Réassurances & Fonds Publics. (2016). *Inondations de mai-juin 2016 en France. Modélisation de l'aléa et et des dommages*. Report. Retrieved from https:// www.ccr.fr/documents/23509/29230/Inondations+de+Seine+et+Loire+mai+2016+avec +Prevention_28062016.pdf/b3456170-8dee-4c3e-b2aa-608172cce8c7

Christoplos, I. (2006, Février 2–3). *The elusive "window of opportunity" for risk reduction in post-disaster recovery*. Intervention in ProVention Consortium Forum, Bangkok, 4 p.

Doorn-Hoekveld, W. J. (2018). *Distributional effects of EU flood risk management and the Law. The Netherlands, Flanders and France as case studies* (PhD thesis). Utrecht University, Utrecht, 231 p.

Duran, P., & Thoenig, J. C. (1996). L'Etat et la gestion publique territoriale. *Revue Française de Science Politique, 46*(4), 580–623.

Grémion, P. (1976). *Le pouvoir périphérique, bureaucrates et notables dans le système politique français*. Paris: le Seuil, 477 p.

Lamond, J., & Penning Rowsell, E. (2014). The robustness of flood insurance regimes given changing risk resulting from climate change. *Climate Risk Management, 2*, 1–10.

Lamond, J., & Proverbs, D. G. (2009). Resilience to flooding: Lessons from international comparison. *Proceedings of the Institution of Civil Engineers - Urban Design and Planning, 162*, 63–70.

Lamond, J., Proverbs, D. G., & Hammonds, F. N. (2009). Accessibility of flood risk insurance in the UK: Confusion, competition and complacency. *Journal of Risk Research, 12*(6), 825–841.

Ledoux, B. (2000). *Retour d'expérience sur la gestion post-catastrophe dans les départements de l'Aude et du Tarn*. Rapport pour le Ministère de l'Aménagement du Territoire et de l'Environnement (MEDD, ex-MATE), Bruno Ledoux Consultants, 70 p.

Lefrou, C., Martin, X., Labarthe, J. P., Varret, J., Mazière, B., Tordjeman, R., & Feunteun, R. (2000). Les crues des 12, 13 et 14 novembre 1999 dans les départements de l'Aude, de l'Hérault, des Pyrénées-Orientales et du Tarn, Rapport au Ministre de l'Aménagement du Territoire et de l'Environnement, 99 p. + annexes.

Maujean, G. (2018, June 25). Pourquoi il faut changer le régime des Cat' Nat'. *Les Echos*. Retrieved from https://www.lesechos.fr/25/06/2018/lesechos.fr/0301876453903_pourquoi-il-faut-changer-le-regime-des—cat-nat-.htm

Mechler, R., Linnerooth-Bayer, J., Hochrainer, S., & Pflug, G. (2006). Assessing financial vulnerability and coping capacity: The IIASA CatSim model. Concepts and methods. In J. Birkmann (Ed.), *Measuring vulnerability and coping capacity to hazards of natural origin* (pp. 380–398). Tokyo: United Nations University Press.

Mission Risques Naturels (MRN). (2018). *Inondations Seine-Loire 2046. Etude sur l'endommagement du bâti causé par un événement 'inondation'*. Case Study Report. Retrieved from https://www.mrn. asso.fr/wp-content/uploads/2018/04/04-06-2018-rapport-mrn_inondations-seine-loire-2016.pdf

Moatty, A. (2017). *Post-flood recovery: An opportunity for disaster risk reduction?* In F. Vinet (Ed.), *Floods* (vol. 2, pp. 349–364). ISTE Press, Elsevier.

Moatty, A., Gaillard, J.-C., & Vinet, F. (2017). Du désastre au développement: Les enjeux de la reconstruction post-catastrophe. *Annales de géographie, 714*, Armand Colin, 169–194.

Moatty, A., & Vinet, F. (2016). *Post disaster recovery: The challenge of anticipation*. Web conference FLOOD risk 2016 - 3rd European Conference on Flood Risk Management. doi:10.1051/e3sconf/ 20160717003

Moatty, A., Vinet, F., Defossez, S., Cherel, J. P., & Grelot, F. (2018). Intégrer une éthique préventive dans le processus de reconstruction post-catastrophe: place des concepts de résilience et d'adaptation dans la reconstruction préventive. *La Houille Blanche*, SHF ed. ISSN: 0018-6368, 11–19.

MRN & et al. (2017, September 8). Conférence B3: exemples d'apports des acteurs de l'assurance française au 'faire et reconstruire mieux'. Retrieved from https://www.mrn.asso.fr/wp-content/uploads/2017/12/2017-presentation-apres-midi-mrn-aqc-elex-saretec-conference-b3.pdf

Quarantelli, E. L. (1999). *The disaster recovery process: What we know and do not know from research.* Preliminary paper n°286. Newark, DE: University of Delaware, Disaster Research Center.

United Nations International Strategy for Disaster Reduction (UNISDR). (2013). *Global assessment report on disaster risk reduction. From shared risk to shared value: The business case for disaster risk reduction.* 288 p.

Vinet, F., Boissier, L., & Defossez, S. (2011). *La mortalité comme expression de la vulnérabilité humaine face aux catastrophes naturelles : deux inondations récentes en France (Xynthia, var, 2010). VertigO* [Online], Volume 11 Numéro 2 | septembre 2011, consulté le 01 août 2018; doi:10.4000/vertigo.11074.

Worms, J. P. (1966, July-September 8–3). Le préfet et ses notables. In *Sociologie du Travail* (pp. 249–275). Paris: Le Seuil.

Stakeholder participation in building resilience to disasters in a changing climate

Paulina Aldunce, Ruth Beilin, John Handmer and Mark Howden

ABSTRACT
The resilience perspective has emerged as a plausible approach to confront the increasingly devastating impacts of disasters; and the challenges and uncertainty climate change poses through an expected rise in frequency and magnitude of hazards. Stakeholder participation is posited as pivotal for building resilience, and resilience is not passive; rather, stakeholders are actively involved in the process of building resilience. Who is involved and how they are involved are crucial aspects for developing resilience in practice. Nevertheless, there are few empirical studies available to inform theory or show how these issues are addressed. This study focuses on revealing how practitioners frame the issue of participation in relation to resilience, its relevance to a changing climate and how, in consequence, they construct practices. Using Hajer's [(1995). *The politics of environmental discourse: Ecological modernization and the policy process*. New York] 'Social-interactive discourse theory', in this interdisciplinary research, we study the frames and subsequent practices developed around a disaster management policy initiative in Australia: the Natural Disaster Resilience Program in Queensland. What emerges from the research findings as critical and requiring urgent attention is stakeholder and especially local government and community participation, and for this to become socially relevant, challenges such as meaningful communication and power structures need to be addressed. What is also critical is to move from experiential learning to social learning. Additionally, the results presented here offer empirical evidence on how broadening the pool of actors can be implemented, and the opportunities that this opens up for building resilience.

Introduction

Stakeholder participation is fundamental for building resilience (Berkes, 2007; O'Brien, O'Keefe, Gadema, & Swords, 2010). In this paper, the arguments that underpin this idea are presented. We focus on *who* must be considered when building resilience to disasters, including aspects associated with the participation of an expanded set of actors and the role of the community and *how* participation is encouraged when partnership is proposed as an appropriate strategy. In doing so the *why* is also revealed.

We begin by considering 'the *who*'. Expanding the participation of actors is valuable for sustaining resilience, because: it brings a diversity of points of view and experiences into the discussion (Folke, 2006; Gero, Méheux, & Dominey-Howes, 2011). It gives opportunities to share knowledge and expertise, recognising that different kind of knowledge can be important (Armitage, Marschke, & Plummer, 2008). It gives space for different needs, and it is an enabler for learning and changing actions (Wildavsky, 1988). Within the resilience literature, uncertainty is conceptualised as part of nature and social systems (Aldunce, Handmer, Beilin, & Howden, in press). Diversity, achieved through expanding the set of actors, is especially relevant in the presence of uncertainty (Aldunce, Beilin, Handmer, & Howden, 2015). On the one hand, it helps to have diverse actors to confront the unknown in a timely manner and to allow for a wider pool of possible ideas and solutions for the development of plans which could not be imagined in a more restricted group of people (Paton, 2006). On the other hand, if the participation of a broader group of actors is sustained over time, this enables iterative processes of evaluation allowing learning from past practices, and inclusion of their experience of successes and failures (Gaillard, 2010; Wildavsky, 1988).

It is widely believed that individuals and local communities are key actors for building resilience and that there is a need for increasing community participation (Aldunce et al., 2015; Gero et al., 2011; Nelson, 2011). This is because they are an invaluable source of knowledge (Bahadur, Ibrahim, & Tanner, 2010; O'Brien et al., 2010); their needs and opinions need to be considered (Twigg, 2007); they are in the 'front line' in experiencing a disaster (Nelson, 2011); and also because their actions affect the overall system. But for community participation to be realised (rather than consultation or mobilisation – see Arnstein's Ladder 1966) what is important is that those in positions of authority and power, for example, agencies and their bureaucrats, have to focus on the needs and priorities of the communities (Chambers & Blackburn, 1996).

Uniqueness is a characteristic of communities that requires consideration in its different manifestations. Every community is unique; hence they have unique experiences which constitute an invaluable source of knowledge (Paton, 2006) and the ways in which communities' vulnerabilities occur are also unique. Building resilience should also be sensitive to the unique social, cultural, economic, political and physical realities in which communities are embedded and in this sense resilience should be conceived of as context dependent (Gaillard, 2010).

The role of community leaders when building resilience is also relevant (Norris, Stevens, Pfefferbaum, Wyche, & Pfefferbaum, 2008; Twigg, 2007). This is because one way of involve communities is through the recognition of organisations, institutions and leaders who naturally emerge in the community. Leaders play a role as brokers, often as the glue that brings together different societal actors, thus enabling not only committed and effective community participation, but also assigning to that participation a meaningful place in governance (Kocho-Schellenberg & Berkes, 2015; Twigg, 2007). Nevertheless, caution is needed when promoting the inclusion of leaders. For example, some members of local *élites* have the power to disrupt community initiatives and some local authorities that want to be involved may have little understanding of the needs and circumstances of those at more risk within the community. As well, leaders may privilege their own ideas and agendas. Further, Twigg et al. (2001) note that deciding which local leaders to include is one of the most difficult challenges in inclusive participation.

Therefore, attention needs to be paid to how participation occurs, who is expected or asked to participate and what that involves. 'Partnership' is one of the most recurrent forms in which stakeholder participation is framed within the resilience literature (Norris et al., 2008; Wildavsky, 1991) and in international guiding documents (UN/ISDR, 2011; US/IOTWS, 2008). A resilience approach acknowledges the relevance of the interaction and interdependency of different social actors (Djalante & Thomalla, 2011). This interaction can be built through partnership initiatives that bring together different sectors or actors in a mutually beneficial relationship (Tompkins & Hurlston, 2010, p. 5). For example, Norris et al. (2008, p. 138) highlight that for building community resilience to disasters it is crucial to have inter-organisational networks which are linked in a reciprocal manner, flexible enough to allow new forms of association to emerge, and where frequent supportive interactions occur in cooperative decision-making processes. This is the opposite to hierarchical systems and can be and crucial to developing strong commitments from different actors when working in collaboration (Gaillard, 2010). Furthermore, some authors (Paton, 2006; Twigg, 2007) make the point that by means of cooperative efforts and collaboration among social actors, a deeper and wider understanding can be achieved of both the issue and its consequent solutions. These authors also stress that cooperation and collaboration among stakeholders leads to a collective expertise, resulting in a collective resilience which exceeds the sum of its parts, similar to the idea of synergy.

An inherent characteristic of disasters is that they constitute complex processes with the presence of high levels of uncertainty; this characteristic has been exacerbated in some situations by a changing climate (Aldunce et al., in press; IPCC, 2012). Promoting collaborative structures is a significant source for building resilience because it gives opportunities for the integration of multiple levels of governance; it opens up channels for the flow and sharing of resources and knowledge; and it facilitates learning, innovation, changing actions and adaptation (Twigg, 2007).

Resilience theory, climate change and disaster risk management (DRM) literature emphasise the importance of stakeholder participation, but because participation has different meanings as conceived in literature and practice, we can expect these different meanings to affect how practitioners operationalise this part of the theory. These differences are addressed in this paper.

The results presented in this paper provide helpful insights to inform policy design and implementation of resilience ideas in DRM and climate change practice.

Methodology

This research uses Hajer's (1995) social-interactive discourse theory to investigate practitioners' frames in relation to stakeholder participation. Within this approch, discourse is defined as 'a specific ensemble of ideas, concepts, and categorisations which are produced, reproduced, and transformed in a particular set of practices and through which meaning is given to physical and social realities' (Hajer, 2000, p. 44). Therefore, discourse analyses (DA) are useful for the examination of multiple and conflicting concepts, ideas and narratives that society holds about an issue (Hajer, 2000). This resonates with this research, as resilience, DRM and climate change are controversial issues.

The methods used were observation, document revision and in-depth interview for the study of the implementation phase of the Natural Disaster Resilience Program (NDRP), at a state level and in two local municipalities (Charleville and Gold Coast) in Queensland, Australia. The 27 practitioners interviewed were personnel of government, non-government, research and private organisations. Ten documents were reviewed, which were texts related to the NDRP implementation and those from the three programmes that the NDRP replaced: the Natural Disaster Mitigation Program (NDMP), the Bushfire Mitigation Program (BMP) and the National Emergency Volunteer Support Fund (NEVSF). A thematic analysis of the data was carried out with the assistance of the software NVivo9, in which data were organised in nodes representing the themes of interest for the research. Within the text, quotes are provided to illustrate participants' perspectives, followed by a code that replaced interviewees' names in order to maintain their confidentiality. Bracketing is used by the researchers to identify assumptions or the researchers' thinking process.

Framing and practice of the role of stakeholder and community participation

This section delves into an analysis of the results, presenting respondents' answers with respect to *who* they consider must be included when building resilience to disasters (with special attention to the role of the community) and *how* they must be included (with a focus on partnership). In exploring the *who* and *how*, the *why* will be revealed.

Who: key actors for building resilience to climate hazards

The study results show that the interviewees frame stakeholder participation as relevant in building resilience. The community and NGOs are most frequently mentioned as central to this process (24 interviewees), followed by government and governmental organisations (22 interviewees) and less frequently mentioned was private and business organisations (8 interviewees). For the 22 respondents who framed governments as essential actors for building resilience, 18 specifically referred to federal and state governments. Within this group, participants claimed that it is necessary to have 'political will' (sr5/lg7[1]) and that it is the government's role to lead the process, to give strategic direction (lg3/sg6/sg11/sg26) and to enable the conditions required by enacting the appropriate legislation, by providing sufficient resources and funding, and by supporting communities (sr2/sn4/sg11/sp13/lr16/lg22). Only one interviewee specifically stressed that it is also a relevant role of government agencies to provide expertise and knowledge to councils and communities (lg22).

> I think there is a role for national and state governments in setting some of the policy and providing some of the resourcing to implement disaster risk resilience programs. (sr2)

Twenty participants, only two more in comparison with those who mentioned the state and federal levels, stated that government agencies at the local level, such as councils and emergency agencies, are central actors in building resilience. Nonetheless, even if the difference in number is insignificant, much more emphasis was given by respondents when describing arguments that support the role of local government. They argued that agencies at the local level are the ones closer to and more interactive with communities.

> I would say the local government would hold the strongest role, clearly because the local governments are the most closely connected to the communities throughout the whole of Australia. (lg12)

Respondents provided further details by pointing out that there is a multifaceted relationship between local government and the community when it comes to building resilience. Firstly, they argued that local government agencies are the ones that most directly support and give services to communities during disasters (sn10/lg17/lg18/lg22/sg24). Secondly, they explained that because building resilience is context dependent, it is the role of local governments to capture contextual realities when translating 'one size fits all' approaches coming from federal governments (sr2/lr16/sg24). Thirdly, they noted that local governments are the ones who talk to and know their communities (sg6/sg8/sn20) and therefore, by means of engagement with different local actors, they are in a position to include different views coming from their communities (lg15). Finally, participants explained that because of local government's position within the administrative government structure, they are the appropriate organisations for linking upper levels of government with grassroots organisations (sg6).

> But then in terms of on the ground rolling it out, talking to people [communities], pulling it together probably local governments are the best placed to do that because they are the ones who are there to pick up the pieces as well. (sg6)

The group of stakeholders least frequently mentioned was the private and business sector, referenced by only eight interviewees. Most of these participants limited their answers to listing this group of stakeholders when mentioning the other groups, but without giving additional details. Only two respondents elaborated further by arguing that businesses, both small scale and large enterprises, have a key role in building resilience by ensuring business continuity and assisting economic recovery after a disaster (lr1/sg27).

With regard to stakeholder participation, we were also interested in exploring whether, within the practices of the NDRP, the importance of broadening the scope of actors from its traditional government emergency response and service delivery model was considered. Some interviewees mentioned that the NDRP opened the possibility for organisations outside those traditionally involved in DRM, to participate for the first time in DRM programmes (sr5/sn9/sg14/lr16/lg15/sn23). One participant, who played a central role in the NDRP design, further emphasised that this has contributed to sharing knowledge (sg14).

> So XX organisation probably could not have applied before [to three former programs that the NDRP replaced], as a non-emergency services volunteer organisation … under the new program [the NDRP], they are invited in. Come on in, give us your ideas. And often, and what I like about it, is they are providing us with non-emergency management ideas. (sg14)

Two sources of information were used to triangulate these findings. These sources were the remaining interviews and a document review. From these interviews, we verified that seven of the participants were affiliated to organisations participating in the NDRP, but that had not taken part in the three programmes replaced by the NDRP. Triangulation, based on document review, verified that a broader list of eligible organisations could apply for funding to the NDRP (NDRP09-10, NDRP10-11), in comparison with the former pro-grammes that the NDRP replaced (NDMP09, BMP09, NEVSF09).

Who: the role of community participation

For interviewees, communities play a pivotal role for building resilience. Nonetheless, one of the first aspects that emerged from the analysis is the inconsistency in how par-ticipants referred to the concept of community. This inconsistency could be attributed to the fact that, even in social sciences, the meaning of community as a concept has been debated (Nisbet, 1970; Studdert, 2006); however, this literature has been insuffi-ciently acknowledged in DRM policies (Phillips et al., 2011). The latter has resulted in the term being ill-defined and used ambiguously in policies worldwide, including in Australia (Phillips et al., 2011). Below, we briefly clarify the definitions that inform this analysis.

From the various definitions available (e.g. of communities with relevance to interests in disasters and emergencies see Handmer & Dovers, 2007, p. 62), we focus on two: *commu-nity as locality* and *community as a social network. Community as locality* encompasses how it is understood by many of the interviewees. This refers to a place-based subset of popu-lation, meaning people sharing a specific area. This is frequently used in the context of hazards because the hazards to which people are exposed are mainly determined by specific locations, and its use also embraces a descriptor to refer to the general public (Cot-trell, 2007; Handmer & Dovers, 2007, p. 60). For example, participants from Charleville described their community as those individuals who live in the area, and that this area's geographical position on a flood plain results in the population being affected frequently by floods.

Community as a social network is the second definition that is used for this theme, and it is understood as people who are connected by interest and therefore join together (Gruzd, Wellman, & Takhteyev, 2011). These associations can originate from or have a stronger presence at the grassroots level, but also and especially in the case of disasters, organis-ations extend their limits beyond specific geographical areas (Cottrell, 2007), such is the case for some national or international NGOs.

In summary, this theme (node) includes an amalgamation of both types of community: community as locality and community as a social network. This is mainly because intervie-wees used them interchangeably; one interviewee may have referred to the two types of communities at the same time, and because to make a clear delineation between them is a very difficult task (Edwards, 2009). Some examples of the trigger words for coding in this node are respondents' mention of 'churches', 'NGOs', 'the community', 'communities' and 'local communities'.

Interviewees gave a diverse range of arguments for *why* communities are one of the central social actors that need to be involved in the development of resilience. These are presented below. One of the stronger arguments was that communities are an invalu-able and unique source of knowledge (sr2/sn4/sg6/sp13/lg17/sn20/lg21/lg22/sn23). They argued that where people have frequently been exposed to disasters they know what the problems are and they have developed different solutions. Some further stated that for these reasons, this knowledge has to be recognised and community feedback needs to be included in a co-learning process.

... the community will be your watchdogs, the communities will identify what works, what doesn't work, how well it is being coordinated because they're the ones that will be affected in the end. (lg22)

One interviewee, for example, used the metaphor of 'old soldiers', to refer to people who have been engaged in NGOs for a long time. S/he suggested that newly enrolled volunteers or young people coming into the organisation should learn from these 'old soldiers' and that this knowledge should be exploited rather than allowed to disappear (lg21). Another participant specifically stressed that it is the community who recognise its vulnerable individuals, and individuals in need of assistance (lg17).

Another argument that arises here is that there is a diversity of points of view amongst individuals within a community, and that all their opinions and concerns are valid and thus should be considered (sg6/sp13/lg15/sn20). A third argument is that individuals are responsible for preparing and responding to disasters (lg12/sp13) and that their communities are the affected ones (lr1/lg22/sn23).

Those who are affected, or will be affected – potentially affected [should be involved in building resilience]. The people, individual households and the community. (lr1)

Another aspect mentioned by respondents is the need to acknowledge and work with community leaders because they know the community members, they can bring people together and they have the ability to mobilise the community (sr5/sg27). One participant explained that these leaders are often not engaged with the DRM system but there is a need for them to be so, in order to elevate the role of community (sg27).

How: co-management and partnership in NDRP practices

Participants recalled the idea of partnership in different ways. For instance, they mentioned that it is particularly important that community groups associate with local government and NGOs; that different groups need to work in partnership, to develop strong links and to have a joint approach and there is a need for cooperative and partnership models in which different social actors are actively involved and engaged. Others said that there is an imperative to connect agencies as well as different communities; that different stakeholders should engage in partnership to form a cluster of groups. Finally, some said there is a necessity for collaborative work between state and local government with local communities.

The ideas of partnership also permeated the practices within the NDRP. One interviewee (sg14) specifically pointed out that the programme has been deliberately implemented as a partnership between different organisations. This was corroborated through observation of some activities during field work, when we were able to confirm that these organisations participated in different phases of the programme's implementation and the applications assessment process, rather than this decision process being carried out by government personnel alone.

Another source of data triangulation was document review and several focused on the importance of forming partnerships. For instance, in the NDRP guidelines for both rounds, partnership applications were encouraged and in the application forms, applicants were required to clearly explain 'How does the project build on partnerships across sectors?' (NDRP09-10 and NDRP10-11). Finally, there were some good examples of partnership applications. For instance, some government organisations applied in partnership with volunteer organisations, while some volunteer organisations applied in partnership with private and research organisations.

Stakeholder participation: a discussion

A broad set of actors, agencies and organisations for building resilience

A broader scope of agencies and organisations was included in the NDRP than in the previous programmes and this constitutes not only one of the most relevant results, but also is a huge step in the praxis of DRM, as producing change in practices is especially difficult to achieve when involving social or governance issues. The implication of widening the set of agencies involved in DRM is discussed next.

Firstly, broadening the range of actors enables resilience building by means of the inclusion of a diversity of ideas, points of view, knowledge and expertise (Comfort, 2005; O'Neill & Handmer, 2012). This happened in this study and resulted in positive outcomes. For example, some of the new agencies involved in the NDRP have a strong social orientation and contributed to the discussion with renewed ideas and social constructions on how to build resilience. Such ideas were included in applications such as the 'Business Roundtable Extend', 'Emergencies Volunteering Community Workshop Project' and 'Volunteer Connect'. These projects aimed not only to create spaces for discussions with different organisation and the community about how they could contribute to resilience, but also to reinforce the networks between different actors.

Secondly, another positive outcome which follows from the first is that a co-creation of knowledge occurred in the process of the case study and this contributed insights into the definition of the issue and the possible solutions required to change the situation. During the field work, we were able to distinguish clearly how ideas proposed by some participants influenced the framing of other participants, or 'reframing' as it is referred to in the DA literature (Fischer, 2003), and thus became part of their own social constructions. This occurred in the NDRP design phase, but also during the applications evaluation meetings. In these meetings, we observed that sometimes arguments presented by some participants rapidly became part of a collective understanding, and also how ideas included in the applications helped contribute specifically to frame solutions. What has been discussed in this paragraph is especially relevant because, as stated by Collins and Ison (2009) under uncertain conditions, such as climate change and DRM, no single group working in isolation will have clear access to the full understanding of issues and their resolutions. This leads to the third implication.

The participation by an array of agencies acquires special relevance in the presence of escalating uncertainty and complexity (Handmer, 2008; Norris, Tracy, & Galea, 2009). To begin with, it facilitates ready access to a wider pool of alternatives when facing the unknown and a deeper pool of resources from which to draw to test these over time (Paton, 2006). Moreover, if participation is sustained over time, it enables a dynamic definition of problems and solutions in order to follow the situational changes as problems and situations emerge and, in most cases, evolve. On the other hand, it also enables learning from past practices through the process of iterative evaluation and learning (Berkes, 2007; Wildavsky, 1988). The latter was observed to some extent within the framework of the NDRP. One example is that some 'applicants' who developed projects in partnership with several organisations also participated in the Assessment Group Expert Advisor group, and this presented an opportunity for them to contribute their field experience gained by working with different actors. Another example, which occurred in the case study but outside of the framework of the NDRP, is that of personnel from Local Government Association Queensland, a government organisation, who ran workshops with stakeholders from the local level to share experiences of past disasters.

While elevating the role of the community, participants focused their attention on enhancing the role of local agencies and governments. This is, for example, because local agencies are responsible for adjusting DRM to fit specific local contexts. This concurs with Collins and Ison (2009, p. 367), in the sense, in the sense that depending on the context different types of participation are required, and that stakeholders need to be engaged in 'context sensitive designs' that consider the specific opportunities, constraints and conflicts of each reality. The NDRP constitutes an advance in this regard, as it allows funding for a wider type of project, giving space to local agencies to apply with ideas that reflected their local realities. Nevertheless, projects that include grassroots organisations were not common or promoted within the programme. This can be explained because as Twigg et al. (2001) argue, people in positions of power, be it political or institutional, may be unwilling to hand over authority to the grass roots, no matter how well educated the electorate may be in terms of responding to collective decision-making (Stoker, 2006). Within the implementation of the NDRP, advances have been made to improve participation but still have not changed the substance of their approach in terms of community participation, for example. This is an opportunity to develop further that sort of practice.

Something worthy of attention is the low importance interviewees attached to the role of the private and business sector, and this contrasts with the literature that assigns a more relevant role to this group in building resilience, by means of business continuity, economic and financial recovery, and as suppliers of basic goods and services (Paton & Hill, 2006). This gives another possible conduit for improving resilience, as even if the guidelines of the NDRP allow private organisations to apply for funding, and this resulted in a couple of these organisations included in applications, this participation is restricted to partnerships with eligible agencies, because private organisations are not eligible by themselves. It would be useful to explore how to promote a wider participation of these kinds of organisations and to raise awareness of their role in building resilience.

Enhancing the role of community and social capital

The rationale in this section is to conduct a combined discussion of community participation and social capital, using community participation as the theoretical link because it is one of the key dimensions of social capital (Norris et al., 2008).

One of the most relevant findings of the study is the strong emphasis given by respondents to elevating the role of the community when building capacity and an understanding of resilience to disasters. This concurs with what is largely stated in the literature (Aldrich & Meyer, 2015; Haque & Etkin, 2007; Pfefferbaum, Reissman, Pfefferbaum, Klomp, & Gurwitch, 2005). For example, respondents explained that communities constitute an invaluable and unique source of knowledge; this is a contextual knowledge which is based on the experience of former disasters contains their myriad practices triggered by specific physical and social contexts. This is not only coherent with the definition of communities from a locality perspective, but also affirm findings in disaster research (O'Brien et al., 2010; Paton, 2006). The metaphor of the 'old soldiers' as knowledge keepers points to the value in experience, and that even if renewed ideas are necessary, establishing ways in which this social memory can be incorporated and maintained is part of the evolution of an approach aimed at improving community resilience. More than just using participation as a slogan, correct platforms and mechanisms are needed for a meaningful participation in which this knowledge has a place to be expressed, and where communication in reciprocal ways occurs. This participation, in which communication takes place, results in the definition of common meanings (Norris et al., 2008), which in turn leads to real possibilities of developing ideas on how to improve and implement resilience, and enable the communities involved to make sense of this process of community building.

Interestingly, interviewees referred more frequently to communities as holders and sources of knowledge than to any other stakeholder group, assigning this group a central role in sharing knowledge and learning. The latter can be explained by the societal value given to the uniqueness of this knowledge. But there may be other explanations. For example, in the Australian context of 'shared responsibility', there is a neo-liberal expectation that as government cannot be there to save all its citizens, there is a need for citizens to become more active in disaster response and resilience (McLennan & Handmer, 2012). Further, because of the predominance of engineered resilience (Cutter et al., 2008, p. 599), there is a historically low rate of participation of communities in formal DRM. However, even if the relevance of community participation is important in the interviewees' framing, this has not sufficiently permeated into their practice as within the scope of the NDRP, communities (except for large NGOs) played a small role and had few effective conduits for participation.

Concerning community participation, there are two additional reflections in regard to dimensions in which resilience theory could further contribute to improving community resilience. These reflections come from a *systemic* understanding of the concept and from the conceptualisation of *community resilience*. From a *systemic* point of view, communities are inherently part of DRM system. Disaster reinforces social and ecological interdependence and interconnectedness through feedback with all other elements of the system (Nelson, 2011) and, as a consequence, actions taken by communities at the local level affect the overall DRM system. Thus, if communities are not fully considered in DRM systems, a crucial part of the system will be missing in the analysis, and opportunities to explore which other elements provide positive and negative feedback, will be lost. One recommended practice that could help in understanding the role of communities in DRM systems and their feedback with other elements, is an incorporation of these communities within disaster rehearsals; to date, this has not always been done.

From the *community resilience* approach, the role of community leaders is crucial for the development of social capital (Norris et al., 2008; Twigg, 2007). The importance of community leaders was rarely mentioned by respondents, and this could decrease the possibility for the community to have a coherent and substantial role within the DRM governance system. Therefore, validating leaders who have emerged in their communities is important, not only because it enables and reinforces community participation, resulting in a proactive and committed community, but also because it also opens up opportunities for diverse forms of leadership, including collaborative rather than hierarchical forms of leadership and networked roles. This allows leaders to play a role as facilitators, bringing people together within the community, as well as binding the community to other stakeholders. Arguably such decentred leadership responds to decentred governance and demands a greater emphasis on civil society than the interviewees in this study may have been contemplating. In considering that all leaders at all levels are capable of also using power to block management initiatives or to pursue their personal agendas, consequently contributing to inequity and injustice (Jentoft, 2007), we note the importance of checks, balances and transparency in defining resilience and its many forms of engagement.

Social networks and social cohesion is another dimension of social capital subject to considerable attention when building resilience to disasters, both within the case study and reflected in the literature (Aldrich & Meyer, 2015; Norris et al., 2008). The interviewees' framing indicated a construction of a modern society that has diminished social connections, less social cohesion, a poor sense of belonging and limited capacity to assist each other. Respondents expected this social malaise was exacerbated in urban areas. Consequently, there is a need for communities to strengthen communities' social networks and to develop group confidence, as this in turn results in people protecting and supporting each other, and developing a shared capability and collective response. How to develop social capital is mainly a community concern, but nevertheless one that can be hindered or reinforced by the governance structures in which communities are embedded. This is discussed next.

Building resilience through co-management and partnership

From the study findings emerged an imperative to have greater and more meaningful participation in response to DRM With regard to community engagement. We agree with others that the epistemologies of participation need critical attention (Collins & Ison, 2009; Handmer, 2008). This is because participation represents one of the major challenges for governments and governance and because in emergent theories it is important to understand how ideas of participation have been applied to practice (Gero et al., 2011; O'Brien et al., 2010).

Interviewees attributed causality to governments for creating public dependency, encouraging the lack of robust delegation of responsibilities, failing to communicate the role of stakeholders and imposing top-down initiatives. They further argued that all these factors have resulted in creating a silo effect with stakeholders acting in isolation at different administrative levels and organisations, and in communities not assuming a more participatory role within the DRM system. Several implications emerge, and two key ones are discussed below.

Firstly, effective and valid participation begins with the creation of a space for a more balanced definition and negotiation of responsibilities and roles. This is relevant because stakeholders' framing of their responsibilities and roles is based on their constructions and interests (Collins & Ison, 2009) and, therefore, including stakeholders' considerations could result in stakeholders becoming empowered by their roles and consequently being more willing to participate. The participation of stakeholders in the definition of their responsibilities does not ensure an effective participation by itself; it is also crucial to consider what

Collins and Ison (2009, p. 364) noted as a reciprocal relationship between 'responsibility' (as enacted) and 'response-ability' (as enabled). As argued by Fisher (2006), in many cases formal arrangements erode the capacity to be 'response-able'. Thus, if devolution of responsibility is passed to non-government stakeholders, including the community, it has to be accompanied not only by sufficient and adequate support to enable 'response-ability', but also by a revision, and consequent redesign, of institutional arrangements, that also distribute power and resources.

Secondly, what is also problematic is that by externally defining the ways in which social actors can contribute to resilience, their participation and ability to consider their input are limited. In contrast to this external top-down approach, what is needed is to jointly define and negotiate the nature of change and the improvements that are necessary. This is because this definition is influenced not only by the physical context, but also by how communities and other stakeholders are organised, the existing relationships between individuals, stakeholder groups, institutions and the state, and their values and constructions about resilience ideas. The relevance of negotiation in cooperative decision-making processes acquires special relevance when talking about resilience, as some disaster-prone systems are considered resilient but this resilience could maintain the systems in a pathological state (Nelson, 2011). The history and context in which a community frequently experiences disasters might push it to build a system frozen in a 'lock-in trap'; and to establish an equilibrium where the stable state is always the same, despite potentially representing a state that has achieved high vulnerability and risk levels (Aldunce, Beilin, Howden, & Handmer, 2014; Wilhite, 1993). We argue that this kind of resilience is negative and agree with Wildavsky (1988) and O'Brien et al. (2010) that resilience ideas must demonstrate their value by helping the system to recognise that to bounce back is still to return to a changed system; and the object becomes to achieve a 'better' position than previously acknowledged or possible. When building resilience, the questions of what a desirable system should look like and how to develop resilience need to be co-produced, and the implementation pathways have to be established; these will depend on whether the platforms for participation and negotiation – by locality and with networks – are in place.

Study findings show that the NDRP has advanced towards participation by means of partnership. The NDRP was successful in working together with various organisations in mutually beneficial relationships. One significant implication of this is that it opened up channels and opportunities for the flow of knowledge and reframing. Nevertheless, the idea that partnership represents an opportunity for social learning, and can be understood as learning and sharing knowledge, is an emergent one in the case study area and, in consequence, is a window of opportunity that can be further explored. Experience indicates that crucial consideration for those actors who are involved in the process include when and how learning takes place, and the power relations among actors (Armitage et al., 2008). Berkes (2007) argue that adaptive co-management combines learning and partnership characteristics and is typically carried out by actors sharing management, power and responsibility. Social learning occurs through participatory processes and benefits from partnership models because they provide different stakeholders with a legitimate platform that allows for their credibility. Moreover, by ensuring different stakeholders engage in learning processes, and by promoting collaborative structures, there is an opportunity for integrating multiple levels of governance and transforming this learning process into actions to achieve a better state. Additionally, the funding rounds of the NDRP, as they take place annually, creates the opportunity to move away from the idea that learning occurs only after mass disasters, and towards learning in 'times of peace' in between big disasters.

Conclusions

Important lessons about how to tackle stakeholder participation for building resilience emerged as a result of witnessing the NDRP implementation process and interviewing participants in the programme. For example, widening the scope of agencies and stakeholder groups, and bringing in new ones created options among municipalities and with collaborators to reconceptualise how to undertake resilience building projects. This was a step change, generating flow on possibilities for change in difficult arenas such as are common in disaster and across social or governance issues.

Building disaster resilience requires governance structures that promote collective and shared efforts, as well as cooperation between practitioners, for instance, through co-management and partnership. For this to be implemented effectively, platforms for dialogue and interaction need to be created and properly funded and facilitated in order to build trust and credibility, to facilitate a more reciprocal relationship among actors, to manage the power structures and to build capacity among disadvantaged groups. This is not, however, a quick process, and it faces challenges; for example, participation in DRM must be part of a larger and coherent approach to restructuring, not just in DRM, but in community engagement and development per se. Interestingly, even if elevating the role of communities was indicated by participants as in need of urgent attention, in the case study the idea of community participation remained mainly in the discourse and did not permeate strongly to the NDRP practice. Thus, what can be suggested with regard to how to elevate community engagement, recognition of the relevance of and need for support for leaders that emerge from within the community is important. They can contribute to the strengthening of social connections and networks, and the inclusion of communities' perspectives. All this gives the opportunity to flourish in a shared capacity, rather than communities' views being interpreted as conflictual or as a threat to government efforts. Moreover, the inclusion of local agencies and governments is a key aspect because they play an essential role, in place, in building resilience, reinforced by their closeness to their communities. They are fundamental in supporting and enabling their participation. Nevertheless, it is crucial to acknowledge that some leaders or local authorities can disrupt management initiatives because they may have little understanding of the needs of the most vulnerable or they may pursue their own agendas, thus creating inequity and injustice. On the other hand, deciding which local leaders to include is one of the most difficult challenges to effective participation in multi-stakeholder processes.

What is salutary in moving towards a resilience approach by means of widening the set of organisations is that it enables the inclusion of a broader range of actors' needs and a diversity of points of view and expertise. Again, it is germane to take into account communities, as they play a key role in constituting an invaluable and unique source of knowledge which is required for enhancing the pool of information, creating meaning, sharing knowledge and supporting learning. If the impetus is a focus on learning, it is pertinent to move from experiential learning (closer to the idea of individual learning) towards social learning (closer to an understanding of group learning and sharing experiences). This is because building resilience is underpinned by the integration of a diversity of frames, which are more likely to be found outside of our immediate networks, and by the inclusion of more diverse social actors. What is needed is an effort to break down the barriers that separate different agents and groups. On the other hand, learning among peers or learning to express ideas in a context of more equal opportunities facilitates a deeper and more sustainable learning compared with a learning that is received from an instructor. Here the challenge lies in actors that are willing to share power and be receptive to new ideas, and listening to people instead of lecturing or attempting to control them.

The power of fostering this kind of social learning is that it provides possibilities for changing practices and for improving institutions, ultimately resulting in decreasing risk. We envisaged this idea as a resilience spiral (see Aldunce et al., 2014). Applying these ideas to the case study, for example, can constitute an opportunity. A low level of resilience framing in which disasters are opportunities for development was found, and even less frames are considered potential for innovation even though, within the jurisdiction of the NDRP, certain levels of social learning were occurring. The programme has created some space for innovation, diminishing a more rigid interpretation of what is possible and more flexible than former programmes; but again more efforts are required to create platforms for learning and innovation, and this requires the active involvement of concerned actors and of practitioners.

Note

1. To protect the anonymity of the interviewee to each one has been assigned by a code.

Acknowledgments

We thank the Department of Community Safety, Queensland and the case study respondents, for their generosity and willingness to participate. This publication is based on Ph.D. thesis study 'Framing resilience: practitioners' views of its meaning and usefulness in disaster risk management practice'.

Disclosure statement

No potential conflict of interest was reported by the authors.

Funding

This work was supported by the 'Becas Bicentenario' from the Government of Chile, the University of Chile, the University of Melbourne and the Commonwealth Scientific and Industrial Research Organization (CSIRO), Australia, for funding the present research. This publication also received the support of and is a contribution to the Center of Resilience and Climate Research (CR)2, FONDAP #1511009.

References

Aldrich, D. P., & Meyer, M. A. (2015). Social capital and community resilience. *American Behavioral Scientist, 59*(2), 254–269. doi:10.1177/0002764214550299
Aldunce, P., Beilin, R., Handmer, J., & Howden, M. (2015). Resilience for disaster risk management in a changing climate: Practitioners' frames and practices. *Global Environmental Change, 30*, 1–11.

Aldunce, P., Beilin, R., Howden, M., & Handmer, J. (2014). Framing disaster resilience: The implications of the diverse conceptualisations of 'bouncing back'. *Disaster Prevention and Management, 23*(3), 252–270.

Aldunce, P., Handmer, J., Beilin, R., & Howden, M. (in press). Is climate change framed as 'business as usual' or as a challenging issue? The practitioners' dilemma. *Environment and Planning C: Government and Policy.*

Armitage, D., Marschke, M., & Plummer, R. (2008). Adaptive co-management and the paradox of learning. *Global Environmental Change, 18*(1), 86–98. doi:10.1016/j.gloenvcha.2007.07.002

Arnstein, S. (1966). A ladder of citizen participation. *Journal of the American Institute of Planners, 35*(4), 216–224.

Bahadur, A., Ibrahim, M., & Tanner, T. (2010). *The resilience renaissance? Unpacking of resilience for tackling climate change and disasters* (Strengthening Climate Resilience Discussion Paper 1). Brighton: Institute of Development Studies.

Berkes, F. (2007). Understanding uncertainty and reducing vulnerability: Lessons from resilience thinking. *Natural Hazards, 41*(2), 283–295.

Chambers, R., & Blackburn, J. (1996). *The power of participation: PRA and policy* (Briefing Paper IDS). Brighton, UK.

Collins, K., & Ison, R. (2009). Jumping off Arnstein's ladder: Social learning as a new policy paradigm for climate change adaptation. *Environmental Policy and Governance, 19*(6), 358–373. doi:10.1002/eet.523

Comfort, L. K. (2005). Risk, security, and disaster management. *Annual Review of Political Science, 8*, 335–356.

Cottrell, A. (2007). What is this thing called community?: An example in far north Queensland. In D. King, & A. Cottrell (Eds.), *Communities living with hazards* (pp. 6–17). Townsville: Centre for Disaster Studies, James Cook University.

Cutter, S. L., Barnes, L., Berry, M., Burton, C., Evans, E., Tate, E., & Webb, J. (2008). A place-based model for understanding community resilience to natural disasters. *Global Environmental Change – Human and Policy Dimensions, 18*(4), 598–606. doi:10.1016/j.gloenvcha.2008.07.013

Djalante, R., & Thomalla, F. (2011). Community resilience to natural hazards and climate change: A review of definitions and operational frameworks. *Asian Journal of Environment and Disaster Management (AJEDM), 3*(3), 339–355.

Edwards, C. (2009). *Resilient nation*. London: Demos.

Fischer, F. (2003). *Reframing public policy: Discursive politics and deliberative practices*. Oxford: Oxford University Press.

Fisher, F. (2006). *Response ability. Environment, health and everyday transcendence*. Melbourne: Vista.

Folke, C. (2006). Resilience: The emergence of a perspective for social-ecological systems analyses. *Global Environmental Change, 16*(3), 253–267. doi:10.1016/j.gloenvcha.2006.04.002

Gaillard, J. C. (2010). Vulnerability, capacity and resilience: Perspectives for climate and development policy. *Journal of International Development, 22*(2), 218–232.

Gero, A., Méheux, K., & Dominey-Howes, D. (2011). Integrating community based disaster risk reduction and climate change adaptation: Examples from the Pacific. *Natural Hazards and Earth System Sciences, 11*, 101–113.

Gruzd, A., Wellman, B., & Takhteyev, Y. (2011). Imagining twitter as an imagined community. *American Behavioral Scientist, 55*(10), 1294–1318. doi:10.1177/0002764211409378

Hajer, M. (1995). *The politics of environmental discourse: Ecological modernization and the policy process*. New York, NY: Oxford University Press.

Hajer, M. (2000). *The politics of environmental discourse: Ecological modernization and the policy process*. Oxford: Clarendon Press.

Handmer, J. (2008). Emergency management thrives on uncertainty. In G. Bammer, & M. Smithson (Eds.), *Uncertainty and risk: Multidisciplinary perspectives* (pp. 231–243). London: Earthscan.

Handmer, J., & Dovers, R. (2007). *Handbook of disaster and emergency policies and institutions*. London: Earthscan.

Haque, C. E., & Etkin, D. (2007). People and community as constituent parts of hazards: The significance of societal dimensions in hazards analysis. *Natural Hazards, 41*(2), 271–282. doi:10.1007/s11069-006-9035-8

IPCC Intergovernmental Panel on Climate Change. (2012). *Managing the risks of extreme events and disasters to advance climate change adaptation* (A special report of Working Groups I and II of the Intergovernmental Panel on Climate Change). Cambridge: Cambridge University Press.

Jentoft, S. (2007). In the power of power: The understated aspect of fisheries and coastal management. *Human Organization, 66*(4), 426–437.

Kocho-Schellenberg, J.-E., & Berkes, F. (2015). Tracking the development of co-management: Using network analysis in a case from the Canadian Arctic. *Polar Record, 51*(4), 422–431. doi:10.1017/s0032247414000436

McLennan, B., & Handmer, J. (2012, August). *Visions of sharing responsibility for disaster resilience: sharing control.* Paper presented at the Proceedings of Bushfire CRC & AFAC 2012 Conference Research Forum, Perth.

Nelson, D. (2011). Adaptation and resilience: Responding to a changing climate. *Wiley Interdisciplinary Reviews Climate Change, 2*(1), 113–120.

Nisbet, R. A. (1970). *The sociological tradition.* London: Heinemann.

Norris, F. H., Stevens, S. P., Pfefferbaum, B., Wyche, K. F., & Pfefferbaum, R. L. (2008). Community resilience as a metaphor, theory, set of capacities, and strategy for disaster readiness. *American Journal of Community Psychology, 41*(1–2), 127–150. doi:10.1007/s10464-007-9156-6

Norris, F. H., Tracy, M., & Galea, S. (2009). Looking for resilience: Understanding the longitudinal trajectories of responses to stress. *Social Science Medicine, 68*(12), 2190–2198. doi:10.1016/j.socscimed.2009.03.043

O'Brien, G., O'Keefe, P., Gadema, Z., & Swords, J. (2010). Approaching disaster management through social learning. *Disaster Prevention and Management, 19*(4), 498–508.

O'Neill, S. J., & Handmer, J. (2012). Responding to bushfire risk: The need for transformative adaptation. *Environmental Research Letters, 7*(1), 014018.

Paton, D. (2006). Disaster resilience: integrating individual, community, institutional and environmental perspectives. In D. Paton, & D. Johnston (Eds.), *Disaster resilience: An integrated approach* (pp. 249–264). Springfield, IL: Charles C. Thomas.

Paton, D., & Hill, R. (2006). Managing company risk and resilience through business continuity management. In D. Paton, & D. Johnston (Eds.), *Disaster resilience: An integrated approach* (pp. 305–316). Springfield, IL: Charles C. Thomas.

Pfefferbaum, B., Reissman, D., Pfefferbaum, R., Klomp, R., & Gurwitch, R. (2005). Building resilience to mass trauma events. In L. Doll, S. Bonzo, J. Mercy, & D. Sleet (Eds.), *Handbook on injury and violence prevention interventions* (pp. 347–358). New York, NY: Kluwer Academic.

Phillips, R., Chaplin, S., Fairbrother, P., Mees, B., Toh, K., & Tyler, M. (2011). Defining community: Debates and implications for bushfire policy. *Fire Note, 88*, 1–4.

Stoker, G. (2006). *Why politics matters. Making democracy work.* Basingstoke: Palgrave Macmillan.

Studdert, D. (2006). *Conceptualising community: Beyond the state and individual.* Basingstoke: Palgrave Macmillan.

Tompkins, E., & Hurlston, L. (2010). *Public-private partnerships for storm risk management in the Cayman Islands.* London. Retrieved from http://www.see.leeds.ac.uk/fileadmin/Documents/research/sri/workingpapers/SRIPs-21_01.pdf

Twigg, J. (2007). *Characteristics of a disaster-resilience community: A guidance note.* Benfield: DFID Disaster Risk Reduction Interagency Coordination Group.

Twigg, J., Bhatt, M., Eyre, A., Jones, R., Luna, E., Murwira, K., ... Wisner, B. (2001). *Guidance notes on participation and accountability*: Benfield Greig Hazard Research Center.

UN/ISDR United Nations International Strategy for Disaster Reduction. (2011). *Global assessment report on disaster risk reduction: revealing risk, redefining development* (U. Nations, Ed.). Geneva.

US/IOTWS Indian Ocean Tsunami Warning System. (2008). *Coastal community resilience (CCR).* Indonesia National Training Report.

Wildavsky, A. (1988). *Searching for safety.* New Brunswick, NJ: Transaction Books.

Wildavsky, A. (1991). *Searching for security.* New Brunswick, NJ: Transaction Books.

Wilhite, D. A. (1993). Chapter 6, Planning for drought: A methodology. In D. A. Wilhite (Ed.), *Drought assessment, management, and planning: Theory and case studies* (pp. 87–108). Boston, MA: Kluwer Academic.

How does social learning facilitate urban disaster resilience? A systematic review

Qingxia Zhang, Junyan Hu, Xuping Song, Zhihong Li, Kehu Yang and Yongzhong Sha

ABSTRACT

Social learning has been considered an emerging policy option for urban disaster risk management. Urban disaster resilience is claimed to be a desirable outcome of many social learning programmes, but little systematic evidence illustrates how urban disaster resilience has been facilitated by social learning programmes. This paper presents a theoretic approach by a systematic review using the thematic content analysis method and finds that all of the social, technological and natural hazards can trigger social learning and the learning type varies according to 'who learns'. It is showed that the mechanisms by which social learning can facilitate urban disaster resilience include: (1) governance capacity; (2) self-organisation; (3) cognitive change; (4) moral or civic responsibility; and (5) open communication and deliberation. Moreover, the leading roles in the risk management practice played by social learning to improve urban disaster resilience are achieved throughout three strategies: (1) meeting the public demands of risk perception; (2) getting more stakeholders involved into a collaborative process; and (3) joint problem-solving. The findings might provide guidelines for the implementation of social learning programme in the future.

1. Introduction

The World Bank and United Nations reports have highlighted that urbanisation will inevitably increase the risk of disasters exposure for cities (Worldbank, 2010). Cities have indeed suffered more from a variety of threats such as social disasters, technological disasters and natural hazards (Alexander, 2002). For example, the 9.11 attacks killed 2996 people, with over 6000 others injured, and caused at least $10 billion in infrastructure and property damage (IAGS, 2004). The 2015 warehouse explosion of hazardous chemicals at Tianjin port of China led to 165 fatalities and almost 1 billion dollars losses (PRC Cabinet, 2016).

The 2015 Hurricane Katrina with caused at least 1836 deaths, also cost a total of 125 billion dollars and affected major U.S. metropolitan area (Zakour, Mock, & Kadetz, 2017). In short, a hazard event occurring in an urban area will lead to major casualties and economic losses for its specific physical and non-physical features (Wamsler, 2014).

In recent years, practice (UNISDR, 2015; ISDR, 2005) and academia (Turner, 1978; May & Williams, 1986; Comfort, 1988; McMillen, Campbell, Svendsen, & Reynolds, 2016; Saja, Teo, Goonetilleke, & Ziyath, 2018) have evolved to focus on the social factors in disaster management, and 'urban resilience' has become a popular concept for disaster risk management (Sanchez, Heijden, & Osmond, 2018; Miller et al., 2010; Godschalk, 2003; Olazabal, Chelleri, & Waters, 2012). Originated from the field of ecology, urban resilience within a disaster context refers to the ability of the whole urban organism to respond and recover from disaster impacts through adaptive processes that facilitate the ability of the social system to re-organise, change, and learn in response to a disaster (Timmermann, 1981; Pelling, 2003; Cutter et al., 2008). In this sense, resilience means more than just responding to and bouncing back after an extreme event (Darkow, 2018). It also incorporates the capacity to change and adapt to environment, which, in turn, requires the essential abilities to cooperate, learn, and apply lessons to continued resilience under future conditions (Cutter et al., 2008). It is clear that management for resilience requires ongoing social learning to increase adaptive capabilities and social learning has become an important policy option for disaster resilience building (Kates, 1962).

Whereas social learning constitutes a central aspect of research and practice on risk management and disaster resilience, no consensus on the definition of social learning has been reached (Reed et al., 2010) (Siebenhüner, Rodela, & Ecker, 2016). Developed by Albert Bandura, social learning is defined as individual learning that takes place in a social context and results in the promotion of individual, team and organisational knowledge acquisition, sharing, and performance improvement (Bandura, 1977). Recently, a more normative concept of social learning has emerged in the context of adaptive management approach and results from these studies show that wider social learning may be picked up by the participation action approach (PAR) (Pahl-Wostl, 2006; Pahl-Wostl et al., 2008; Cundill et al., 2014). PAR project, which emphasises on 'learning by doing' and 'feedback learning' actually promotes changes in thinking and learning (Ballard & Belsky, 2013). However, Reed et al. (2010) pointed out that frequently confusion between the concept itself and its potential outcome makes it difficult to identify and measure. To address this problem, we adopted a definition combining both Bandura's theory and adaptive management approach and social learning was considered to be a process, in which individual and collective frame alignment situated in communication action. From the definition, social learning can occur through observation and interaction with others as Bandura argues, it is also like shifts in thinking that occur through collaborative action focused on problem resolving (Smith, DuBois, & Krasny, 2016, p. 442).

In general, social learning is both an individual and a collective process and its processes have some cyclical processes or feedback loops (Argyris & Schon, 1974; Flood & Romm, 1996). The process dynamics of social learning are explained by the learning loop theory (Argyris & Schon, 1974; Flood & Romm, 1996; Ernst, 2019). According to the theory, social learning programmes in the context of risk management, start with single-loop learning, that is learning about event itself, then to double-loop learning, which learns to manage disasters generally and finally triple-loop learning, which means a deeper learning about socio-technical frames or cultures (Bateson, 1972; Pahl-Wostl, Becker, Knieper, & Sendzimir, 2013). All three types of social learning imply something else that accumulating data and instead refer to what the relationship between the experience and subsequent action is (Broto, Glendinning, Dewberry, Walsh, & Powell, 2014, p. 189). Since the cyclical processes or feedback loops are not the content of this review, therefore they would not be considered in this paper. While many existing studies on urban resilience focused on a particular type of social learning such as collaborative learning, organisational learning, and collective learning (Blackmore, 2007), the types of social learning based on the criteria of 'who learns' can also generate new insights and which can be seen in Section 3.3.

Social learning for urban-level resilience involves a variety of social processes and takes an adaptive management approach (Klinke & Renn, 2010; Rist, Felton, Samuelsson, Sandström, & Rosvall, 2013; Cook, Casagrande, Hope, Groffman, & Collins, 2004; Pahl-Wostl et al., 2013). While resilience is thought both as a set of capacities and as an adaptive process, therefore it is a dynamic phenomenon and it is changing with time and different sets of actors. The process of adaptive resilience, which plays out under varying circumstances, entails some form of social learning (Ross, 2013). To further examine the role of social learning, the adaptive management approach involves flexible institutional and organisational arrangements that encourage reflection and innovative responses (Armitage, Marschke, & Plummer, 2008), knowledge flows and cognitive change (Boyd, Ensor, Broto, & Juhola, 2014; Toubin, Laganier, Diab, & Serre, 2015; Smith et al., 2016), networks of mutual exchange (Orleans Reed et al., 2013), civil participation of all social components (Folke, 2006; Lorenz, 2013; Smith et al., 2016; Schauppenlehner-Kloyber & Penker, 2016). Collectively, these studies provide some reasonable evidence of an association between social learning and disaster resilience.

While a considerable amount of literature has been published on urban disaster resilience facilitated by the means of social learning, however, researches of the mechanisms by which urban resilience are achieved in a systematic way are very rare. Given exclusively focus on the original studies with a clear objective of urban resilience improvement, this paper tries to provide an answer to the question that 'How does social learning facilitate urban disaster resilience?' by means of a systematic review. This theory-based qualitative systematic review moves beyond the question of 'what works' to 'how and why' and provides a conjectured explanation of the causal chain, which may benefit policymakers to devote to enhancing urban disaster resilience.

2. Method

This study is a kind of qualitative systematic review (Higgins & Green, 2011) and mainly focuses on qualitative evidence of the original studies. Scholars have developed various qualitative synthesis methods, including meta-ethnography, critical interpretive synthesis, and thematic synthesis (Thomas, 2012). We utilised thematic synthesis approach in accordance with the question to be answered. The systematic review was conducted according to the Cochrane qualitative and implementation methods group guidance series (Harris et al., 2018; Noyes et al., 2018; Harden et al., 2018), and the identification and selection process was guided by the Prisma statement (Moher et al., 2015; Tian, Zhang, Ge, Yang, & Song, 2017; Ge et al., 2018). The Prisma flow diagram is shown in Figure 1.

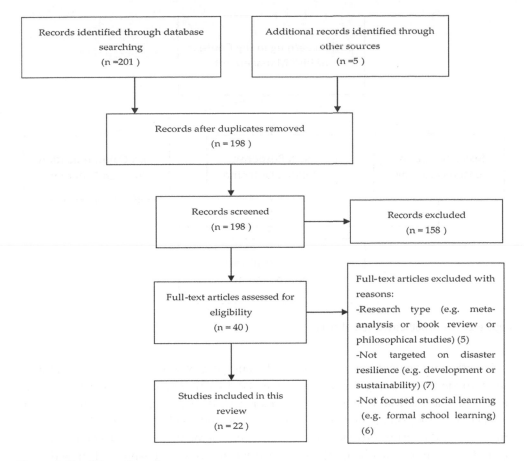

Figure 1. The Prisma flow diagram.

2.1. The hypothesised causal chain

To explore the mechanisms through which social learning programme may facilitate urban disaster resilience, this review combines scholarship on activity theory, social learning, and risk management. Social learning is a kind of activity as Vygotsky (1978) pointed out that human beings deeply understand the things around them and acquire knowledge through their meaningful actions, such as collaborative dialogue, interaction and other social activities. The framework of activity theory developed by Leont'ev (1981) consists of six elements, and they are subject, object, tools, community, rules and divisions of labour (Chung, Hwang, & Lai, 2019). In the context of risk management, the objects of social learning programme may be the direct effects, which may mediate the relationships between social learning and urban disaster resilience. Combined with the three categories of learning outcomes by Kraiger, Ford, and Salas (1993): cognitive, skill-based, affective and risk management process theory by Staveren (2008) and Fone and Young (2005), we would illustrate mechanisms from three aspects: risk perception-based mechanism; risk response related mechanism and risk communication related mechanism. The hypothesised causal chain between social learning and urban disaster resilience is presented in Figure 2.

Figure 2. The hypothesised causal chain.

For risk management, the meaning of different types of mechanisms can be summarised through the following example questions: (1) risk perception-based mechanism: how do changes in the risk awareness through knowledge acquiring and sharing affect the behaviour of risk management; (2) risk response related mechanisms: which behavioural mechanisms underlines processes of resilience improvement; and (3) risk communication related mechanisms: which mechanisms explain the indirect changes of individual or macro-level changes like trust or respect and so on.

2.2. Search strategy

The search of the literature concentrated on electronic databases with a broad-based search term strategy to identify related researches. Keywords used in the primary search include: (Learn* OR stud*) AND (resilien*) AND (urban OR cit* OR municipal). The search was run in Social Sciences Citation Index (SSCI) (2000–2018.06); Arts & Humanities Citation Index (A&HCI) (1990–2018.06); Conference Proceedings Citation Index- Social Science & Humanities (CPCI-SSH) – (2000–2018.06); SAGE Journals Online (1965–2018.06); and Wiley InterScience (1997–2018.06)

2.3. Inclusion and exclusion criteria

Original studies that investigated the relationship between social learning and urban disaster resilience were included in this review. The exclusion criteria were: (1) studies with a clear objective of urban resilience improvement are included, while the ones with barely improved outcome were excluded; (2) studies of disaster resilience focus on the whole urban organism are included, while the ones of national or international resilience were excluded; (3) the participants of social learning including governments, experts, the media and the public were included, while the ones with a discussion for formal school learning on urban resilience were excluded; (4) empirical studies that focused on data from cases, interviews, observations, historical materials or numerical measurements to explore urban disaster resilience throughout social learning were included, while philosophical studies were excluded; and (5) studies presented in other language other than English or that were not primary studies (meta-analyses or book reviews) were excluded. Two of the authors independently screened the abstracts and titles, and discussed disagreements with the leading author to achieve consensus. The authors also screened all the full text to ensure that no important documents were missed.

2.4. Data extraction and synthesis

Included studies were analysed qualitatively using the method of content analysis along the causal chain to pool the evidence. The hypothesised causal chain of social learning programme was used for analysis. The outcomes and discussion content in the literature were thus extracted and grouped according to the hypothesis presented in Figure 2. The processes of analysis were: (1) data extraction according to the initial questions ('to whom', 'by whom' and 'how'). We categorised the types of urban hazard and the types of social learning as background information and listed changes following social learning of each study as mechanisms in Table 4. Sometimes authors explicitly refer to 'mechanisms', more often they do not, though. By reading between the lines, one is able to detect statements alluding to mechanisms. (2) themes classification and generation according to hypothesised causal chain and detailed mechanisms were detected using the method of theme synthesis. It should be added that the lines between different mechanisms are not clear-up all the time and our research team discussed it and decision was made at last by the leading author. (3) Some moderator variables based on intervention and learner characteristics were identified through clarifying the review of qualitative literature.

3. Results

3.1. Study descriptions

Our search identified 206 articles, among which 166 were excluded after duplication and title-abstract review. Among the remained 40 potentially studies, 18 were excluded because of the research type (5), or because they did not target on disaster resilience (7), or because they did not focus on social learning (6), and 22 (Albright & Crow, 2015; Ashley et al., 2012; Boyd et al., 2014; Boyd & Osbahr, 2010; Brody, Zahran, Highfield, Bernhardt, & Vedlitz, 2009; Dawes, Cresswell, & Cahan, 2004; Dieleman, 2013; Evers et al., 2012; Frantzeskaki & Kabisch, 2016; Goldstein, Wessells, Lejano, & Butler, 2013; Hagemeier-Klose, Beichler, Davidse, & Deppisch, 2014; Hooli, 2015; Johannessen & Wamsler, 2017; Kroepsch et al., 2018; Lassa & Nugraha, 2014; McEwen, Holmes, Quinn, & Cobbing, 2018; Orleans Reed et al., 2013; Slack & McEwen, 2013; Smith et al., 2016; Tidball, Krasny, Svendsen, Campbell, & Helphand, 2010; Toubin et al., 2015; Witting, 2017) studies were included in our review (Figure 1). The quality of the selected studies was ensured by inclusion of

Table 4. Identification of salient mechanisms and outcomes of the included studies.

Study	Changes following social learning	Indicators of urban disaster resilience	Others made the interventions better
Smith et al., 2016	Individual and collective cognitive change, consideration of others, exchange information and knowledge, deliberation and reflection, working towards collective goals, inclusiveness in defining goals, social-ecological systems thinking	Form the basis for possible future action, improve the sense of well-being, negative emotions disappear	Class, race, age, social identity
Dawes et al., 2004	Public involvement, open communication	Build public trust and strong ties in community life	Usable data, versatile technologies
Reed et al., 2013	Deliberation, formation of new relationship and formal and informal networks, building capacity for analysis and self-representation	Transformative change is equitable and socially just	Access to knowledge and information
Dieleman, 2013	Citizen participation, strengthening of communal networks, local self-reliance	Mitigation and adaption strategies	Small-scale technology
Lassa & Nugraha, 2014	Stakeholder participation, building champions and encouraging action	City involved in planning and city budget	Incentives to City Team
Witting, 2017	Rules that facilitate communication and reciprocity, collective learning and collective action, institutional change in policy-making	Trust, reciprocity	Fitness between rules and decision process
Toubin et al., 2015	Raise manager awareness, construct shared vision, effective collaboration	Protection, adaptation or recovery	-
Tidball et al., 2010	Green practice (collective action), members' interaction, transmit memories of greening as a source of healing	Individual, social and environmental well-being	Organised by NGOs or government
Slack & McEwen, 2013	Awareness creation, community collaboration, community professionals' participation	Human, social and economic capital	-
McEwen et al., 2018	Co-working, participation process, management emotion, awareness, effective citizenship or stewardship, knowledge sharing, deliberation and care for others	Community capital	Stronger groups in proximal area
Kroepsch et al., 2018	Remember disasters and apply memories to make changes, discuss policy problems, race to rebuild	Individual and institutional changes	Pathway for learning
Johannessen & Wamsler, 2017	Risk perception and action capacity, recognise a development pathway, supporting reorganisation	Transition, governance arrangement	-
Hooli, 2015	Cope with impacts, tuning the seasonal weather calendars to local circumstances	Prediction, prevention, preparation, coping, social and political conditions	-
Frantzeskaki & Kabisch, 2016	Build governance capacity for adaptation, engagement in scenario work, dialogues and discussion between scientists	Relationships and trust in both cities	Create cities ties
Hagemeier-Klose et al., 2014	Improve capacity for self-organisation, informal networks and informal exchange and operation	New governance and adaptive capacity building	Science and practice corporation
Evers et al., 2012	Increase knowledge, personal responsibility, risk awareness, position regarding, participatory governance	Transparency	Data and time
Boyd et al., 2014	Understand government support, cooperative action, balance between formal and informal institutions	Creating and delivering visions of future city, state-based institutions Adaptation of local policies	

(Continued)

Table 4. Continued.

Study	Changes following social learning	Indicators of urban disaster resilience	Others made the interventions better
Albright & Crow, 2015	Stakeholders and the public engagement, changes of beliefs, attitudes, behaviours and goals, in-depth deliberation		Multi-sector initiatives
Ashley et al., 2012	Stakeholder engagement, mutual development of ideas, act upon potential "quick wins"	Changes to policy, practice and culture, change of top-down governance	-
Goldstein et al., 2013	Bridge knowledge practices, shared understanding, diversity in perspectives for going forward, self-organisation	Coherent logic	-
Boyd & Osbahr, 2010	Sharing information between international organisations, network development for information sharing, changes in leadership	Framing of organisational objectives, culture of sharing, foster innovation	-
Brody et al., 2009	Public information, maps and regulation, damage reduction and flood preparedness	Directly benefit from mitigation efforts	Incentives to the stakeholder

Note: (-): unclear or unable to ascertain.

only the published ones. Table 1 summarises the characteristics of the included studies; Among the 22 studies, only 2 discussed human-made disasters and 24 concerned with natural and technological hazards (with 4 overlapped). Most of the studies were conducted in the United States and Europe, and only three studies were transnational studies.

3.2. Type of urban hazard involved in the analysis

We classified urban hazard into three types: natural hazards, technological hazards, and social (or man-made) hazards using types in the disaster literature (Alexander, 2002). As reported in Table 2, the types of hazards involved in these studies are mainly technical hazards and natural hazards, and only 2 study is about social hazards.

3.3. Type of social learning

Social learning type varies based on the criteria of 'who learns', and disaster reduction activities mainly include two major actors: governmental and non-governmental actors such as private and civil society participants (Bakema, Parra, & McCann, 2019). Accordingly, this review pointed out that social learning can be divided into the types of management knowledge learning, common-sense knowledge learning, expert knowledge learning, media (interested bystander) knowledge learning and exchanged knowledge learning among stakeholders. Table 3 shows that the learning of common-sense knowledge and management knowledge are explored by more than half of the included studies, while media learning and expert learning only constitute a very small percentage (4.55% each). There are also eight studies about exchanged knowledge learning.

In addition, it can be seen from the included studies that the majority of papers focused on social learning interventions involving citizens in many kinds of government programmes (Dawes et al., 2004), education programmes (Smith et al., 2016), board meetings

Table 1. Characteristics of the included studies.

Study	Study objective	Study sites	Disaster type	Data collection	Methodology	Analysis
Smith et al., 2016	Three youth restoration programmes	New York& Boulder	Floods	Interview, observations, participation	Case study	Edge list analysis
Dawes, Cresswell, & Cahan, 2004	New York City's Emergency Operations Center	New York	World Trade Center attack	Interview, observations, participation	Phenomenology	Content analysis
Reed et al., 2013	Asian Cities Climate Change Resilience Network	10 cities in Asian	Climate change	Observations	Case study	Thematic analysis
Dieleman, 2013	Mexico City repair project	Mexico	Climate change	Participation	Case study	Inductive
Lassa & Nugraha, 2014	Asian Cities Climate Change Resilience Network	Bandar Lampung	Solid waste, flood	Interview, observations, second-hand info	Case study	Thematic analysis
Witting, 2017	The process of planning infrastructure investments	Denver Metropolitan	Urban Services	Interview	Case study	Content analysis
Toubin et al., 2015	Paris urban services	Paris	Urban Services	Interview	Case study	-
Tidball et al., 2010	Memories of trees and other living things	New York, Katrina	9.11/Hurricane	Participation	Case study	-
Slack & McEwen, 2013	An interprofessional education programme	Arizona	All kind of	Focus group	Case study	Thematic analysis
McEwen et al., 2018	Community Action Groups	England	Flood	Interview, observations, focus group	Case comparison	Thematic analysis
Kroepsch et al., 2018	Community Learning and Adaptation Discourse	Colorado	Wildfire	Investigate, second-hand info	Case study	Content analysis
Johannessen & Wamsler, 2017	Water experts	Four cities	WASH	Interview	Case comparison	Thematic analysis
Hooli, 2015	Resilience of the poorest	Northern Namibia	Flood	Interview, focus group	Case study	Content analysis
Frantzeskaki & Kabisch, 2016	Urban environmental governance	Berlin, Rotterdam	Environmental governance	Participation	Case comparison	Content analysis
Hagemeier-Klose et al., 2014	Inter- and Transdisciplinary Cooperation	Rostock	Climate change	Workshops, focus group, participatory, interview	Case study	Content analysis
Evers et al., 2012	Collaborative modelling of stakeholders	Restork, Stockholm	Flood	Interview, workshop	Case study	Content analysis

(Continued)

Table 1. Continued.

Study	Study objective	Study sites	Disaster type	Data collection	Methodology	Analysis
Boyd et al., 2014	Environmentalities of urban climate governance	Maputo, Mozambique	Climate change	Interview, participation, second-hand info	Case study	Content analysis
Albright & Crow, 2015	Policy learning of stakeholders	Colorado	Flood	Interview	Case study	Content analysis
Ashley et al., 2012	Learning and Action Alliances	England	Flood	-	Case study	-
Goldstein et al., 2013	Collaborative storytelling	Santa Ana, Los Angeles	Flood/Fire/River restoration	Participation	Case study	-
Boyd & Osbahr, 2010	Internationally networked organisations	International	Climate change	Interview, participation, second-hand info	Case study	Content analysis
Brody et al., 2009	FEMA CRS programme	Florida	Flood	Second-hand data	Statistics	Panel regression

(-): unclear or unable to ascertain.
Definitions: WASH: water sanitation and hygiene system; thematic analysis: concepts or theories are inductively derived from the data; content analysis: codes are identified before searching for their occurrence in the data

Table 2. Types of the hazard and related examples.

Origin	Broad type	Examples
Man-made (Social) hazard	Terrorist	9.11 attack
Technological hazard	Solid waste, Fire, Urban service, WASH (water, sanitation and hygiene) system, Environmental issues, Urban development	Urban water service of Paris/ South Africa/India/ Sweden/Philippines, Denver drainage facilities, Environmental issues of Rotterdam and Berlin, Renaissance Plan of Santa Ana, Los Angeles River Restoration
Natural hazard	Hurricane, Flood, Climate change, Wildfire, Storm, Rainfall event	Hurricane Katrina, Hurricane Sandy, Rainfall event in Colorado, Climate change in some cities or areas, UK Summer 2007 floods, Wildfires in Colorado/U.S. wildfires in the early 2000s, Northern-Namibia Floods, Alster and Cranbrook Flood, Yorkshire flood, Florida flood

Table 3. Types and characteristics of social learning.

Major Actor	Learning Content	Learning Characteristics	Frequency	Percent
Manager (Government)	Management knowledge	Lessons for organisational management and emergency management for the government	5	22.72%
Expert	Expert knowledge	Experts acting as managers of social learning programme	1	4.55%
The public	Common-sense knowledge	Knowledge acquisition from the general public by way of indigenous knowledge or storytelling	7	31.8%
Media	Media knowledge	Move services and guidance to people to overcome uncertainty and improve the way they deal with disasters by social media	1	4.55%
Among Stakeholders	Exchanged knowledge	Knowledge exchange: expert–expert/expert–manager/ expert–public/internationally	8	36.3%

(Witting, 2017), forums (McEwen et al., 2018), workshops (Frantzeskaki & Kabisch, 2016), green practices (Tidball et al., 2010) or service supplies (Toubin et al., 2015; Slack & McEwen, 2013). For example, Smith et al. (2016) described a 5–6-week summer restoration programme investigating the role of civic ecology education programmes played in shaping youths' capacity to understand and respond to environmental disturbance. In addition, Tidball et al. (2010) explored how green practice and green memorialisation fostered resilience after 9.11 attack and Hurricane Katrina.

3.4. How is social learning operationalised to urban disaster resilience?

In this section, we synthesise evidence in the causal pathways through which social learning may have an effect in urban disaster resilience outcomes. The desirable outcome of social learning programme of each included studies is listed in Table 4, and they can be grouped into five types, and the detailed information can be seen in Figure 3 regarding the link between social learning and urban disaster resilience. In general terms, the detailed mechanisms of the effect of social learning on urban resilience can be grouped into (1) governance capacity; (2) self-organisation; (3) cognitive change; (4) open communication and deliberation; and (5) moral or civic responsibility.

The rationale for the effects of social learning programmes on urban disaster resilience development depends on how they enhance knowledge of the risk by sharing and

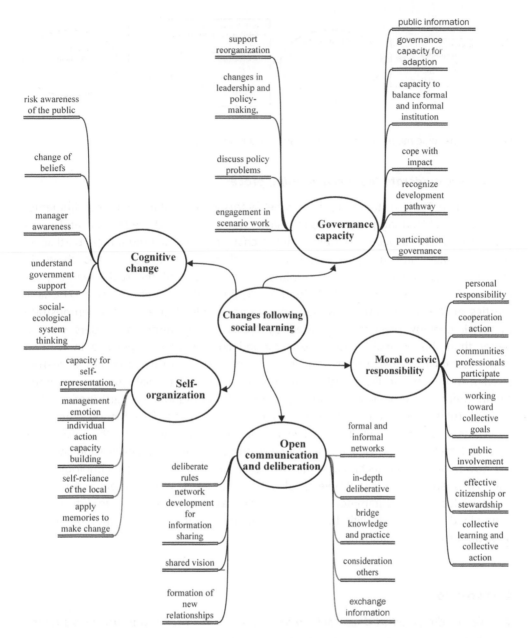

Figure 3. The mechanisms between social learning and urban disaster resilience.

communication, and how they improve the ability to cope with disaster and develop new policies. As a result of these benefits, social learning is thought to be an enabler of disaster resilience by increasing governance capacity, leading to cognitive change and moral and civic responsibility, and promoting open communication and deliberation, and self-organisation. Combined with the three aspects of hypothesised causal chain in Figure 2, the conjectured logic model of this process can be seen in Figure 4.

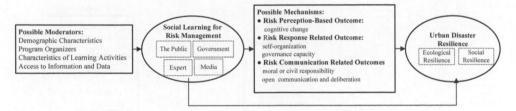

Figure 4. The logic model for the effects of social learning on urban disaster resilience.

3.5. Others moderate the social learning process

According to Table 4, there are also some factors that facilitate or hinder implementation of social learning programme in disaster management. These factors can be categorised into four aspects: (1) Demographic characteristics of programme participants like class, race, age, marginal population, social identity. Several studies show that the objectives of these actors may influence the results of the social learning programme (Orleans Reed et al., 2013; Smith et al., 2016; McEwen et al., 2018). (2) Characteristics of the programme organiser. For example, Tidball et al. (2010) demonstrated that civil ecology practices, organised by NGOs and government leaders were far better than the original spontaneously formed communities of practice in fostering urban sustainability because they have more resources. (3) Characteristics of learning activity, more specifically, participation intensity (referring to dialogue and communication intensity), participation frequency, participation duration and participation (deliberation) rules. Collaborative decisions could not be reached in limited number of participants within a limited time (Evers et al., 2012). (4) Access to the information, data and technologies used in the learning process. Other evidences suggested that the power of versatile technologies (Dawes et al., 2004), small-scale technology (Dieleman, 2013) and usable data (Dawes et al., 2004; Evers et al., 2012) can act as a facilitator to social learning programme. In the first two cases, studies found that technology maintained a close link with the extension of participation and promoted efficiency of social learning programme.

4. Discussion

4.1. Key findings: what does social learning mean for facilitating urban disaster resilience?

This review applied a thematic synthesis to explore how social learning facilitates urban resilience. We developed a logic framework and found that all of social, technological and natural hazards can trigger social learning. The low proportion of social hazards studies included in the review may be the result of our keyword searches. In terms of social learning types, both exchanged knowledge learning among stakeholders and common-sense learning of the public are far more than expert knowledge and media knowledge learning. Media learning, as a popular means of information transmission, which could be used to create a 'superordinate' geographical community identity, should be given more attention in the future (Feng, Hossain, & Paton, 2018).

As Van der Knaap, Leeuw, Bogaerts, and Nijssen (2008), Vaessen et al. (2014) and Molina, Carella, Pacheco, Cruces, and Gasparini (2017) discussed that systematic review, which focuses on the mechanisms underlying processes of change, synthesises high internal validity studies. This paper only pays attention to original studies with a clear objective of urban resilience improvement and the studies with barely improved outcome were not our focus. To conclude, we found five interesting mechanisms that work in the context of causal relationships between social learning and urban resilience improvement on individual or collective level. Our conclusions on mechanisms are also in line with those by (Berkes, Colding, & Folke, 2008; Trujillo, Suárez, Lema, & Londoño, 2015; Toubin et al., 2015; O'Donnell, Lamond, & Thorne, 2018) who reported that knowledge growth and public participation throughout social learning have an important impact on risk management performance, which is mainly realised through risk perception, risk communication and risk response. Although our five categories are not entirely exhaustive, they form a framework to better understand the social learning processes. According to the included literature, we further discuss the following mechanisms. The discussion of reorganised mechanisms thus concerns some kinds of outcomes that a social learning approach could but not necessarily engender.

4.1.1. Governance capacity and self-organisation

Social learning in the risk management context means a transition from learning together to managing together (Cernesson, Echavarren, & Enserink, 2005) and elaborated design and creation of social learning programmes during the learning process can navigate policy and governance reform agenda and support management policy and risk response (Murti & Mathez-Stiefel, 2019). Previous studies in England, Restock, Stockholm, Berlin and Rotterdam reported themes of governance capacity that related to social learning programme (Ashley et al., 2012; Evers et al., 2012; Frantzeskaki & Kabisch, 2016). Among which, some studies evaluated the participatory model involving top-down, bottom-up or horizontal support approach (Ashley et al., 2012; McEwen et al., 2018). A good example of 'bottom-up' governance can be found in the Yorkshire learning and action alliance, which included different stakeholders into the participation process and may lead to horizontal and bottom-up knowledge mobilisation and exchange (Ashley et al., 2012). Besides, the interviews with flood group members and flood risk management agencies in the UK outlined how Stakeholder groups emerge from transient and disconnected communities. Furthermore, the heterogeneous composition of the group had possibility to influence the process and the results of collaborative modelling (Evers et al., 2012). The case of learning and action alliance indicated that changes in policy, practice and culture are also a desirable result of social learning (Ashley et al., 2012). This was similar to what Antje Witting (2017) and Boyd et al. (2014) described that proper institution and policy-making to balance the formal and informal institution. Johannessen and Wamsler (2017) demonstrated that reorganising of failing structures (such as organisations) is necessary for a transition into something better.

Whereas the improvement of local governmental capacities for policy-making and creative enforcement and changes in leadership was detected, strong evidence of self-organisation capacity was found when a city or community take ways to maintain and re-create its identity after impacts or disasters (Lebel et al., 2006). In line with Hagemeier-Klose et al. (2014), Goldstein et al. (2013) pointed out that a new, relaxing participate network that characterised by informal networks and informal knowledge exchanges and cooperation can serve as a starting point for self-organisation capacity improvement, and it might also incubate new models of governance. The workshop and focus group in Vietnam indicated that the inclusion of representatives from different areas or social economic groups (farmers, fishermen, residents of informal settlements) facilitate to map the root causes of vulnerabilities (Dieleman, 2013). Besides, City repair projects, innovation can be found all around the world, seemed to teach citizens to collaborate in networks and to become more self-reliant (Orleans Reed et al., 2013).

4.1.2. Cognitive change

Social learning facilitates individual and group learning and provides insights into how stakeholder perceive risk management (Jones, Ross, Lynam, Perez, & Leitch, 2011). While the evidences indicated knowledge acquisition, it also suggested that active participation during social learning programme lead to risk cognitive change on individual- and group-level. This was the case in New York and Boulder where, during three youth restoration programmes, an elevation or refinement in thinking was detected. The three youth restoration programmes revealed that after implementing the programmes, there were more ideas about the impacts emerged while the number of concepts related to the causes of the disturbance has decreased. Moreover, results presented that the participants' focus also changed, at the beginning, their focus was the problem they have experienced, but after the programme, they focused more on solution and actions that they could take in response to future disturbances (Smith et al., 2016). There is also an evidence suggesting the importance of manager awareness change. The manager of service provision in the City of Paris found (1) the existence of service interdependencies (between manager and manager, between manager and other urban stakeholders), and (2) the importance of the manager's capacity for continuous operation in case of disturbance, and both of the service interdependencies and the capacities will change the way they handle the shock from passive crisis management to active risk management (Toubin et al., 2015). Risk perception is believed to change people's preparedness for, responses to and recovery from disasters, and is very important for effective risk communication (Xu, Zhang, & Xue, 2014). However, White, Kates, and Burton (2001) and Weichselgartner and Pigeon (2015) have argued that the role of knowledge and participation to resilience improvement may be limited, and more effects are needed to improve our understanding of disaster risk, knowledge and learning.

4.1.3. Open communication and deliberation

There are also some evidences suggesting the communication and deliberation acted as mechanism linked social learning to urban disaster resilience. In both cases of Los Angeles river restoration and U.S. Fire Learning Network, participants integrated their diverse experiences, enhanced their ability to embrace multiple voices and help communities operate effectively (Goldstein, et al., 2013). In other cases, the existing rules facilitated communication and reciprocity between all the players in the action arena (Witting, 2017). According to Dawes et al. (2004), trust is a condition under which collective learning and collective action can be achieved in response to changes. Reflecting on the co-production process in both Berlin and Rotterdam, we are aware that the European research projects has given openness to trust relations cultivation (Frantzeskaki & Kabisch, 2016;

McEwen et al., 2018) However, some scholars claimed that the co-production of knowledge and tools for disaster management is a long process. Unless the expert and the government are willing to participate in that process and get local stakeholders involved, effective risk reduction is unlikely to occur (Scolobig & Pelling, 2016). The approach of deliberation and dialogue were grounded in a set of rules and communal networks' development of information sharing among different actors such as the public, the experts, the government and the NGOs (Dieleman, 2013; Witting, 2017). Social learning in the aftermath of the 2013 Colorado flood, presented a link between rules and district's capacity from a recurring empirical situation. It is concluded that the fitness between rules and decision processes as well as the fitness between rules and the social-ecological context can accelerate the achievement of urban disaster resilience (Witting, 2017).

4.1.4. Moral or civic responsibility

Moral or civic responsibility was described as non-market values-mercy, justice, solidarity, care and service (West, 2002). Several studies suggested that moral or civic responsibility, as a spontaneous social force, can be achieved by participation. It manifests throughout the establishment of civic awareness and responsibility; thus, small love can be turned into big love (Orleans Reed et al., 2013; Smith et al., 2016; Witting, 2017; McEwen et al., 2018). In general, studies suggest that social cohesion can be formed and the resilience of society can be improved by the ways of volunteer activities (McEwen et al., 2018; Tidball et al., 2010), personal responsibility (Evers et al., 2012), and collective action (such as green practice from Tidball et al., 2010), effective citizenship or stewardship (McEwen et al., 2018) and so on.

4.2. Policy and management implications

This article provides some valuable knowledge by a systematic review. The activity theory is perfected and expanded in the risk management context, and the understanding of social learning activity is deepened. Furthermore, our review provides scientific and reliable evidences for risk management decision-making and practical application from three aspects based on our logic model connecting social learning and urban disaster resilience.

First, it is important to meet the public demands of risk perception by social learning. Utilisation of adequate and efficient messages will improve transparency and reduce the 'knowledge gap' between experts, government and the public (Garschagen & Kraas, 2011). Social learning provides a channel that allows a better understanding of the situation so that everyone involved in is familiar with the process of crisis management and enhances the trust of the public to government (Han, Lu, Hörhager, & Yan, 2017). Of course, honest and open principles of information dissemination will reaffirm government credibility and legitimacy (Sellnow, Ulmer, Seeger, & Littlefield, 2008).

Second, collaborative processes based on mutual learning, which get more stakeholders involved into communication about disasters during the entire social learning process and build an audience-centred communication mechanism by public participants and in-depth deliberation, are the main driving force of risk communication strategy (Smillie & Blissett, 2010; Kapucu, Hawkins, & Rivera, 2013). Additionally, the social learning process enhances moral or civil responsibility. Groups and individuals who are not political elites or experts, could influence political discourse by a process of deliberation and dia-logue, during which reasonable policies could possibly be enacted by combing the results of past policies and new information.

Third, greater attention needs to be given to action capacity. What has become increas-ingly clear, however, is that awareness cannot be equal to action. The willingness and the capacity of individuals and organisations to take action, the uncertainty of the action result and the resource are maybe the obstacles to take action. Government, private entities and the public are the key players to disaster management and only when they work together and shared sense of responsibility can they play their part and achieve the final goals (Kapucu et al., 2013).

4.3. Strengths and limitations of the review

Social learning is a complex intervention because of its variation on context and standard-isation (Rogers, 2008). There are many studies on quality evaluation, methodologies and report guidelines about systematic reviews in other subjects that could guide this review (Li et al., 2015; Li et al., 2018; Wang et al., 2018). This synthesis focuses on the causal links between contexts, intervention mechanisms and observed outcomes, which can be achieved by using evidences to develop explanations of how, for whom and in what circumstances an intervention may be successful.

There are several inevitable limitations of this review. First, difficulties in urban resilience measurement because of inherent relationships between urban resilience, community resilience and individual resilience, may limit the strength of explanation of this review for urban resilience (Sharifi et al., 2017). Second, the meaning of resilience remains con-tested for its numerous definitions (Bhamra, Dani, & Burnard, 2011), and it might limit the opportunity for understanding social learning roles in urban disasters because we were unable to synthesise some related studies using database searching. We synthesised only the findings of the studies that met our criteria. The potential loss of other relevant studies was compensated by the reference search. Thirdly, this review is limited by the lack of information from non-English language papers, this means that study findings still need to be interpreted cautiously. Finally, social learning programmes may exist in some government websites or reports that we did not synthesise in this review, which should be considered in the future.

5. Conclusion

Efforts to develop urban disaster resilience stress the capability facilitation of all actors and their organisations to learn from disaster for effective mitigation, recovery and transform-ation. Complex urban disaster risks often have domino effects (Renn & Walker, 2008). Many countries, organisations and communities have implemented a number of responses, including social learning by governmental or non-governmental programmes. Although significant variations have been designed and implemented in some countries, the avail-able evidence on the role of social learning in urban resilience is still limited. Few evi-dences on social learning programme are from the developing countries or undeveloped countries.

In this systematic review, we documented the significant impact of social learning on urban disaster resilience improvement. It is important to realise that social learning can play a crucial role in urban resilience throughout the mechanisms of governance capacity improvement, cognitive change, self-organisation, moral and civic responsibility and open communication and deliberation. However, our conclusions cannot be generalised to all cities because of the limitation of this review and to what extent social learning facilitates urban disaster resilience still need more researches in the future.

Acknowledgements

The authors would like to thank all of the unknown reviewers for their helpful comments on the article.

Disclosure statement

No potential conflict of interest was reported by the authors.

Funding

This work was supported by the Key Project of China Ministry of Education for Philosophy and Social Science under Big Data Driven Risk Research on City's Public Safety [grant number 16JZD023]; the Fundamental Research Funds for the Central Universities under Big Data Driven Risk Pre-Warning Research on City's Public Safety [grant number 17LZUJBWZD012] and Lanzhou university under Evidence-based Social Science Research [grant number 16LZUJBWTD013, 18LZUJBWZX006].

References

Albright, E. A., & Crow, D. A. (2015). Learning processes, public and stakeholder engagement: Analyzing responses to Colorado's extreme flood events of 2013. *Urban Climate*, *14*, 79–93. doi:10.1016/j.uclim.2015.06.008

Alexander, D. E. (2002). *Principles of emergency planning and management*. Edinburgh: Dunedin Academic Press.

Argyris, C., & Schon, D. A. (1974). *Theory in practice: Increasing professional effectiveness*. San Francisco: Jossey-Bass.

Armitage, D., Marschke, M., & Plummer, R. (2008). Adaptive co-management and the paradox of learning. *Global Environmental Change*, *18*(1), 86–98.

Ashley, R. M., Blanskby, J., Newman, R., Gersonius, B., Poole, A., Lindley, G., ... Nowell, R. (2012). Learning and action alliances to build capacity for flood resilience. *Journal of Flood Risk Management*, *5*(1), 14–22. doi:10.1111/j.1753-318X.2011.01108.x

Bakema, M. M., Parra, C., & McCann, P. (2019). Learning from the rubble: The case of Christchurch, New Zealand, after the 2010 and 2011 earthquakes. *Disasters*, *43*(2), 431–455. doi:10.1111/disa.12322

Ballard, H. L., & Belsky, J. M. (2013). Participatory action research and environmental learning: Implications for resilient forests and communities. *Environmental Education Research*, *16*(5), 152–168.

Bandura, A. (1977). *Social learning theory*. Englewood Cliffs, NJ: Prentice-Hall Series in Social Learning Theory: Prentice Hall.

Bateson, G. (1972). The logical categories of learning and communication. In *Steps to an ecology of mind*. New York: Ballantine Books.

Berkes, F., Colding, J., & Folke, C. (2008). *Navigating social-ecological systems: Building resilience for complexity and change*. New York: Cambridge University Press.

Bhamra, R., Dani, S., & Burnard, K. (2011). Resilience: The concept, a literature review and future directions. *International Journal of Production Research*, *49*(18), 5375–5393.

Blackmore, C. (2007). What kinds of knowledge, knowing and learning are required for addressing resource dilemmas?: A theoretical overview. *Environmental Science & Policy*, *10*(6), 512–525.

Boyd, E., Ensor, J., Broto, V. C., & Juhola, S. (2014). Environmentalities of urban climate governance in Maputo, Mozambique. *Global Environmental Change*, *26*, 140–151. doi:10.1016/j.gloenvcha.2014.03.012

Boyd, E., & Osbahr, H. (2010). Responses to climate change: Exploring organisational learning across internationally networked organisations for development. *Environmental Education Research*, *16*(5–6), 629–643. doi:10.1080/13504622.2010.505444

Brody, S. D., Zahran, S., Highfield, W. E., Bernhardt, S. P., & Vedlitz, A. (2009). Policy learning for flood mitigation: A longitudinal assessment of the community rating system in Florida. *Risk Analysis*, *29*(6), 912–929. doi:10.1111/j.1539-6924.2009.01210.x

Broto, V. C., Glendinning, S., Dewberry, E., Walsh, C., & Powell, M. (2014). What can we learn about transitions for sustainability from infrastructure shocks? *Technological Forecasting and Social Change*, *84*, 186–196.

Cernesson, F., Echavarren, J. M., & Enserink, B. (2005). *Learning together to manage together: Improving participation in water management*. Germany: University of Osnabrück.

Chung, C.-J., Hwang, G.-J., & Lai, C.-L. (2019). A review of experimental mobile learning research in 2010–2016 based on the activity theory framework. *Computers & Education*, *129*, 1–13.

Comfort, L. K. (1988). *Managing disaster: Strategies and policy perspectives*. Durham: Duke University Press.

Cook, W. M., Casagrande, D. G., Hope, D., Groffman, P. M., & Collins, S. L. (2004). Learning to roll with the punches: Adaptive experimentation in human-dominated systems. *Frontiers in Ecology and the Environment*, *2*(9), 467–474.

Cundill, G., Lotz-Sisitka, H., Mukute, M., Belay, M., Shackleton, S., & Kulundu, I. (2014). A reflection on the use of case studies as a methodology for social learning research in sub Saharan Africa. *Njas-Wageningen Journal of Life Sciences*, *69*, 39–47.

Cutter, S. L., Barnes, L., Berry, M., Burton, C., Evans, E., Tate, E., & Webb, J. (2008). A place-based model for understanding community resilience to natural disasters. *Global Environmental Change*, *18*(4), 598–606.

Darkow, P. M. (2018). Beyond "bouncing back": Towards an integral, capability-based understanding of organizational resilience. *Journal of Contingencies and Crisis Management*, *27*(2), 145–156.

Dawes, S. S., Cresswell, A. M., & Cahan, B. B. (2004). Learning from crisis. *Social Science Computer Review*, *22*(1), 52–66. doi:10.1177/0894439303259887

Dieleman, H. (2013). Organizational learning for resilient cities, through realizing eco-cultural innovations. *Journal of Cleaner Production*, *50*, 171–180. doi:10.1016/j.jclepro.2012.11.027

Ernst, A. (2019). Review of factors influencing social learning within participatory environmental governance. *Ecology and Society*, *24*(1), 3.

Evers, M., Jonoski, A., Maksimovič, Č, Lange, L., Ochoa Rodriguez, S., Teklesadik, A., … Makropoulos, C. (2012). Collaborative modelling for active involvement of stakeholders in urban flood risk management. *Natural Hazards and Earth System Sciences*, *12*(9), 2821–2842. doi:10.5194/nhess-12-2821-2012

Feng, S., Hossain, L., & Paton, D. (2018). Harnessing informal education for community resilience. *Disaster Prevention and Management*, *27*(1), 43–59.

Flood, R. L., & Romm, N. R. (1996). Contours of diversity management and triple loop learning. *Kybernetes*, *25*(7/8), 154–163.

Folke, C. (2006). Resilience: The emergence of a perspective for social–ecological systems analyses. *Global Environmental Change*, *16*(3), 253–267.

Fone, M., & Young, P. C. (2005). *Managing risks in public organizations* (pp. 49–67). San Francisco: Jossey-Bass.

Frantzeskaki, N., & Kabisch, N. (2016). Designing a knowledge co-production operating space for urban environmental governance – lessons from Rotterdam, Netherlands and Berlin, Germany. *Environmental Science & Policy*, *62*, 90–98. doi:10.1016/j.envsci.2016.01.010

Garschagen, M., & Kraas, F. (2011). *Urban climate change adaptation in the context of transformation: Lessons from Vietnam*. Resilient cities: Springer. 131–139.

Ge, L., Tian, J., Li, Y., Pan, J., Li, G., Wei, D., ... Song, F. (2018). Association between prospective registration and overall reporting and methodological quality of systematic reviews: A meta-epidemiological study. *Journal of Clinical Epidemiology*, 93, 45–55.

Godschalk, D. R. (2003). Urban hazard mitigation: Creating resilient cities. *Natural Hazards Review*, 4(3), 136–143.

Goldstein, B. E., Wessells, A. T., Lejano, R., & Butler, W. (2013). Narrating resilience: Transforming urban systems through collaborative storytelling. *Urban Studies*, *52*(7), 1285–1303. doi:10.1177/0042098013505653

Hagemeier-Klose, M., Beichler, S. A., Davidse, B. J., & Deppisch, S. (2014). The dynamic knowledge loop: Inter- and transdisciplinary cooperation and adaptation of climate change knowledge. *International Journal of Disaster Risk Science*, *5*(1), 21–32. doi:10.1007/s13753-014-0015-4

Han, Z., Lu, X., Hörhager, E. I., & Yan, J. (2017). The effects of trust in government on earthquake survivors' risk perception and preparedness in China. *Natural Hazards*, *86*(1), 437–452.

Harden, A., Thomas, J., Cargo, M., Harris, J., Pantoja, T., Flemming, K., ... Noyes, J. (2018). Cochrane qualitative and implementation methods group guidance series – paper 5: Methods for integrating qualitative and implementation evidence within intervention effectiveness reviews. *Journal of Clinical Epidemiology*, 97, 70–78.

Harris, J. L., Booth, A., Cargo, M., Hannes, K., Harden, A., Flemming, K., ... Noyes, J. (2018). Cochrane qualitative and implementation methods group guidance series – paper 2: Methods for question formulation, searching, and protocol development for qualitative evidence synthesis. *Journal of Clinical Epidemiology*, 97, 39–48.

Higgins, J. P. T., & Green, S. (Eds). (2011). *Cochrane handbook for systematic reviews of interventions*. England: John Wiley & Sons.

Hooli, L. J. (2015). Resilience of the poorest: Coping strategies and indigenous knowledge of living with the floods in Northern Namibia. *Regional Environmental Change*, *16*(3), 695–707. doi:10.1007/s10113-015-0782-5

IAGS. (2004). How much did the September 11 terrorist attack cost America? Retrieved from http://www.iags.org/costof911.html (accessed on 13 August 2019)

ISDR, U. (2005). *Hyogo framework for action 2005-2015: Building the resilience of nations and communities to disasters*. Paper presented at the Extract from the final report of the World Conference on Disaster Reduction (A/CONF. 206/6).

Johannessen, Å, & Wamsler, C. (2017). What does resilience mean for urban water services? *Ecology and Society*, *22*(1), 18. doi:10.5751/es-08870-220101

Jones, N. A., Ross, H., Lynam, T., Perez, P., & Leitch, A. (2011). Mental models: an interdisciplinary synthesis of theory and methods. *Ecology and Society*, *16*(1), 46.

Kapucu, N., Hawkins, C. V., & Rivera, F. I. (2013). *Disaster resiliency: Interdisciplinary perspectives*. Disaster Resiliency: Routledge. 121–142.

Kates, R. W. (1962). *Hazard and choice perception in flood plain management* (Research, Paper No. 78). University of Chicago, Department of Geography.

Klinke, A., & Renn, O. (2010). *Risk governance: Contemporary and future challenges*, 9–27. doi:10.1007/978-90-481-9428-5_2

Kraiger, K., Ford, J. K., & Salas, E. (1993). Application of cognitive, skill-based, and affective theories of learning outcomes to new methods of training evaluation. *Journal of Applied Psychology*, *78*(2), 311–328.

Kroepsch, A., Koebele, E. A., Crow, D. A., Berggren, J., Huda, J., & Lawhon, L. A. (2018). Remembering the past, anticipating the future: Community learning and adaptation discourse in media commemorations of catastrophic wildfires in Colorado. *Environmental Communication*, *12*(1), 132–147. doi:10.1080/17524032.2017.1371053

Lassa, J. A., & Nugraha, E. (2014). From shared learning to shared action in building resilience in the city of Bandar Lampung, Indonesia. *Environment and Urbanization*, *27*(1), 161–180. doi:10.1177/0956247814552233

Lebel, L., Anderies, J. M., Campbell, B., Folke, C., Hatfield-Dodds, S., Hughes, T. P., & Wilson, J. (2006). Governance and the capacity to manage resilience in regional social-ecological systems.

Leont'ev, A. N. (1981). *Problems of the development of the mind.* Moscow: Progress Publishers.

Li, X., Wei, L., Shang, W., Xing, X., Yin, M., Ling, J., … Yang, K. (2018). Trace and evaluation systems for health services quality in rural and remote areas: A systematic review. *Journal of Public Health, 26* (2), 127–135.

Li, X. X., Zheng, Y., Chen, Y. L., Yang, K. H., & Zhang, Z. J. (2015). The reporting characteristics and methodological quality of Cochrane reviews about health policy research. *Health Policy, 119*(4), 503–510.

Lorenz, D. F. (2013). The diversity of resilience: Contributions from a social science perspective. *Natural Hazards, 67*(1), 7–24.

May, P. J., & Williams, W. (1986). *Disaster policy implementation: Managing programs under shared governance.* New York: Plenum Press.

McEwen, L., Holmes, A., Quinn, N., & Cobbing, P. (2018). 'Learning for resilience': Developing community capital through flood action groups in urban flood risk settings with lower social capital. *International Journal of Disaster Risk Reduction, 27*, 329–342. doi:10.1016/j.ijdrr.2017.10.018

McMillen, H., Campbell, L. K., Svendsen, E. S., & Reynolds, R. (2016). Recognizing stewardship practices as indicators of social resilience: In living memorials and in a community garden. *Sustainability, 8* (8), 26. doi:10.3390/su8080775

Miller, F., Osbahr, H., Boyd, E., Thomalla, F., Bharwani, S., Ziervogel, G., … Rockström, J. (2010). Resilience and vulnerability: Complementary or conflicting concepts? *Ecology and Society, 15*(3), 11.

Moher, D., Shamseer, L., Clarke, M., Ghersi, D., Liberati, A., Petticrew, M., … Group, P.-P. (2015). Preferred reporting items for systematic review and meta-analysis protocols (PRISMA-P) 2015 statement. *Systematic Reviews, 4*(1), 1. doi:10.1186/2046-4053-4-1

Molina, E., Carella, L., Pacheco, A., Cruces, G., & Gasparini, L. (2017). Community monitoring interventions to curb corruption and increase access and quality in service delivery: A systematic review. *Journal of Development Effectiveness, 9*(4), 462–499.

Murti, R., & Mathez-Stiefel, S.-I. (2019). Social learning approaches for ecosystem-based disaster risk reduction. *International Journal of Disaster Risk Reduction, 33*, 433–440.

Noyes, J., Booth, A., Flemming, K., Garside, R., Harden, A., Lewin, S., … Thomas, J. (2018). Cochrane qualitative and implementation methods group guidance series – paper 3: Methods for assessing methodological limitations, data extraction and synthesis, and confidence in synthesized qualitative findings. *Journal of Clinical Epidemiology, 97*, 49–58.

O'Donnell, E. C., Lamond, J. E., & Thorne, C. (2018). Learning and action alliance framework to facilitate stakeholder collaboration and social learning in urban flood risk management. *Environmental Science & Policy, 80*, 1–8.

Olazabal, M., Chelleri, L., & Waters, J. (2012). Why urban resilience? In L. Chelleri & M. Olazabal (Eds.), *Report of multidisciplinary perspectives on urban resilience.* Bilbao: Basque Centre for Climate Change.

Orleans Reed, S., Friend, R., Toan, V. C., Thinphanga, P., Sutarto, R., & Singh, D. (2013). "Shared learning" for building urban climate resilience – experiences from Asian cities. *Environment and Urbanization, 25*(2), 393–412. doi:10.1177/0956247813501136

Pahl-Wostl, C. (2006). The importance of social learning in restoring the multifunctionality of rivers and floodplains. *Ecology and Society, 11*(1), 10.

Pahl-Wostl, C., Becker, G., Knieper, C., & Sendzimir, J. (2013). How multilevel societal learning processes facilitate transformative change: A comparative case study analysis on flood management. *Ecology and Society, 18*(4), 58.

Pahl-Wostl, C., Sendzimir, J., Jeffrey, P., Aerts, J., Berkamp, G., & Cross, K. (2008). Managing change toward adaptive water management through social learning. *Ecology and Society, 12*(2), 30.

Pelling, M. (2003). *Natural disaster and development in a globalizing world.* Londen & New York: Routledge.

PRC Cabinet. (2016). Tianjin port 8·12 Ruihai dangerous goods warehouse special major fire and explosion accident investigation report. Retrieved from http://www.gov.cn/foot/2016-02/05/5039788/files/460731d8cb4c4488be3bb0c218f8b527.pdf (accessed on 10 March 2019)

Reed, M., Evely, A. C., Cundill, G., Fazey, I. R. A., Glass, J., Laing, A., … Raymond, C. (2010). What is social learning? *Ecology and Society, 15*(4), 1.

Renn, O., & Walker, K. (2008). *Global risk governance: Concept and practice using the IRGC framework.* The Netherlands: Springer.

Rist, L., Felton, A., Samuelsson, L., Sandström, C., & Rosvall, O. (2013). A new paradigm for adaptive management. *Ecology and Society, 18*(4), 63.

Rogers, P. J. (2008). Using programme theory to evaluate complicated and complex aspects of interventions. *Evaluation, 14*(1), 29–48.

Ross, A. D. (2013). *Local disaster resilience: Administrative and political perspectives.* New York: Routledge.

Saja, A. A., Teo, M., Goonetilleke, A., & Ziyath, A. M. (2018). An inclusive and adaptive framework for measuring social resilience to disasters. *International Journal of Disaster Risk Reduction, 28,* 862–873.

Sanchez, A., Heijden, J., & Osmond, P. (2018). The city politics of an urban age: Urban resilience conceptualisations and policies. *Palgrave Communications, 4*(1), 25.

Schauppenlehner-Kloyber, E., & Penker, M. (2016). Between participation and collective action – from occasional liaisons towards long-term co-management for urban resilience. *Sustainability, 8*(7), 18. doi:10.3390/su8070664

Scolobig, A., & Pelling, M. (2016). The co-production of risk from a natural hazards perspective: Science and policy interaction for landslide risk management in Italy. *Natural Hazards, 81*(1), 7–25.

Sellnow, T. L., Ulmer, R. R., Seeger, M. W., & Littlefield, R. (2008). *Effective risk communication: A message-centered approach.* New York: Springer Science & Business Media.

Sharifi, A., Chelleri, L., Fox-Lent, C., Grafakos, S., Pathak, M., Olazabal, M., & Yamagata, Y. (2017). Conceptualizing dimensions and characteristics of urban resilience: insights from a co-design process. . *Sustainability, 9*(6), 1032.

Siebenhüner, B., Rodela, R., & Ecker, F. (2016). Social learning research in ecological economics: A survey. *Environmental Science & Policy, 55,* 116–126.

Slack, M. K., & McEwen, M. M. (2013). Perceived impact of an interprofessional education program on community resilience: An exploratory study. *Journal of Interprofessional Care, 27*(5), 408–412. doi:10.3109/13561820.2013.785501

Smillie, L., & Blissett, A. (2010). A model for developing risk communication strategy. *Journal of Risk Research, 13*(1), 115–134.

Smith, J. G., DuBois, B., & Krasny, M. E. (2016). Framing for resilience through social learning: Impacts of environmental stewardship on youth in post-disturbance communities. *Sustainability Science, 11*(3), 441–453. doi:10.1007/s11625-015-0348-y

Staveren, M. V. (Producer). (2008, 2019, August 13). Risk, innovation & change: design propositions for implementing risk management in organizations. Retrieved from https://core.ac.uk/download/pdf/11464730.pdf

Thomas, J. (2012). Methods for the synthesis of qualitative research. In *NCRM research methods festival 2012,* 2nd–5th July 2012, St. Catherine's College, Oxford.

Tian, JH, Zhang, J., Ge, L., Yang, KH, & Song, FJ. (2017). The methodological and reporting quality of systematic reviews from China and the USA are similar. *Journal of Clinical Epidemiology, 85,* 50–58. doi:10.1016/j.jclinepi.2016.12.004.

Tidball, K. G., Krasny, M. E., Svendsen, E., Campbell, L., & Helphand, K. (2010). Stewardship, learning, and memory in disaster resilience. *Environmental Education Research, 16*(5-6), 591–609. doi:10.1080/13504622.2010.505437

Timmermann, P. (1981). Vulnerability, resilience and the collapse of society. *Environmental Monograph, 1,* 1–42.

Toubin, M., Laganier, R., Diab, Y., & Serre, D. (2015). Improving the conditions for urban resilience through collaborative learning of Parisian urban services. *Journal of Urban Planning and Development, 141*(4), 05014021. doi:10.1061/(asce)up.1943-5444.0000229

Trujillo, E. M., Suárez, D. E., Lema, M., & Londoño, A. (2015). How adolescents learn about risk perception and behavior in regards to alcohol use in light of social learning theory: A qualitative study in Bogotá, Colombia. *International Journal of Adolescent Medicine and Health, 27*(1), 3–9.

Turner, B. A. (1978). *Man-made disasters*: Wykeham.

UNISDR, U. (2015). *Sendai framework for disaster risk reduction 2015–2030*. Paper presented at the 3rd United Nations world Conference on DRR.

Vaessen, J., Rivas, A., Duvendack, M., Palmer Jones, R., Leeuw, F. L., Van Gils, G., … Hombrados, J. G. (2014). The effects of microcredit on women's control over household spending in developing countries: A systematic review and meta-analysis. *Campbell Systematic Reviews, 10*(8), 1–205.

Van der Knaap, L. M., Leeuw, F. L., Bogaerts, S., & Nijssen, L. T. (2008). Combining Campbell standards and the realist evaluation approach: The best of two worlds? *American Journal of Evaluation, 29*(1), 48–57.

Vygotsky, L. (1978). Interaction between learning and development. In *Mind and society*. Cambridge, MA: Harvard University Press.

Wamsler, C. (2014). *Cities, disaster risk and adaptation*. Londan: Routledge.

Wang, X., Chen, Y., Yao, L., Zhou, Q., Wu, Q., Estill, J., … Norris, S. L. (2018). Reporting of declarations and conflicts of interest in WHO guidelines can be further improved. *Journal of Clinical Epidemiology, 98*, 1–8. doi:10.1016/j.jclinepi.2017.12.021

Weichselgartner, J., & Pigeon, P. (2015). The role of knowledge in disaster risk reduction. *International Journal of Disaster Risk Science, 6*(2), 107–116.

West, C. (2002). The moral obligations of living in a democratic society. In L. Martin Alcoff, D. Batstone, R. N. Bellab, J. Butler, B. Christian, M. Lerner, … C. west (Eds.), *The good citizen* (pp. 5–12). New York: Routledge.

White, G. F., Kates, R. W., & Burton, I. (2001). Knowing better and losing even more: The use of knowledge in hazards management. *Global Environmental Change Part B: Environmental Hazards, 3*(3), 81–92.

Witting, A. (2017). Ruling out learning and change? Lessons from urban flood mitigation. *Policy and Society, 36*(2), 251–269. doi:10.1080/14494035.2017.1322772

Worldbank, & U.N. (2010). *Natural hazards, unnatural disasters: The economics of effective prevention*. Washington, DC: The World Bank.

Xu, J., Zhang, Y., Liu, B., & Xue, L. (2014). *Risk perception in natural disaster management*. Paper presented at the 2014 Tech4Dev International Conference.

Zakour, M. J., Mock, N., & Kadetz, P. (2017). *Creating Katrina, rebuilding resilience: Lessons from New Orleans on vulnerability and resiliency*. Oxford: Butterworth-Heinemann.

Local government, political decentralisation and resilience to natural hazard-associated disasters

Vassilis Tselios and Emma Tompkins

ABSTRACT
Natural hazards affect development and can cause significant and long-term suffering for those affected. Research has shown that sustained long-term disaster preparedness combined with appropriate response and recovery are needed to deliver effective risk reductions. However, as the newly agreed Sendai framework recognises, this knowledge has not been translated into action. This research aims to contribute to our understanding of how to deliver longer term and sustained risk reduction by evaluating the role of political decentralisation in disaster outcomes. Specifically, we investigate whether countries which devolve power to the local level experience reduced numbers of people affected by storms and earthquakes, and have lower economic damage. Using regression analysis and cross-country data from 1950 to 2006, we find that, in relation to both storms and earthquakes, greater transfers of political power to subnational tiers of government reduce hazard impacts on the population. The downside is that more politically decentralised countries, which are usually wealthier countries, can increase the direct economic losses associated with a natural hazard impact after the storm or earthquake than those which are more centralised. However, overall, it seems advantageous to give subnational governments more authority and autonomy in storm and earthquake risk planning.

1. Introduction

Disasters, which are the result of natural hazards, can act as a confounding factor in development.[1] Estimates suggest that from 1992 to 2012, disasters killed 1.3 million people, and caused US$2 trillion of damage – far exceeding the amount provided in development assistance over the same period (Foresight, 2012). In a 20-year study from 1995 to 2015, the Centre for Research on the Epidemiology of Disasters (CRED) and the United Nations Office for Disaster Risk Reduction estimated that, on average, 30,000 deaths were attributed to weather-related disasters (UNISDR, 2015). The fact that natural hazard-associated disasters delay development and cause significant harm is undisputed, and theories of how to reduce this level of harm are well developed (Wisner, Blaikie,

Cannon, & Davis, 2004). There is widespread agreement internationally that action is needed, seen most recently through the adoption of the Sendai Framework on Disaster Risk Reduction (UNISDR, 2015). This framework espouses improved understanding of disaster risk; strengthening disaster risk governance to manage disaster risk; public and private investments in disaster risk reduction for widespread societal resilience and to deliver better recovery, rehabilitation and reconstruction through enhancing disaster preparedness. These principles build on previous international efforts to reduce the damage from natural hazards, notably, the 'Hyogo Framework for Action 2005–2015: Building the Resilience of Nations and Communities to Disasters' and the 'International Decade for Natural Disaster Reduction Programme Forum 1999'. Although reports indicate some success in declining disaster fatalities, this contrasts with slow progress in reducing damage and loss. This slow progress stresses the importance of the role of governance, especially at local scales, in reducing disaster risks.

This paper aims to investigate whether political decentralisation, especially more locally representative democracy, can minimise disaster losses, after controlling for factors recognised as key determinants of human and economic disaster outcomes. Political decentralisation here refers to the distribution of power in public decision-making from the central government to citizens or their elected representatives. It is usually present where there is representative government and pluralism in politics. There is an argument that decentralisation can improve the provision of local public goods and services, as there is more local accountability and representation (Bardhan, 2002; Faguet, 2014; Marks & Lebel, 2016; Oates, 1972; Tiebout, 1956). Our central hypothesis in this paper is that more politically decentralised countries fare better, when affected by natural hazards, in terms of damage to the population and the economy. If this hypothesis proves correct, we will be able to provide concrete guidance on governance improvements that can reduce disaster losses and deliver on the Sendai framework.

Our analysis focuses specifically on disasters triggered by storms and earthquakes due to differences in their predictability (i.e. earthquakes are less predictable than storms) and hence the ability of governance bodies to prepare for them. In the 40 years from 1972 to 2012, earthquakes, storms and droughts were the three largest causes of disaster mortality (Foresight, 2012), despite some advances in development and participatory application of early warning systems (Basher, 2006; Twigg, 2003).

Significant and widespread advances can be seen in the science of short term and seasonal forecasting of storms (DeMaria, Sampson, Knaff, & Musgrave, 2014; Kahn, 2005). Earthquakes remain less predictable than storms, although in some countries there is more effective integration of earthquake risk into planning than for storms. We argue that there is a lack of political decentralisation and, more specifically, a lack of local representation in decision-making about information needs for disaster preparedness. Hence, weather forecasts and early warning systems are not necessarily developed with the users' needs in mind. Delegating power may help to reduce disaster risks. In other words, giving citizens or their elected representatives more power in public decision-making may improve community resilience to disasters. This is associated with pluralistic politics, representative government and democracy.

Political decentralisation, which is a top-down process, can occur in multiple economic and political contexts. Hence, we also explore the role of national wealth, political orientation and local representation in the relationship between political decentralisation and damage associated with natural hazards. More specifically, we identify whether there are differences in this relationship: (a) between middle-income and high-income countries, (b) between countries with left, centre and right-wing politics and (c) between countries with different types of local representation. Thus, national wealth, political orientation and local representation are examined as mediating variables. Previous research has shown that meditating factors play a key role in the effect of decentralisation on the reduction of damage associated with natural hazards. Yamamura (2012), for instance, shows that decentralisation makes a greater contribution to mitigating damage in countries with higher quality institutions.

Various studies have been undertaken which allude to the importance of governance mechanisms in disaster outcomes (e.g. Brooks, Adger, & Kelly, 2005), but no macro-level study has been undertaken on the relationship between political decentralisation and natural hazard-associated disaster outcomes while analysing the mediating role of national wealth, political orientation and local representation. Cross-national studies in this area either have dwelt on fiscal decentralisation (Ahmed & Iqbal, 2009; Escaleras & Register, 2012; Skidmore & Toya, 2013; Yamamura, 2012) or are case studies (Ainuddin, Aldrich, Routray, Ainuddin, & Achkazai, 2013; Blackburn, 2014; Marks & Lebel, 2016). Using cross-country data over the 1950–2006 period, our study reveals that political decentralisation is correlated with disaster outcomes, and hence indicates the importance of focusing new disaster mitigation initiatives and future disaster risk reduction research on this area.

2. Literature review

2.1. Disaster risk reduction as an evolving field

A rich body of literature on natural hazards and disasters, dating back over 60 years, highlights the complexity of interactions between politics, economics, society, hazards and the resulting human and economic disaster outcomes. Vulnerability theory, for example, recognises both the role of exposure to hazards and the impact of social and economic deprivation on disaster outcomes (Adger, 2006). Hazard theory catalogues the root causes of vulnerability that create worsening disaster outcomes, pointing to power structures, political ideologies and economic systems that contribute to people living in unsafe conditions (Wisner et al., 2004). Sadly, the catalogue of recent major disasters (e.g. Hurricane Katrina in New Orleans in 2005; Cyclone Sidr in Bangladesh in 2007; the Haiti earthquake in 2010; Typhoon Haiyan in the Philippines in 2013; the Amatrice earthquake in Italy in 2016 to name a few) indicates that despite improved theoretical explanation of why disasters happen, appropriate mitigating actions are not being taken, so tragic disasters still happen.

Globally, it is now recognised that disaster losses are most effectively lowered by reducing disaster risks. In this context, government plays a vital role in both the preparation for and response to disaster events. Recent research has started to unravel the components of community resilience to hazards and disasters to clarify the role of the governance context. For example, Norris, Stevens, Pfefferbaum, Wyche, and Pfefferbaum (2008) recognise the importance of the equitable distribution of post-disaster resources. Twigg (2007) identifies the importance of integrated multi-scale decision-making, accountability and community participation. Djalante, Holley, and Thomalla (2011) recognise the importance of polycentric and multi-layered institutions, and following Tompkins, Lemos, and Boyd (2008), the central role of participation and collaboration, self-organisation and effective local networks in post-disaster recovery. Ainuddin et al. (2013) identify the usefulness of efficient preparedness and coordination of provisional and national level agencies to enhance community awareness and preparedness. According to Wisner et al. (2004) and Marks and Lebel (2016), greater local participation in decisions is expected to lead to more appropriate and sustainable disaster risk reduction interventions. All of these areas are linked to the extent to which the political system is decentralised, as this determines whether resources can flow to those who need them quickly.

2.2. Local government and political decentralisation

Although the vulnerability and hazard literature explains the causes of natural hazards, this paper takes a step forward by examining whether political decentralisation, whereby power and resources flow to the local level, can reduce the human and economic damage associated with natural hazards. Both natural hazards and political decentralisation have a local or regional dimension. Natural hazards strike a local or regional part of a country (Escaleras & Register, 2012), they rarely affect entire nations. Political decentralisation refers to the degree to which central government allows local or regional government entities to undertake the political functions of governance (Pike, Rodriguez-Pose, Tomaney, Torrisi, & Tselios, 2012), and hence to prepare for region-specific risks and respond effectively to regional crises. Moreover, international organisations, such as the World Bank, recognise the importance of decentralisation, by including it as a requirement in their development assistance programmes (Escaleras & Register, 2012). However, local government and political decentralisation have both pros and cons for humans and the economy.

On the one hand, delivered effectively decentralisation can bring government closer to the people by improving the provision of local public goods and services (Bardhan, 2002; Faguet, 2014; Marks & Lebel, 2016; Oates, 1972; Tiebout, 1956). Greater transfers of political powers to local and regional bodies can promote a better matching of public policies to local needs. Subnational governments may have an information advantage over central governments when it comes to responding to the heterogeneous needs of local citizens and especially when these needs arise from disasters. Although central government may be better placed to respond due to greater access to resources, locally elected governments are better placed geographically and politically to respond to local needs, due to their proximity to those affected (Adger, Hughes, Folke, Carpenter, & Rockstrom, 2005; Chhotray & Few, 2012; Norris et al., 2008; Rodríguez-Pose & Ezcurra, 2011). Relatively, more decentralised countries may fare better when disasters occur in terms of their effects on the population and on the economy, because local and regional officials are better able to set the optimal mix of local and regional policies prior to the disaster event than bureaucrats in distant central governments (de Mello, 2011; Lessmann, 2009; Tselios, Rodríguez-Pose, Pike, Tomaney, & Torrisi, 2012). It has been argued that decentralisation increases social capital, which has been shown to improve community resilience to natural hazard-associated disasters (e.g. see Brouwer & Nhassengo, 2006; Murphy, 2007; Paton, Millar, & Johnston, 2001; Tompkins, Hurlston, & Poortinga, 2009). When delivered effectively, decentralisation promotes greater voice and participation and limits the oppor-tunities for corruption (Brenner, 2004; Le Galès, 2002; Weingast, 2009). It empowers under-represented groups in society, including marginalised groups (e.g. the less well-off, the socially excluded and immigrants) by giving them a local voice. Such empowerment may lead to the adoption of local disaster risk reduction policies involving a wider range of actors and which are thus more sensitive to the local needs to prepare for and manage a natural hazard. Overall, subnational governments are perceived to have a com-parative advantage over national governments in the management of land use, economic development, safety and other local-based policies that affect disaster risk (Skidmore & Toya, 2013, p. 45).

On the other hand, political decentralisation can create negative effects for humans and the economy. First of all, there is a role for national governments in setting certain disaster management policies (Skidmore & Toya, 2013). Decentralisation may reduce the capacity of central government to transfer post-disaster resources to those most badly affected, from the less-damaged to the more-damaged localities. In general, the differences in socioeconomic endowment and institutional capacities among regions within any given country may undermine the potential benefits associated with the better matching of pol-icies to local needs (Kamel, 2012; Rodríguez-Pose & Ezcurra, 2010). For example, in regions with extreme poverty and loss of economies of scale, decentralisation may not mean a better matching of the provision of local public goods and services, because these regions face capacity constraints (Pelling, 2011; Prud'homme, 1995; Rodríguez-Pose & Ezcurra, 2011). In some contexts, local government is not capable of interacting effectively with the international disaster relief agencies (Holloway, 2003). Further, subnational gov-ernments may attract fewer skilled and capable officials and decision-makers (Prud'-homme, 1995), implying that subnational governments can end-up being less efficient at delivering disaster risk reduction. Moreover, some countries often devolve responsibil-ities but not skills or human and financial resources (Marks & Lebel, 2016). Finally, with higher degrees of decentralisation, there are costs associated with establishing multiple layers of government (Escaleras & Register, 2012). This can create some redundancy in government, which can deliver more flexible responses in times of stress. Poor countries may not be able to afford this cost of providing institutional redundancy. Where there are multiple tiers of government, accountability can be weak as citizens may be less able to identify responsible individuals to attribute blame for failures and credit for successes (Fisman & Gatti, 2002). Local traps can also arise because a narrow focus on local percep-tions, knowledge and interests may become a barrier to recognising solutions from other locations or that are available at other levels (Brown & Purcell, 2005). All these factors may offset the assumed benefits of political decentralisation. Overall, there is an optimal mix of responsibility between national and subnational governments in disaster management risk which requires coordination and the sharing of costs between national and subna-tional governments (Marks & Lebel, 2016; Skidmore & Toya, 2013).

2.3. The role of economic development, political institutions and local representation

Up to this point, we have focused on the question of whether political decentralisation can improve disaster outcomes. However, the distribution of the benefits and costs of decentralisation is dependent on national economic and political factors. In this subsection, we examine whether economic development, political institutions and local representation affect the possible association between disaster outcomes and political decentralisation.

First, we expect that the relationship between disaster outcomes and political decentralisation varies with levels of *economic development*. Specifically, the negative effects of decentralisation may most affect the poorer regions of the less-developed countries, due to their lack of resources and capacity to address natural hazards. Poorer regions face greater budget constraints than richer ones, while the latter rely on their own revenues, meaning that they are often in a better shape to address problems (Tselios et al., 2012). The positive effects of decentralisation are less likely to occur in less-developed countries, due to the absence of strong local accountability and higher levels of corruption, nepotism and clientelism. Examples of good practice do exist, for example, in Ceará, Brazil, where locally active leaders managed to push for greater accountability in drought management (Lemos, 2007). Cheshire and Gordon (1998) argue that the transfer of powers and resources to subnational tiers of government benefits the most prosperous regions, such as those with better socioeconomic endowments. Richer regions can generally extract greater resources, not only through the taxation of their own citizens but also through a greater political leverage to negotiate with the central government (Rodríguez-Pose & Ezcurra, 2011; Rodriguez-Pose & Gill, 2004). This argument is often reinforced through the literature on social capital and disaster recovery, which highlights the fact that those communities which are better linked to policy-makers, at any level of government, tend to experience more rapid rates of post-disaster recovery (Aldrich, 2012). Hence, the degree to which economic agents benefit from and are able to comply with and employ their own established safety standards depends on the level of economic development (Skidmore & Toya, 2013, p. 45). Despite our assumptions that richer nations will recover more quickly (for example, due to effective building regulations, careful development planning, provision of higher quality infrastructure and resource availability to provide high-quality emergency care (Kahn, 2005)), rich nations are not always the site of best practice. Cuba, a poor country with a centrally planned economy in the Caribbean, is often cited as an exemplar of disaster risk reduction (Sims & Vogelmann, 2002; Thompson & Gaviria, 2004). The United States of America, in contrast, delivered very poor quality disaster risk reduction in response to Hurricane Katrina in New Orleans in 2005 (Colten, Kates, & Laska, 2008; Laska & Morrow, 2006). While the political context of Hurricane Katrina was highly significant, this example serves to show that national wealth is not a straightforward indicator of successful disaster risk reduction.

Second, we expect that the relationship between disaster outcomes and political decentralisation varies with *political orientation*. Notably, the effects of decentralisation and thus the opportunities for citizens to take interest in public affairs may differ in countries with more dominant left-wing or right-wing politics, because of differences in political strategies towards local authorities and central government. Left-wing politics support socioeconomic equality and egalitarianism, often in opposition to the socioeconomic hierarchy and inequality, advocate strong government intervention in the economy and prefer local control of the economy, while right-wing politics support the socioeconomic order, stratification, hierarchy and inequality, advocate little government intervention in the economy and prefer a decentralised economy based on economic freedom.

Third, we anticipate that the effects of decentralisation will differ according to the extent of *local representation*. There are countries where both the legislature and executive are locally elected, countries where the executive is appointed but the legislature is locally elected, and countries where neither the executive nor the legislature are locally elected, when it comes to the municipal governments or the state/province governments. We argue that countries with different types of local representation are likely to perform differently in disaster recovery, due to differences in approaches to post-disaster macroeconomic stabilisation (Oates, 1999; Tselios et al., 2012) and in regional redistribution of post-disaster resources. The role of local representation is a key factor because an appropriate political setting for downward accountability requires a suitable environment for local elected leaders to act independently and responsively (Yilmaz, Beris, & Serrano-Berthet, 2008). The local leadership will be influenced by at least three sets of factors: the institutional arrangements for the separation of powers among the executive, legislative and judicial bodies; the election laws and the electoral system and the existence and functioning of a party system and political party laws (Yilmaz et al., 2008). Overall, local representation is likely to mediate the relationship between political decentralisation and disaster outcomes.

3. Data, variables and empirical functions

3.1. Data sources

We now explore the relationship between political decentralisation and disaster occurrence associated with natural hazards, after controlling for some economic and natural characteristics. The control variables are used not only to capture some structural characteristics of the countries and to deal with some important sources of heterogeneity but also as statistical controls in order to make the estimated parameters of the main explanatory variable (i.e. political decentralisation) more precise.

Data on the *disaster outcomes* were obtained from the Emergency Events Database (EM-DAT).[2] This database contains essential core data on the occurrence and effects of over 18,000 mass disasters in the world from 1900 to the present. Thus, this is a panel database. In order for a disaster to be entered into the database, at least 1 of the following 4 criteria (known as CRED criteria) has to be fulfilled: 10 or more people reported killed, 100 or more people reported affected, a call for international assistance and a declaration of a state of emergency. The EM-DAT database is an unbalanced panel database, because (a) if a country–year has experienced a disaster, it must satisfy one or more of the above four criteria and must have data available on all the following variables: number of people killed, injured, homeless and affected and estimated economic damage (Escaleras & Register, 2012), (b) a country–year has experienced a disaster but it has not been observed (especially before 1988) and (c) a country–year has not experienced a disaster.[3] Thus, some countries might experience multiple disasters associated with natural hazards in the same year, while others experience none.

Political decentralisation is proxied by regional authority indices defined by Hooghe, Marks, and Schakel (2008b). These indices cover 42 middle-income and high-income countries (but not low-income countries), and cover 8 dimensions of regional authority, for the years 1950 to 2006. Countries are coded on an annual basis for years in which a country is independent and (semi-)democratic (Hooghe, Marks, & Schakel, 2008a, p. 259). Hooghe et al. (2008b) consider two domains of regional authority: 'Self-rule' and 'Shared-rule'.[4] The aggregate 'Self-rule' and 'Shared-rule' score constitutes the overall regional authority: the Regional Authority Index, or 'RAI-total' score. We recognise that the level of political decentralisation varies between countries. For instance, in 2006, Bosnia and Herzegovina (score = 30.5), Germany (score = 29.3), Belgium (score = 29), United States (score = 23), Canada (score = 22.7) and Italy (score = 22.7) have the highest 'RAI-total' score. All these political decentralisation indices have several advantages over rival ones, such as the Schneider (2003) index, as they measure political decentralisation along a multitude of dimensions and allow for some change over time (Tselios et al., 2012). For example, the 'RAI-total' score for Belgium is 14 from 1950 to 1969, 22.9 from 1970 to 1979, 25.8 from 1980 to 1988, 32.1 from 1989 to 1994 and 29 from 1995 to 2006. Regional Authority Indices have been used in many recent studies (Rodríguez-Pose & Ezcurra, 2010, 2011; Tselios et al., 2012); however, as with the EM-DAT database, this is an unbalanced database, because it does not provide regional authority scores for all of the years 1950–2006.[5]

Data on *controls* were obtained from the Penn World Table (PWT)[6] and the EM-DAT database. The PWT provides economic data for 189 countries for some or all of the years 1950–2010. This database is produced by researchers at the University of Pennsylvania and is based on the so-called benchmark comparisons of the International Comparison Programme. It allows for real quantity comparisons both between countries and over time and it has been used in many comparative studies. However, the PWT database is also an unbalanced database.[7]

We combined EM-DAT, Hooghe et al. (2008b) and PWT, to generate a database of 41 countries from 1950 and 2006 (i.e. Albania, Australia, Austria, Belgium, Bosnia and Herzegovina, Bulgaria, Canada, Croatia, Cyprus, Czech Republic, Denmark, Estonia, Finland, France, Germany, Greece, Hungary, Iceland, Ireland, Italy, Japan, Latvia, Lithuania, Luxembourg, Macedonia, Malta, Netherlands, New Zealand, Norway, Poland, Portugal, Romania, Russia, Slovak Republic, Slovenia, Spain, Sweden, Switzerland, Turkey, United Kingdom and United States). This resulting database contains only high- and upper middle-income countries and is amenable to estimation methods that manage potential heterogeneity bias (Rodríguez-Pose & Tselios, 2009). The missing year–country observations of political decentralisation for those countries which have experienced a storm or earthquake according to the CRED specific criteria is low and thus unlikely to affect the association between disaster outcomes and political decentralisation (see Appendix).

3.2. Variables and descriptive statistics

For the purposes of this project, we consider two types of disaster associated with natural hazards: storms[8] and earthquakes (seismic activities). In this study, we also use three indices (variables) to measure the impacts of these two natural hazards.

(A) The number of total people affected. This includes three categories:
   - number of people injured (i.e. people suffering from physical injuries, trauma or an illness requiring medical treatment as a direct results of a disaster);
   - number of people made homeless (i.e. people needing immediate shelter); and
   - number of people affected (i.e. people requiring immediate assistance during a period of emergency; this can also include displaced or evacuated people).
(B) The number of people killed, i.e. persons confirmed as dead and persons missing and presumed dead.
(C) The disaster-related economic damage (in 000' US$).

Several methodologies exist to quantify disaster losses, but there is no standard procedure to determine a global figure for economic impact. Economic damage is defined as the direct losses associated with a natural hazard impact after the event, but the indirect damage and longer term macroeconomic effects are not considered (Pielke, Rubiera, Landsea, Fernandez, & Klein, 2003). The first two indices (i.e. the number of total people affected and the number of people killed) provide information on the *human* impact of disasters, while the last index provides information on the *economic* impact of disasters.

Figure 1 shows the number of observations for the 41 countries over 1950–2006 which have experienced a storm or earthquake according to the CRED specific criteria. This figure clearly shows that the number of observations has risen. This increase has been discussed and is attributed to societal change and economic development (Bouwer, Crompton, Faust, Höppe, & Pielke , 2007; Escaleras & Register, 2012), as well as improvements in data collection and reporting in the EM-DAT database (Guha-Sapir, Hargitt, & Hoyois, 2004).

Figure 2 shows the number of people killed as a result of storms or earthquakes for 41 countries over 1950–2006.[9] We observe that there are two spikes in the number of people killed as a result of storms between 1950 and 1960 and two spikes in the number of people killed as a result of earthquakes between 1988 and 2000. The spikes for storms may reflect the improvements in the prediction of storms (through technology, such as computer modelling in storms) and in health care, while the spikes for earthquakes may depict the improvements in the EM-DAT database.

Noting the wide variation in the observations on all disaster outcome variables (i.e. the number of total people affected, the number of people killed and the economic damage), we take the log form (ln) of these variables.[10] Since some disasters in the sample resulted in no deaths, we add one death to each observation (Escaleras & Register, 2012; Kahn, 2005).[11]

For the political decentralisation variables, we use all three Regional Authority Indices developed by Hooghe et al. (2008b): Self-rule, Shared-rule and RAI-total. We also transform these variables taking the natural logarithm, to make them more normally distributed. The logarithm transformation also helps the interpretation of the results, as the natural disaster outcome variables are also in natural logarithmic form.

For the controls, we use three variables obtained from the PWT: (a) government consumption share of GDP per capita (%), (b) openness (%) and (c) annual growth rate in GDP per capita ($\ln(\text{GDP}pc_t/\text{GDP}pc_{t-1})$). We do not include other controls, such as the population, economic development (GDP per capita),[12] consumption share of GDP per capita and investment share of GDP per capita, due to their high correlation (above 0.5) with the three control variables or the political decentralisation variables. Finally, we

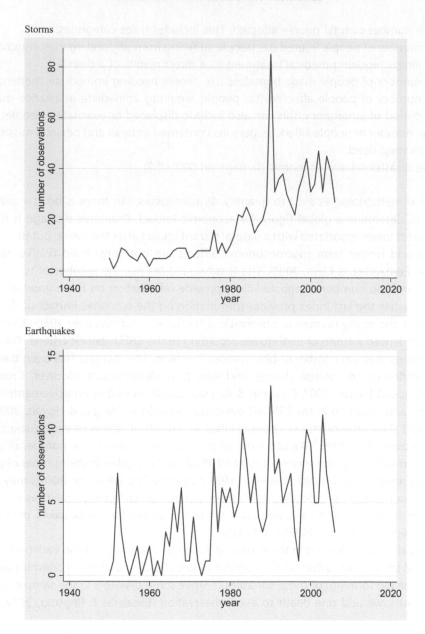

Figure 1. Number of observations by year.

include the number of disasters in log form (ln) (due to the wide variation in the observations) as a control variable (source: EM-DAT).[13] We believe that these control variables capture some features of the nations and take into account some important sources of heterogeneity.

The mean, standard deviation and minimum and maximum value for the above variables are reported in Table 1. It should be noted that the descriptive statistics refer to those observations where there are data for all variables (i.e. 392 observations for storms and 144 observations for earthquakes).

Storms

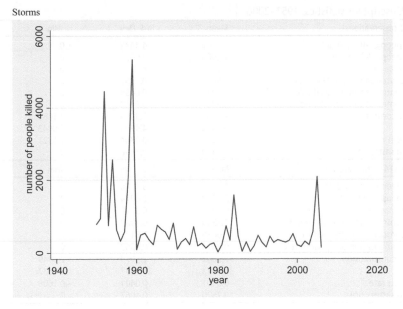

Earthquakes

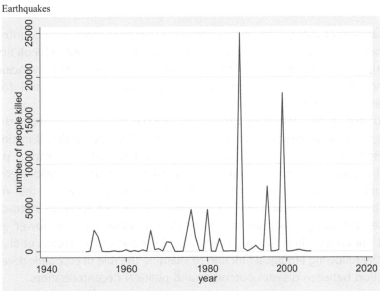

Figure 2. Number of people killed by year.

3.3. Empirical functions

We use the following empirical function to examine the association between disaster outcomes and political decentralisation

$$\text{Dis}_{it} = f(\text{PolDec}_{it},\ \text{Control}_{it},\ u_t) \tag{1}$$

where Dis_{it} is the disaster outcomes from storms or earthquakes for country i ($i = 1, 2, 3, \dots, 41$) at time (year) t ($t = 1951, 1952, \dots, 2006$), PolDec_{it} is a proxy for the degree of political

Table 1. Descriptive statistics, 1951–2006.

Storms (392 observations)	Mean	Std. Dev.	Min	Max
Number of people affected (ln)	3.3309	4.3889	0	15.4408
Number of people killed (ln)	2.2903	2.0053	0	8.5688
Economic damage (ln)	7.1825	6.1143	0	18.8796
Political decentralisation (ln) (RAI-total)	2.5881	0.7350	0	3.4995
Political decentralisation (ln) (self-rule)	2.4275	0.6434	0	3.1987
Political decentralisation (ln) (shared-rule)	0.9957	0.9448	0	2.5649
Government consumption	7.8707	2.3090	3.1681	19.2755
Openness	46.1956	34.4722	5.0043	212.8417
Annual growth rate	0.0258	0.0304	−0.1321	0.1238
Number of disasters (ln)	0.5549	0.7498	0	3.2958
Earthquakes (144 observations)	Mean	Std. Dev.	Min	Max
Number of people affected (ln)	7.1196	3.7138	0	14.2801
Number of people killed (ln)	2.3399	2.4074	0	9.7972
Economic damage (ln)	6.8706	6.0728	0	18.4207
Political decentralisation (ln) (RAI-total)	2.1549	0.8989	0	3.4995
Political decentralisation (ln) (self-rule)	2.0636	0.8125	0	3.0910
Political decentralisation (ln) (shared-rule)	0.5824	0.8562	0	2.5649
Government consumption	7.2802	2.5979	2.6949	15.9205
Openness	30.4823	23.1837	5.0687	130.9326
Annual growth rate	0.0301	0.0401	−0.1608	0.1900
Number of disasters (ln)	0.2620	0.4067	0	1.7918

decentralisation for country i at time t, Control$_{it}$ is a vector of controls for country i at time t and u_t represents time-dummies. We include time-dummies to control for all time-specific space-invariant variables (Baltagi, 2005). In this study, these variables may include climate change and improvements in technology. Time-dummies also control for possible improvements in the EM-DAT database after 1988.

As part of our analysis, we also examine whether the association between disaster outcomes and political decentralisation differs by national wealth, political institutions and local representation. In other words, we examine whether GDP per capita, party orientation with respect to economic policy, electoral level of the municipal governments and electoral level of the state/provincial governments influence the relationship between disaster outcomes and political decentralisation. We apply an interaction analysis, because comparing subgroup-based correlation coefficients has lower explanatory capacity as the division into subgroups reduces the sample size (Tselios et al., 2012).

We use the following empirical function to examine the role of *economic development* in the association between disaster outcomes and political decentralisation.

$$\text{Dis}_{it} = f(\text{PolDec}_{it} x D_{\lambda it},\ \text{Control}_{it},\ u_t) \tag{2}$$

where $D_{\lambda it}$ is a vector of dummy variables[14] for economic development with $\lambda = 1, 2$, where $\lambda = 1$ is the dummy variable for the medium-income countries and $\lambda = 2$ is the dummy variable for the high-income countries. We classify the countries into medium-income and high-income using the World Bank's definition of development (Escaleras & Register, 2012).

We use the following empirical function to examine the role of *political institutions* and *local representation* in the association between disaster outcomes and political decentralisation.

$$\text{Dis}_{it} = f(\text{PolDec}_{it} x D_{\mu it},\ \text{Control}_{it},\ u_t) \tag{3}$$

where $D_{\mu it}$ is a vector of dummy variables for political institutions or local representation. Data were obtained from the Keefer (2012) database. This database is available from 1975 to the present. More specifically,

- if μ denotes the party orientation with respect to economic policy, $\mu = 1, 2, 3$, where $\mu = 1$ is the dummy variable for the left-party countries (i.e. for parties that are defined as communist, socialist, social democratic or left-wing), $\mu = 2$ is the dummy variable for the centre-party countries (i.e. for parties that are defined as centrist or when party position can best be described as centrist such as the party advocating strengthening private enterprise in a social-liberal context) and $\mu = 3$ is the dummy variable for the right-party countries (i.e. for parties that are defined as conservative, Christian Democratic or right-wing); and
- if μ denotes the electoral level of either the municipal governments or the state/province governments, $\mu = 1, 2, 3$, where $\mu = 1$ is the dummy variable for the countries where both the local legislature and the local executive are locally elected, $\mu = 2$ is the dummy variable for the countries where the executive is appointed, but the legislature is locally elected, and $\mu = 3$ is the dummy variable for the countries where neither local executive nor legislature are locally elected.

4. Results and discussion

4.1. Political decentralisation and storm disaster outcomes

Table 2 displays the regression results for empirical function (1) above, using the number of people affected (ln) (Regression 1), the number of people killed (ln) (Regression 2), and the economic damage (ln) (Regression 3) as dependent variables. Political decentralisation (ln) constitutes the overall regional authority score ('RAI-total' score). Regression 1 and 2 show that there is a negative and statistically significant association between political decentralisation and human outcomes caused by storms. The coefficient on the number of people affected (−1.1705) is higher than that on the number of people killed (−0.1847), i.e. decentralised countries experience lower numbers of people affected, and to some degree numbers killed as a result of storms. Thus, there seems to be a

Table 2. Political decentralisation and storms' disaster outcomes.

	Number of people affected (ln) (1)	Number of people killed (ln) (2)	Economic damage (ln) (3)
Political decentralisation (ln)	−1.1705***	−0.1847*	0.9842**
Government consumption	−0.1746**	0.0120	−0.1293
Openness	−0.0423***	−0.0180***	−0.0008
Annual growth rate	2.4677	2.5543	−31.0671***
Number of disasters (ln)	2.6504***	1.2477***	3.9439***
Time-dummies	Yes	Yes	Yes
Constant	4.5497	6.9612***	2.3343
Observations	392	392	392
R-squared	0.4689	0.6471	0.4617

Note: Standard errors are not reported.
*$p < 0.1$.
**$p < 0.05$.
***$p < 0.01$.

need to further decentralise disaster risk reduction, as local government can be more responsive to the needs of local residents (Marks & Lebel, 2016). Regression 3 shows that there is a positive and statistically significant association between political decentralisation and economic outcomes, that is, more decentralised countries experience higher economic losses with storms. This finding is on the surface counter-intuitive, that is, one would assume that economic losses and human impact go hand in hand. There are two possible causes of this anomaly. One comes from the work of Pielke et al. (2003), who point out that economic losses from storms in the USA are related to rising wealth and population density in storm affected areas. Their work hints that even in decentralised countries with some effective preparedness, the economic impact of storms is rising due to increasing wealth in affected areas. The second argument is that historically disaster risk reduction has focussed on protecting lives, rather than making livelihoods and homes resilient to shocks (UNISDR, 2005, 2009). Both point to the need for a focus on the resilience of livelihoods and economies in the face of storm shocks.

For the control variables, the results are discussed only briefly. First, we find that the relationship between government consumption share and the number of people affected is negative and statistically significant (Regression 1), that is, where government spending in the economy is relatively large, fewer people are affected by storms. This is perhaps due to relatively more spending on storm protection, early warning or disaster recovery. The coefficient on government consumption for Regression 2 and 3 is statistically insignificant. Second, the coefficient on openness is negative and statistically significant for both human outcomes (Regression 1 and 2), that is, the more economically open, the fewer people killed or affected by storms, but it is not statistically significant for economic outcomes (Regression 3). This was a difficult finding to explain, and may simply be an artefact of the data, that is, the European countries which are not affected by major cyclones are part of a shared economic area, whereas the USA and Australia, which are affected by major cyclones, are less open due to their size and physical location. However, this is an interesting finding that requires more investigation. Third, there is a negative association between the annual GDP growth rate and the economic damage (Regression 3), that is, countries with faster growing GDP per capita tend to experience lower levels of economic loss from storms. However, the coefficient on growth for Regression 1 and 2 is statistically insignificant. Fourth, there is a positive (and unsurprising) association between the number of disasters and the human and economic outcomes (Regression 1, 2 and 3).

We now explore whether *national wealth*, *political orientation* and *local representation* intervene in the relationship between political decentralisation and storm disaster outcomes according to empirical functions (2) and (3). Table 3 displays the coefficients on political decentralisation by levels of national wealth, by political orientation and by local representation. We find evidence supporting our main finding (as we saw in Table 2) that relatively more politically decentralised countries fare better when storms strike in terms of the effect on the population (i.e. number of people affected and number of people killed), but with higher economic losses than less politically decentralised countries, though the effect appears much more robust (a) in high-income countries (Regression 1, 2 and 3); and, (b) in countries where the executive of municipal governments is appointed, but the legislature is locally elected (Regression 7, 8 and 9). Table 3 shows that there are no differences in the impact of political decentralisation when

Table 3. Political decentralisation and storms' disaster outcomes by economic development, political institution and local representation.

	Number of people affected (ln)	Number of people killed (ln)	Economic damage (ln)
	(1)	(2)	(3)
Political decentralisation (ln)			
• Medium-income countries *(62)*	−1.1333***	−0.0833	−0.7748
• High-income countries *(330)*	−1.1674***	−0.1761*	0.8360**
Controls	Yes	Yes	Yes
Time-dummies	Yes	Yes	Yes
Constant	4.5026	6.8328***	4.5608
Observations	392	392	392
R-squared	0.4689	0.6481	0.4922
	(4)	(5)	(6)
Political decentralisation (ln)			
• Left-party countries *(122)*	−0.9871***	−0.1541	0.9399**
• Center-party countries *(14)*	−1.2970**	−0.1166	1.1366
• Right-party countries *(145)*	−0.8290**	−0.1795	0.9246**
Controls	Yes	Yes	Yes
Time-dummies	Yes	Yes	Yes
Constant	8.4944***	3.3652***	9.2107***
Observations	281	281	281
R-squared	0.5154	0.5226	0.4592
	(7)	(8)	(9)
Political decentralisation (ln)			
• Municipal governments: legislature and executive are locally elected *(219)*	−0.9370**	−0.0273	1.1213**
• Municipal governments: legislature is locally elected *(41)*	−0.8813**	−0.5391***	1.4128**
• Municipal governments: no local elections *(0)*	No data	No data	No data
Controls	Yes	Yes	Yes
Time-dummies	Yes	Yes	Yes
Constant	4.9053*	1.8793**	8.6225**
Observations	260	260	260
R-squared	0.5315	0.6242	0.4872
	(10)	(11)	(12)
Political decentralisation (ln)			
• State/province governments: legislature and executive are locally elected *(214)*	−1.1411***	−0.1780	1.1997***
• State/province governments: legislature is locally elected *(67)*	−1.3200***	−0.1630	0.2901
• State/province governments: no local elections *(17)*	−1.8772***	0.2442	−0.6188
Controls	Yes	Yes	Yes
Time-dummies	Yes	Yes	Yes
Constant	6.9011***	3.2212***	8.1585***
Observations	298	298	298
R-squared	0.4989	0.5379	0.4890

Note: Parentheses in italics show the number of observations for each classification; standard errors are not reported.
*$p < 0.1$.
**$p < 0.05$.
***$p < 0.01$.

there is right-wing or left-wing dominant politics (Regressions 4, 5 and 6). Finally, countries where neither the legislature nor the executive of state/province governments are locally elected manage better storm outcomes on the population (i.e. number of people affected) (Regression 10), while the impact of decentralisation on economic damage is statistically significant only for countries where both the legislature and executive are locally elected (Regression 12).

4.2. Political decentralisation and earthquake disaster outcomes

Table 4 presents the relationship between political decentralisation and earthquake disaster outcomes following empirical function (1). We find decentralisation to be negatively and statistically significant associated with the number of people affected (Regression 1), that is, more decentralised countries have fewer people affected in earthquakes. However, decentralisation is positively and statistically significant associated with the economic damage (Regression 3), that is, more decentralised countries have more economic damage. This mirrors the findings under storms. While the population outcomes from earthquakes refer only to the number of people affected, the population outcomes from storms refer to both the number of people affected and the number of people killed. The conclusion from this is that greater transfers of political power to subnational tiers of governments could be justified as a means to reduce earthquake disaster impacts on the population, but again with a high economic cost.

As for the control variables, the coefficients on the three economic variables are statistically significant for the population outcomes only (Regression 1 and 2). The coefficients on both government consumption and openness are negative, which is consistent with the findings for storms (Table 2). It is remarkable that the coefficient on annual growth rate is positive, that is, the faster the rates of economic growth the higher the economic costs of an earthquake disaster. This supports earlier work noting that 'rapid growth comes at the expense of weak or ignored building codes, poor land zoning controls, and the like' (Escaleras & Register, 2012, p. 171). Finally, as expected, we find that there is a positive association between the number of disasters and the human and economic losses from earthquakes (Regression 1, 2 and 3).

We consider whether political decentralisation influences the earthquakes' outcomes differently across countries with different levels of national wealth and with different political orientation and local representation following empirical functions (2) and (3). Table 5 shows that the negative impacts on the population appear more robust (a) in high-income countries (Regression 1), (b) in right-leaning countries (Regression 4: as the magnitude of the coefficient on political decentralisation is higher for right-party countries than for left-party countries) and (c) in countries where both the legislature and executive of municipal or state/province governments are locally elected (Regression 7, 8 and 11). The positive economic damage effects appear more robust in right-party countries (Regression 6).

Table 4. Political decentralisation and earthquakes' disaster outcomes.

	Number of people affected (ln)	Number of people killed (ln)	Economic damage (ln)
	(1)	(2)	(3)
Political decentralisation (ln)	−0.8624**	−0.1403	1.2615*
Government consumption	−0.3379**	−0.1878**	−0.0436
Openness	−0.0423**	−0.0207*	−0.0453
Annual growth rate	25.6556**	13.5087**	19.2058
Number of disasters (ln)	1.8927**	1.7279***	3.2098**
Time-dummies	Yes	Yes	Yes
Constant	2.1062	2.2694	−3.9603
Observations	144	144	144
R-squared	0.5578	0.6155	0.4128

Note: Standard errors are not reported.
*$p < 0.1$
**$p < 0.05$.
***$p < 0.01$.

Table 5. Political decentralisation and earthquakes' disaster outcomes by economic development, political institution and local representation.

	Number of people affected (ln)	Number of people killed (ln)	Economic damage (ln)
	(1)	(2)	(3)
Political decentralisation (ln)			
• Medium-income countries *(51)*	−0.2581	0.1910	−0.7306
• High-income countries *(93)*	−0.8055**	−0.1091	1.0737
Controls	Yes	Yes	Yes
Time-dummies	Yes	Yes	Yes
Constant	2.2089	2.3258	−4.2990
Observations	144	144	144
R-squared	0.5698	0.6241	0.4618
	(4)	(5)	(6)
Political decentralisation (ln)			
• Left-party countries *(28)*	−0.8187*	−0.2594	0.7970
• Center-party countries *(14)*	0.4964	0.0742	1.2444
• Right-party countries *(54)*	−1.1979**	−0.3463	2.0991**
Controls	Yes	Yes	Yes
Time-dummies	Yes	Yes	Yes
Constant	12.8831***	8.1904***	7.5139
Observations	96	96	96
R-squared	0.6139	0.6356	0.4867
	(7)	(8)	(9)
Political decentralisation (ln)			
• Municipal governments: legislature and executive are locally elected *(73)*	−1.0777**	−0.5919**	1.4917
• Municipal governments: legislature is locally elected *(5)*	−0.5542	−0.6630	−0.2164
• Municipal governments: no local elections *(2)*	2.5709	3.1199**	4.5276
Controls	Yes	Yes	Yes
Time-dummies	Yes	Yes	Yes
Constant	13.0200***	8.5250***	7.7685
Observations	80	80	80
R-squared	0.7197	0.7285	0.5274
	(10)	(11)	(12)
Political decentralisation (ln)			
• State/province governments: legislature and executive are locally elected *(56)*	−0.4596	−0.5133*	0.9901
• State/province governments: legislature is locally elected *(21)*	0.0055	−0.0581	−0.1197
• State/province governments: no local elections *(23)*	1.1602	−0.3287	−1.1476
Controls	Yes	Yes	Yes
Time-dummies	Yes	Yes	Yes
Constant	8.4622**	8.4620***	13.1836
Observations	100	100	100
R-squared	0.5690	0.6419	0.4273

Note: Parentheses in italics show the number of observations for each classification; standard errors are not reported.
*$p < 0.1$.
**$p < 0.05$.
***$p < 0.01$.

4.3. Sensitivity analysis

Tables 2 and 4 showed that there is a negative and statistically significant relation between a country's degree of political decentralisation, measured by 'RAI-total' score, and the number of people affected by disasters triggered by either storms or earthquakes. The same is true when we decompose political decentralisation into 'Self-rule' and 'Shared-rule' scores (Table 6: Regressions 1, 4, 7 and 10). However, the magnitude of the coefficient

Table 6. Decomposition of political decentralisation and disaster outcomes.

	Number of people affected (ln)	Number of people killed (ln)	Economic damage (ln)
Storms			
	(1)	(2)	(3)
Political decentralisation (ln) (self-rule)	−1.3147***	−0.0862	1.0698**
Controls	Yes	Yes	Yes
Time-dummies	Yes	Yes	Yes
Constant	4.9953	6.6785***	2.0634
Observations	392	392	392
R-squared	0.4672	0.6441	0.4604
	(4)	(5)	(6)
Political decentralisation (ln) (shared-rule)	−0.8037***	−0.3486***	0.8636***
Controls	Yes	Yes	Yes
Time-dummies	Yes	Yes	Yes
Constant	1.3521	6.5080***	4.9794
Observations	392	392	392
R-squared	0.4628	0.6650	0.4650
Earthquakes			
	(7)	(8)	(9)
Political decentralisation (ln) (self-rule)	−0.9101**	−0.1092	1.3632*
Controls	Yes	Yes	Yes
Time-dummies	Yes	Yes	Yes
Constant	2.1667	2.2712	−4.0556
Observations	144	144	144
R-squared	0.5548	0.6146	0.4113
	(10)	(11)	(12)
Political decentralisation (ln) (shared-rule)	−0.9694**	−0.4230*	1.5030**
Controls	Yes	Yes	Yes
Time-dummies	Yes	Yes	Yes
Constant	1.6754	2.1170	−3.3039
Observations	144	144	144
R-squared	0.5639	0.6289	0.4210

Note: Standard errors are not reported.
*$p < 0.1$.
**$p < 0.05$.
***$p < 0.01$.

of political decentralisation measured by 'Self-rule' for storms (Regression 1) is higher than that measured as 'Shared-rule' (Regression 4). Table 6 also shows that the negative association between political decentralisation and deaths associated with storms or earthquakes is negative and statistically significant, but only when political decentralisation is measured as 'Shared-rule' (Regression 5 and 11). Finally, no matter how political decentralisation is measured, there is a positive and statistically significant relation between a country's degree of political decentralisation and the economic losses from disasters (Regressions 3, 6, 9 and 12). Overall, both the authority of a regional government over those living in the region ('Self-rule' score) and the authority a regional government co-exercises in the country as a whole ('Shared-rule' score) matter when it comes to the number of people affected and the economic damage, while only the authority a regional government co-exercises in the country as a whole matters for the number of people killed.

We finally replicate the baseline empirical function above [empirical function (1)] through estimation with fixed effects. In other words, we add unobservable national specific effects into empirical function (1). These are time-invariant nationally omitted variables, such as physical endowments which may affect exposure to hazards (e.g. presence of mountains, rivers and coastal proximity). This estimator controls for the omitted variables that are peculiar to each country, accommodating some national heterogeneity.

Table 7 shows a negative and statistically significant association between decentralisation and the number of storm-related fatalities (Regression 2), as well as a negative and statistically significant association between decentralisation and the number of people affected by earthquakes (Regression 4).

Consequently, we find evidence that greater transfers of political power to subnational tiers of governments are justified as a means to reducing the impacts of storms and earthquakes on the population (Regression 3 and 6). These findings must be interpreted with some caution, because although the fixed effects estimator wipes out all the space-specific time-invariant variables, reducing the risk of obtaining biased estimation results (Baltagi, 2005), this reduction in bias comes at a significant cost as it removes cross-national variation from the data, affecting the efficiency of the parameter estimates, especially when the cross-national variation is high (Higgins & Williamson, 1999; Rodríguez-Pose, Psycharis, & Tselios, 2012; Rodríguez-Pose & Tselios, 2010). Finally, we prefer the pooled ordinary least squared (see Tables 2–6) to the fixed effect (Table 7) coefficients, because the latter are interpreted as time-series effects, or short/medium-run effects, as they reflect within-country time-series variation, whereas the former reflect long run effects (Durlauf & Quah, 1999; Mairesse, 1990; Partridge, 2005).

Table 7. Political decentralisation and disaster outcomes: adding fixed effects.

	Number of people affected (ln)	Number of people killed (ln)	Economic damage (ln)
Storms			
	(1)	(2)	(3)
Political decentralisation (ln)	−0.8643	−0.8400**	0.3555
Government consumption	−0.3707*	0.0502	0.0762
Openness	−0.0329	0.0220**	0.0255
Annual growth rate	−9.8797	−1.9404	−26.4381**
Number of disasters (ln)	2.5009***	0.9943***	2.8898***
Time-dummies	Yes	Yes	Yes
Fixed effects	Yes	Yes	Yes
Constant	5.0725	6.3778***	−0.0171
Observations	392	392	392
R-squared	0.4651	0.4008	0.4097
Earthquakes			
	(4)	(5)	(6)
Political decentralisation (ln)	−2.4468**	−0.5551	−3.4492
Government consumption	−0.0376	0.1105	0.3476
Openness	−0.1210	−0.0043	−0.2300
Annual growth rate	41.3474***	13.4828*	46.2483*
Number of disasters (ln)	0.6487	0.8146	3.2856*
Time-dummies	Yes	Yes	Yes
Fixed effects	Yes	Yes	Yes
Constant	−0.4953	−0.0568	−0.6761
Observations	144	144	144
R-squared	0.6295	0.6013	0.4867

*$p < 0.1$; standard errors are not reported.
**$p < 0.05$.
***$p < 0.01$.

5. Conclusions

Hazards happen and due to a variety of factors, some hazards result in disasters. Our analysis builds on previous work that highlights the importance of political institutions in disaster outcomes, and reveals that political decentralisation is an important factor in determining the impact of storms and earthquakes. Thus, our analysis underlines the importance of local capability in managing disaster risk. We are aware that this study is limited by both its focus on only two types of hazards and on only high-income and upper middle-income countries. Nonetheless, important conclusions can be drawn for those countries affected by storms and earthquakes.

In relation to both storms and earthquakes, our results suggest that greater transfers of political power to subnational tiers of governments reduce hazard impacts on the population. Hence, not only do countries with more fiscally decentralised governments experience fewer disaster-induced fatalities (Escaleras & Register, 2012; Skidmore & Toya, 2013; Yamamura, 2012) but also countries with more politically decentralised governments. There are several reasons why more effective local government may be able to influence disaster outcomes. It could be argued that local governments can tailor disaster risk reduction resources more closely to the needs of local citizens than national government at all stages of the disaster. The downside is that more politically decentralised countries, which are usually wealthier countries (Rodríguez-Pose & Ezcurra, 2010; Tselios et al., 2012), can increase the direct economic losses associated with a disaster impact after the storm or earthquake more than the less politically decentralised ones. Another explanation is that local officials might be better informed about local needs before, during and after a storm and, hence, are better able to set the optimal mix of local policies than are central bureaucrats. Or it may simply be that local actors, faced by the recurrent risk, can be involved in longer term risk reduction. However, our research generates a clear conclusion: from a policy point of view, it seems advantageous to give subnational governments more authority and autonomy in storm and earthquake risk planning. However, other factors need to be considered to reduce the economic damage following disasters triggered by storms or earthquakes. Finally, our results show that national wealth, political orientation and local representation can moderate the effect of political decentralisation on disaster outcomes.

This leaves us with the difficult task of identifying a way forward for research and policy. Our findings support much of the community resilience literature which argues that more political decentralisation can reduce the human cost of disasters, but we recognise that this analysis has its limits. It is often the collapse of markets, the destruction of stock and the lack of access to capital that creates the longer term damage for poor or fragile economies after a disaster. The next step is to identify the optimal mix of decentralisation and growth policies to enable economies to bounce back more quickly after disasters.

Notes

1. A natural hazard is a natural event (e.g. earthquake, storm, flood, hurricane, volcanic eruption, tsunami) where there is a threat to humans, society and the economy. A hazard has the potential to cause widespread destruction and loss of lives and thus the potential to cause disaster.
2. See www.emdat.be

3. On the one hand, the set of data where some country–year data are not observed due to factors a and c will not produce bias because it has nothing to do with missing observations; and on the other hand, the set of data where some country–year data are not observed due to factor b may produce bias because it has to do with missing observations. Nevertheless, we control for factor b using time-dummy variables.

4. 'Self-rule' refers to the authority of a regional government over those living in the region and considers: *regional authority over institutional depth* (i.e. the extent to which a regional government is autonomous rather than deconcentrated), *policy scope* (i.e. the range of policies for which a regional government is responsible) *fiscal autonomy* (i.e. the extent to which a regional government can independently tax its population) and *representation* (i.e. the extent to which a region is endowed with an independent legislature and executive). 'Shared-rule' refers to the authority a regional government co-exercises in the country as a whole and considers: *regional authority over law making* (i.e. the extent to which regional representatives co-determine national legislation), *executive control* (i.e. the extent to which a regional government co-determines national policy in intergovernmental meetings), *fiscal control* (i.e. the extent to which regional representatives co-determine the distribution of national tax revenues) and *constitutional reform* (i.e. the extent to which regional representatives co-determine constitutional change).

5. Here, the set of data where some country–year data are not observed is due to the missing observations. However, evaluating and understanding the distribution of missing data, we observe that the missing data will produce little or no bias in the conclusions drawn about the population. Moreover, the 'RAI-total' index has been used in many empirical studies.

6. https://pwt.sas.upenn.edu/php_site/pwt_index.php

7. Again, we are confident the missing data will produce little or no bias and the data on controls is still representative of the population as the PWT database has been used in many empirical studies.

8. The disaster type of 'storm' includes tropical storms, extra-tropical cyclones (winter storms) and local/convective storms [thunderstorm/lightening, snowstorm/blizzard, sandstorm/dust storm, generic (severe) storm, tornado and orographic storm (strong winds)].

9. The number of (total) people affected by storms or earthquakes by year and the economic damage from storms or earthquakes by year can be provided by the authors upon request.

10. The distribution of the disaster outcome variables are asymmetric and skewed, so most of the mass is either on the left or on the right, while most of the mass of the logarithmic transformation of these variables is nearly symmetrical.

11. For those countries where there were no resulting deaths (0), we cannot take the natural logarithm because the logarithm of zero is not defined. Adding one death to each observation means the natural logarithm of one is zero, which represents countries with zero deaths.

12. We prefer to use the government consumption share of GDP per capita rather than the GDP per capita because the latter 'no doubt reflects factors such as public desire for enhanced building codes, the existence of early warning systems, and more disaster sensitive land zoning and use decisions' (Escaleras & Register, 2012, 171). All these factors are better captured by the government consumption share of GDP per capita.

13. We do not include the magnitude scale and value of storms (in kph, speed of wind) and earthquakes (Richter scale) because some countries have experienced multiple natural hazards in the same year. Instead, we use the number of disasters each year, which has been used in empirical studies on disasters (Ahmed & Iqbal, 2009; Escaleras & Register, 2012).

14. A dummy variable takes the value of 0 or 1 to indicate the absence or presence of some categorical effect that may be expected to shift the outcome. It is a proxy variable for qualitative facts in regression analysis. A vector of dummy variables includes a set of dummy variables. In other words, it represents levels within multiple variables.

Disclosure statement

No potential conflict of interest was reported by the authors.

References

Adger, W. N. (2006). Vulnerability. *Global Environmental Change, 16*, 268–281.

Adger, W. N., Hughes, T. P., Folke, C., Carpenter, S. R., & Rockstrom, J. (2005). Social-ecological resilience to coastal disasters. *Science, 309*, 1036–1039.

Ahmed, M., & Iqbal, K. (2009). *Disaster and decentralisation* (Working Paper Series No. 2009-03). Carleton College: Department of Economics.

Ainuddin, S., Aldrich, D. P., Routray, J. K., Ainuddin, S., & Achkazai, A. (2013). The need for local involvement: Decentralization of disaster management institutions in Baluchistan, Pakistan. *International Journal of Disaster Risk Reduction, 6*, 50–58.

Aldrich, D. P. (2012). *Building resilience: Social capital in post-disaster recovery*. Chicago: University of Chicago Press.

Baltagi, B. H. (2005). *Econometric analysis of panel data*. Chichester: John Wiley.

Bardhan, P. (2002). Decentralization of governance and development. *Journal of Economic Perspectives, 16*, 185–205.

Basher, R. (2006). Global early warning systems for natural hazards: Systematic and people-centred. *Philosophical Transactions of the Royal Society A: Mathematical, Physical and Engineering Sciences, 364*, 2167–2182.

Blackburn, S. (2014). The politics of scale and disaster risk governance: Barriers to decentralisation in Portland, Jamaica. *Geoforum, 52*, 101–112.

Bouwer, L. M., Crompton, R. P., Faust, E., Höppe, P., & Pielke Jr R. A. (2007). Disaster management: Confronting disaster losses. *Science, 318*, 753.

Brenner, N. (2004). *New state spaces: Urban governance and the rescaling of statehood*. Oxford: Oxford University Press.

Brooks, N., Adger, W. N., & Kelly, P. M. (2005). The determinants of vulnerability and adaptive capacity at the national level and the implications for adaptation. *Global Environmental Change, 15*, 151–163.

Brouwer, R., & Nhassengo, J. (2006). About bridges and bonds: Community responses to the 2000 floods in Mabalane District, Mozambique. *Disasters, 30*, 234–255.

Brown, J. C., & Purcell, M. (2005). There's nothing inherent about scale: Political ecology, the local trap, and the politics of development in the Brazilian Amazon. *Geoforum, 36*, 607–624.

Cheshire, P. C., & Gordon, I. R. (1998). Territorial competition: Some lessons for policy. *The Annals of Regional Science, 32*, 321–346.

Chhotray, V., & Few, R. (2012). Post-disaster recovery and ongoing vulnerability: Ten years after the super-cyclone of 1999 in Orissa, India. *Global Environmental Change, 22*, 695–702.

Colten, C. E., Kates, R. W., & Laska, S. B. (2008). Three years after Katrina: Lessons for community resilience. *Environment: Science and Policy for Sustainable Development, 50*, 36–47.

DeMaria, M., Sampson, C. R., Knaff, J. A., & Musgrave, K. D. (2014). Is tropical cyclone intensity guidance improving? *Bulletin of the American Meteorological Society, 95*, 387–398.

Djalante, R., Holley, C., & Thomalla, F. (2011). Adaptive governance and managing resilience to natural hazards. *International Journal of Disaster Risk Science, 2*, 1–14.

Durlauf, S. N., & Quah, D. T. (1999). The new empirics of economic growth. In J. B. Taylor & M. Woodford (Eds.), *Handbook of macroeconomics* (pp. 555–677). Amsterdam: North-Holland: Elsevier.

Escaleras, M., & Register, C. A. (2012). Fiscal decentralization and natural hazard risks. *Public Choice, 151*, 165–183.

Faguet, J. P. (2014). Decentralization and governance. *World Development, 53*, 2–13.

Fisman, R., & Gatti, R. (2002). Decentralization and corruption: Evidence across countries. *Journal of Public Economics, 83*, 325–345.

Foresight. (2012). Reducing the risks of future disasters for decision makers. Final project report. In Foresight and UK Government Office of Science (Eds.), *Foresight reports* (p. 1–132). London: Government Office of Science.

Guha-Sapir, D., Hargitt, D., & Hoyois, P. (2004). *Thirty years of natural disasters 1974–2003: The numbers*. Brussels: Presses univ. de Louvain.

Higgins, M., & Williamson, J. (1999). Explaining inequality the world round: cohort size, Kuznets curves, and openness (NBER Working Paper No. 7224, National Bureau of Economic Research).

Holloway, A. (2003). Disaster risk reduction in southern Africa. *African Security Review, 12*, 29–38.

Hooghe, L., Marks, G., & Schakel, A. H. (2008a). Appendix B: Country and regional scores. *Regional and Federal Studies, 18*, 259–274.

Hooghe, L., Marks, G., & Schakel, A. H. (2008b). Operationalizing regional authority: A coding scheme for 42 countries, 1950–2006. *Regional and Federal Studies, 18*, 123–142.

Kahn, M. E. (2005). The death toll from natural disasters: The role of income, geography, and institutions. *Review of Economics and Statistics, 87*, 271–284.

Kamel, N. (2012). Social marginalisation, federal assistance and repopulation patterns in the New Orleans Metropolitan Area following Hurricane Katrina. *Urban Studies, 49*(14), 3211–3231.

Keefer, P. (2012, December). *Database of political institutions: Changes and variable definitions*. Washington: Development Research Group, The World Bank.

Laska, S., & Morrow, B. H. (2006). Social vulnerabilities and hurricane Katrina: An unnatural disaster in New Orleans. *Marine Technology Society Journal, 40*, 16–26.

Le Galès, P. (2002). *European cities: Social conflicts and governance*. New York: Oxford University Press.

Lemos, M. C. (2007). Drought, governance and adaptive capacity in North East Brazil: A case study of Ceará. In UNDP for Human Development Report Office Occasional Paper (Eds.), *Fighting climate change: Human solidarity in a divided world* (p. 1–15). New York: UNDP Human Development Report 2007/2008.

Lessmann, C. (2009). Fiscal decentralization and regional disparity: Evidence from cross-section and panel data. *Environment and Planning A, 41*, 2455–2473.

Mairesse, J. (1990). Time-series and cross-sectional estimates on panel data: Why are they different and why they should be equal. In J. Hartog, G. Ridder, & J. Theeuwes (Eds.), *Panel data and labor market studies* (pp. 81–95). New York: North Holland.

Marks, D., & Lebel, L. (2016). Disaster governance and the scalar politics of incomplete decentralization: Fragmented and contested responses to the 2011 floods in Central Thailand. *Habitat International, 52*, 57–66.

de Mello, L. (2011). Does fiscal decentralisation strengthen social capital? Cross-country evidence and the experiences of Brazil and Indonesia. *Environment and Planning C: Government and Policy, 29*, 281–296.

Murphy, B. L. (2007). Locating social capital in resilient community-level emergency management. *Natural Hazards, 41*, 297–315.

Norris, F. H., Stevens, S. P., Pfefferbaum, B., Wyche, K. F., & Pfefferbaum, R. L. (2008). Community resilience as a metaphor, theory, set of capacities, and strategy for disaster readiness. *American Journal of Community Psychology, 41*, 127–150.

Oates, W. E. (1972). *Fiscal federalism*. New York: Harcourt Brace Jovanovich.

Oates, W. E. (1999). An essay on fiscal federalism. *Journal of Economic Literature, 37*, 1120–1149.

Partridge, M. D. (2005). Does income distribution affect U.S state economic growth? *Journal of Regional Science, 45*, 363–394.

Paton, D., Millar, M., & Johnston, D. (2001). Community resilience to volcanic hazard consequences. *Natural Hazards, 24*, 157–169.

Pelling, M. (2011). Urban governance and disaster risk reduction in the Caribbean: The experiences of Oxfam GB. *Environment and Urbanization, 23*, 383–400.

Pielke, R. A., Jr., Rubiera, J., Landsea, C., Fernandez, M. L., & Klein, R. (2003). Hurricane vulnerability in Latin America and The Caribbean: Normalized damage and loss potentials. *Natural Hazards Review, 4*, 101–114.

Pike, A., Rodriguez-Pose, A., Tomaney, J., Torrisi, G., & Tselios, V. (2012). In search of the 'economic dividend' of devolution: Spatial disparities, spatial economic policy, and decentralisation in the UK. *Environment and Planning C: Government and Policy, 30,* 10–28.

Prud'homme, R. (1995). The dangers of decentralization. *The World Bank Research Observer, 10,* 201–220.

Rodríguez-Pose, A., & Ezcurra, R. (2010). Does decentralization matter for regional disparities? A cross-country analysis. *Journal of Economic Geography, 10,* 619–644.

Rodríguez-Pose, A., & Ezcurra, R. (2011). Is fiscal decentralization harmful for economic growth? Evidence from the OECD countries. *Journal of Economic Geography, 11,* 619–643.

Rodriguez-Pose, A., & Gill, N. (2004). Is there a global link between regional disparities and devolution? *Environment and Planning A, 36,* 2097–2117.

Rodríguez-Pose, A., Psycharis, Y., & Tselios, V. (2012). Public investment and regional growth and convergence: Evidence from Greece. *Papers in Regional Science, 91,* 543–568.

Rodríguez-Pose, A., & Tselios, V. (2009). Education and income inequality in the regions of the European Union. *Journal of Regional Science, 49,* 411–437.

Rodríguez-Pose, A., & Tselios, V. (2010). Returns to migration, education and externalities in the European Union. *Papers in Regional Science, 89,* 411–434.

Schneider, A. (2003). Decentralization: Conceptualization and measurement. *Studies in Comparative International Development, 38,* 32–56.

Sims, H., & Vogelmann, K. (2002). Popular mobilization and disaster management in Cuba. *Public Administration and Development, 22,* 389–400.

Skidmore, M., & Toya, H. (2013). Natural disaster impacts and fiscal decentralization. *Land Economics, 89,* 101–117.

Thompson, M., & Gaviria, I. (2004). *Weathering the storm: Lessons in risk reduction from Cuba.* Boston, MA: Oxfam America.

Tiebout, C. M. (1956). A pure theory of local expenditures. *Journal of Political Economy, 64,* 416–424.

Tompkins, E. L., Hurlston, L. A., & Poortinga, W. (2009). *Disaster resilience: Fear, friends and foreignness as determinants of risk mitigating behaviour in small islands.* (SRI working paper 18).

Tompkins, E. L., Lemos, M. C., & Boyd, E. (2008). A less disastrous disaster: Managing response to climate-driven hazards in the Cayman Islands and NE Brazil. *Global Environmental Change, 18,* 736–745.

Tselios, V., Rodríguez-Pose, A., Pike, A., Tomaney, J., & Torrisi, G. (2012). Income inequality, decentralisation, and regional development in Western Europe. *Environment and Planning A, 44,* 1278–1301.

Twigg, J. (2003). The human factor in early warnings: Risk perception and appropriate communications. In J. Zschau & A. Küppers (Eds.), *Early warning systems for natural disaster reduction* (pp. 19–26). Berlin: Springer.

Twigg, J. (2007). *Characteristics of a disaster-resilient community. A guidance note.* London: Benfield UCL Hazard Research Centre and Interagency Coordination Group.

UNISDR. (2005). *Hyogo framework for action 2005-2015: Building the resilience of nations and communities to disasters.* Geneva: International Strategy for Disaster Reduction, p. 22. Retrieved from www.unisdr.org

UNISDR. (2009). *Global assessment report on disaster risk reduction. Risk and poverty in a changing climate.* Geneva: Author.

UNISDR. (2015). *The human cost of weather-related disasters 1995-2015.* Geneva: CRED and UNISDR.

Weingast, B. R. (2009). Second generation fiscal federalism: The implications of fiscal incentives. *Journal of Urban Economics, 65,* 279–293.

Wisner, B., Blaikie, P., Cannon, T., & Davis, I. (2004). *At risk. Natural hazards, people's vulnerability and disasters.* London: Routledge.

Yamamura, E. (2012). *Death tolls from natural disasters: Influence of interactions among fiscal decentralization, institutions and economic development.* (EERI Research paper Series No 08/2012). Brussels: Economics and Econometrics Research Institute.

Yilmaz, S., Beris, Y., & Serrano-Berthet, R. (2008). *Local government discretion and accountability: A diagnostic framework for local governance* (Paper No. 113 / July 2008, Local Governance and Accountability Series, Social Development Working Papers). Washington: The World Bank.

Appendix. Number of observations for storms and earthquakes.

Country	Year	Storms Obs	Obs (1950–2006)	Diff	% of obs	Earthquakes Obs	Obs (1950–2006)	Diff	% of obs
Albania	1992–2006	2	2	0	0.005	1	5	−4	0.007
Australia	1950–2006	31	31	0	0.078	4	4	0	0.028
Austria	1955–2006	9	9	0	0.023	1	1	0	0.007
Belgium	1950–2006	12	12	0	0.030	2	2	0	0.014
Bosnia and Herzegovina	1995–2006	2	2	0	0.005	0	0	0	0.000
Bulgaria	1991–2006	4	5	−1	0.010	1	4	−3	0.007
Canada	1950–2006	26	26	0	0.065	0	0	0	0.000
Croatia	1991–2006	1	1	0	0.003	1	1	0	0.007
Cyprus	1960–2006	4	4	0	0.010	1	2	−1	0.007
Czech Republic	1993–2006	2	2	0	0.005	0	0	0	0.000
Denmark	1950–2006	9	9	0	0.023	0	0	0	0.000
Estonia	1992–2006	1	1	0	0.003	0	0	0	0.000
Finland	1950–2006	1	1	0	0.003	0	0	0	0.000
France	1950–2006	22	22	0	0.055	1	1	0	0.007
Germany	1950–2006	22	22	0	0.055	3	3	0	0.021
Greece	1950–2006	6	6	0	0.015	22	22	0	0.153
Hungary	1990–2006	4	5	−1	0.010	0	0	0	0.000
Iceland	1950–2006	0	0	0	0.000	2	2	0	0.014
Ireland	1950–2006	10	10	0	0.025	0	0	0	0.000
Italy	1950–2006	11	11	0	0.028	16	16	0	0.111
Japan	1950–2006	53	53	0	0.133	22	22	0	0.153
Latvia	1990–2006	2	2	0	0.005	0	0	0	0.000
Lithuania	1992–2006	3	3	0	0.008	0	0	0	0.000
Luxembourg	1950–2006	2	2	0	0.005	0	0	0	0.000
Macedonia	1991–2006	1	1	0	0.003	0	0	0	0.000
Malta	1964–2006	0	0	0	0.000	0	0	0	0.000
Netherlands	1950–2006	12	12	0	0.030	1	1	0	0.007
New Zealand	1950–2006	8	8	0	0.020	3	3	0	0.021
Norway	1950–2006	4	4	0	0.010	0	0	0	0.000
Poland	1990–2006	5	6	−1	0.013	0	1	−1	0.000
Portugal	1976–2006	3	4	−1	0.008	0	0	0	0.000
Romania	1991–2006	6	6	0	0.015	1	3	−2	0.007
Russia	1993–2006	10	14	−4	0.025	7	19	−12	0.049
Slovak Republic	1993–2006	1	1	0	0.003	0	0	0	0.000
Slovenia	1990–2006	0	0	0	0.000	2	2	0	0.014
Spain	1978–2006	10	12	−2	0.025	1	2	−1	0.007
Sweden	1950–2006	5	5	0	0.013	0	0	0	0.000
Switzerland	1950–2006	15	15	0	0.038	0	0	0	0.000
Turkey	1950–2006	6	6	0	0.015	31	31	0	0.215
United Kingdom	1950–2006	19	19	0	0.048	1	1	0	0.007
United States	1950–2006	55	55	0	0.138	20	20	0	0.139
Sum		399	409	−10		144	168	−24	

Index

Note: Figures are indicated by *italics* and tables by **bold** type. Endnotes are indicated by the page number followed by "n" and the endnote number e.g., 57n50 refers to endnote 50 on page 57.